NATURAL HISTORY
UNIVERSAL LIBRARY

西方博物学大系

主编：江晓原

THE QUADRUPEDS OF NORTH AMERICA

北美四足动物志

[美] 约翰·詹姆斯·奥杜邦 著

华东师范大学出版社

图书在版编目(CIP)数据

北美四足动物志 = The quadrupeds of North America : 英文 / (美) 约翰 · 詹姆斯 · 奥杜邦著. — 上海 : 华东师范大学出版社, 2018
(寰宇文献)
ISBN 978-7-5675-7724-4

Ⅰ. ①北… Ⅱ. ①约… Ⅲ. ①陆栖 - 动物志 - 北美洲-英文 Ⅳ. ①Q959

中国版本图书馆CIP数据核字(2018)第096597号

北美四足动物志
The quadrupeds of North America
(美) 约翰 · 詹姆斯 · 奥杜邦著

特约策划 黄曙辉 徐 辰
责任编辑 庞 坚
特约编辑 许 倩
装帧设计 刘怡霖

出版发行 华东师范大学出版社
社 址 上海市中山北路3663号 邮编 200062
网 址 www.ecnupress.com.cn
电 话 021-60821666 行政传真 021-62572105
客服电话 021-62865537
门市(邮购)电话 021-62869887
地 址 上海市中山北路3663号华东师范大学校内先锋路口
网 店 http://hdsdcbs.tmall.com/

印 刷 者 虎彩印艺股份有限公司
开 本 16开
印 张 88.5
版 次 2018年6月第1版
印 次 2018年6月第1次
书 号 ISBN 978-7-5675-7724-4
定 价 1880.00元(精装全三册)

出 版 人 王 焰

《西方博物学大系》总序

江晓原

《西方博物学大系》收录博物学著作超过一百种，时间跨度为15世纪至1919年，作者分布于16个国家，写作语种有英语、法语、拉丁语、德语、弗莱芒语等，涉及对象包括植物、昆虫、软体动物、两栖动物、爬行动物、哺乳动物、鸟类和人类等，西方博物学史上的经典著作大备于此编。

中西方"博物"传统及观念之异同

今天中文里的"博物学"一词，学者们认为对应的英语词汇是Natural History，考其本义，在中国传统文化中并无现成对应词汇。在中国传统文化中原有"博物"一词，与"自然史"当然并不精确相同，甚至还有着相当大的区别，但是在"搜集自然界的物品"这种最原始的意义上，两者确实也大有相通之处，故以"博物学"对译Natural History一词，大体仍属可取，而且已被广泛接受。

已故科学史前辈刘祖慰教授尝言：古代中国人处理知识，如开中药铺，有数十上百小抽屉，将百药分门别类放入其中，即心安矣。刘教授言此，其辞若有憾焉——认为中国人不致力于寻求世界"所以然之理"，故不如西方之分析传统优越。然而古代中国人这种处理知识的风格，正与西方的博物学相通。

与此相对，西方的分析传统致力于探求各种现象和物体之间的相互关系，试图以此解释宇宙运行的原因。自古希腊开始，西方哲人即孜孜不倦建构各种几何模型，欲用以说明宇宙如何运行，其中最典型的代表，即为托勒密（Ptolemy）的宇宙体系。

比较两者，差别即在于：古代中国人主要关心外部世界"如何"运行，而以希腊为源头的西方知识传统（西方并非没有别的知识传统，只是未能光大而已）更关心世界"为何"如此运行。在线性发展无限进步的科学主义观念体系中，我们习惯于认为“为何”是在解决了“如何”之后的更高境界，故西方的分析传统比中国的传统更高明。

然而考之古代实际情形，如此简单的优劣结论未必能够成立。例如以天文学言之，古代东西方世界天文学的终极问题是共同的：给定任意地点和时刻，计算出太阳、月亮和五大行星（七政）的位置。古代中国人虽不致力于建立几何模型去解释七政“为何”如此运行，但他们用抽象的周期叠加（古代巴比伦也使用类似方法），同样能在足够高的精度上计算并预报任意给定地点和时刻的七政位置。而通过持续观察天象变化以统计、收集各种天象周期，同样可视之为富有博物学色彩的活动。

还有一点需要注意：虽然我们已经接受了用“博物学”来对译 Natural History，但中国的博物传统，确实和西方的博物学有一个重大差别——即中国的博物传统是可以容纳怪力乱神的，而西方的博物学基本上没有怪力乱神的位置。

古代中国人的博物传统不限于“多识于鸟兽草木之名”。体现此种传统的典型著作，首推晋代张华《博物志》一书。书名“博物”，其义尽显。此书从内容到分类，无不充分体现它作为中国博物传统的代表资格。

《博物志》中内容，大致可分为五类：一、山川地理知识；二、奇禽异兽描述；三、古代神话材料；四、历史人物传说；五、神仙方伎故事。这五大类，完全符合中国文化中的博物传统，深合中国古代博物传统之旨。第一类，其中涉及宇宙学说，甚至还有“地动”思想，故为科学史家所重视。第二类，其中甚至出现了中国古代长期流传的“守宫砂”传说的早期文献：相传守宫砂点在处女胳膊上，永不褪色，只有性交之后才会自动消失。第三类，古代神话传说，其中甚至包括可猜想为现代“连体人”的记载。第四类，各种著名历史人物，比如三位著名刺客的传说，此三名刺客及所刺对象，历史上皆实有其人。第五类，包括各种古代方术传说，比如中国古代房中养生学说，房中术史上的传说人物之一“青牛道士封君达”等等。前两类与西方的博物学较为接近，但每一类都会带怪力乱神色彩。

“所有的科学不是物理学就是集邮”

在许多人心目中，画画花草图案，做做昆虫标本，拍拍植物照片，这类博物学活动，和精密的数理科学，比如天文学、物理学等等，那是无法同日而语的。博物学显得那么的初级、简单，甚至幼稚。这种观念，实际上是将“数理程度”作为唯一的标尺，用来衡量一切知识。但凡能够使用数学工具来描述的，或能够进行物理实验的，那就是“硬”科学。使用的数学工具越高深越复杂，似乎就越“硬”；物理实验设备越庞大，花费的金钱越多，似乎就越“高端”、越“先进”……

这样的观念，当然带着浓厚的“物理学沙文主义”色彩，在很多情况下是不正确的。而实际上，即使我们暂且同意上述“物理学沙文主义”的观念，博物学的“科学地位”也仍然可以保住。作为一个学天体物理专业出身，因而经常徜徉在“物理学沙文主义”幻影之下的人，我很乐意指出这样一个事实：现代天文学家们的研究工作中，仍然有绘制星图，编制星表，以及为此进行的巡天观测等等活动，这些活动和博物学家“寻花问柳”，绘制植物或昆虫图谱，本质上是完全一致的。

这里我们不妨重温物理学家卢瑟福（Ernest Rutherford）的金句：“所有的科学不是物理学就是集邮（All science is either physics or stamp collecting）。”卢瑟福的这个金句堪称“物理学沙文主义”的极致，连天文学也没被他放在眼里。不过，按照中国传统的“博物”理念，集邮毫无疑问应该是博物学的一部分——尽管古代并没有邮票。卢瑟福的金句也可以从另一个角度来解读：既然在卢瑟福眼里天文学和博物学都只是“集邮”，那岂不就可以将博物学和天文学相提并论了？

如果我们摆脱了科学主义的语境，则西方模式的优越性将进一步被消解。例如，按照霍金（Stephen Hawking）在《大设计》（*The Grand Design*）中的意见，他所认同的是一种“依赖模型的实在论（model-dependent realism）”，即“不存在与图像或理论无关的实在性概念（There is no picture- or theory-independent concept of reality）”。在这样的认识中，我们以前所坚信的外部世界的客观性，已经不复存在。既然几何模型只不过是对外部世界图像的人为建构，则古代中国人干脆放弃这种建构直奔应用（毕竟在实际应用中我们只需要知道七政“如何”运行），又有何不可？

传说中的“神农尝百草”故事，也可以在类似意义下得到新的解读：“尝百草”当然是富有博物学色彩的活动，神农通过这一活动，得知哪些草能够治病，哪些不能，然而在这个传说中，神农显然没有致力于解释“为何”某些草能够治病而另一些则不能，更不会去建立“模型”以说明之。

“帝国科学”的原罪

今日学者有倡言“博物学复兴”者，用意可有多种，诸如缓解压力、亲近自然、保护环境、绿色生活、可持续发展、科学主义解毒剂等等，皆属美善。编印《西方博物学大系》也是意欲为“博物学复兴”添一助力。

然而，对于这些博物学著作，有一点似乎从未见学者指出过，而鄙意以为，当我们披阅把玩欣赏这些著作时，意识到这一点是必须的。

这百余种著作的时间跨度为15世纪至1919年，注意这个时间跨度，正是西方列强“帝国科学”大行其道的时代。遥想当年，帝国的科学家们乘上帝国的军舰——达尔文在皇家海军“小猎犬号”上就是这样的场景之一，前往那些已经成为帝国的殖民地或还未成为殖民地的“未开化”的遥远地方，通常都是踌躇满志、充满优越感的。

作为一个典型的例子，英国学者法拉在（Patricia Fara）《性、植物学与帝国：林奈与班克斯》（*Sex, Botany and Empire, The Story of Carl Linnaeus and Joseph Banks*）一书中讲述了英国植物学家班克斯（Joseph Banks）的故事。1768年8月15日，班克斯告别未婚妻，登上了澳大利亚军舰“奋进号”。此次“奋进号”的远航是受英国海军部和皇家学会资助，目的是前往南太平洋的塔希提岛（Tahiti，法属海外自治领，另一个常见的译名是“大溪地”）观测一次比较罕见的金星凌日。舰长库克（James Cook）是西方殖民史上最著名的舰长之一，多次远航探险，开拓海外殖民地。他还被认为是澳大利亚和夏威夷群岛的“发现”者，如今以他命名的群岛、海峡、山峰等不胜枚举。

当“奋进号”停靠塔希提岛时，班克斯一下就被当地美丽的土著女性迷昏了，他在她们的温柔乡里纵情狂欢，连库克舰长都看不下去了，“道德愤怒情绪偷偷溜进了他的日志当中，他发现自己根本不可能不去批评所见到的滥交行为”，而班克斯纵欲到了“连嫖妓都毫无激情”的地步——这是别人讽刺班克斯的说法，因为对于那时常年航行于茫茫大海上的男性来说，上岸嫖妓通常是一项能够唤起“激情”的活动。

而在“帝国科学”的宏大叙事中，科学家的私德是无关紧要的，人们关注的是科学家做出的科学发现。所以，尽管一面是班克斯在塔希提岛纵欲滥交，一面是他留在故乡的未婚妻正泪眼婆娑地“为远去的心上人绣织背心”，这样典型的“渣男”行径要是放在今天，非被互联网上的口水淹死不可，但是“班克斯很快从他们的分离之苦中走了出来，在外近三年，他活得倒十分滋润”。

法拉不无讽刺地指出了“帝国科学”的实质：“班克斯接管了当地的女性和植物，而库克则保护了大英帝国在太平洋上的殖民地。”甚至对班克斯的植物学本身也调侃了一番：“即使是植物学方面的科学术语也充满了性指涉。……这个体系主要依靠花朵之中雌雄生殖器官的数量来进行分类。”据说“要保护年轻妇女不受植物学教育的浸染，他们严令禁止各种各样的植物采集探险活动。”这简直就是将植物学看成一种“涉黄”的淫秽色情活动了。

在意识形态强烈影响着我们学术话语的时代，上面的故事通常是这样被描述的：库克舰长的“奋进号”军舰对殖民地和尚未成为殖民地的那些地方的所谓“访问”，其实是殖民者耀武扬威的侵略，搭载着达尔文的“小猎犬号”军舰也是同样行径；班克斯和当地女性的纵欲狂欢，当然是殖民者对土著妇女令人发指的蹂躏；即使是他采集当地植物标本的“科学考察”，也可以视为殖民者“窃取当地经济情报”的罪恶行为。

后来改革开放，上面那种意识形态话语被抛弃了，但似乎又走向了另一个极端，完全忘记或有意回避殖民者和帝国主义这个层面，只歌颂这些军舰上的科学家的伟大发现和成就，例如达尔文随着“小猎犬号”的航行，早已成为一曲祥和优美的科学颂歌。

其实达尔文也未能免俗，他在远航中也乐意与土著女性打打交道，当然他没有像班克斯那样滥情纵欲。在达尔文为“小猎犬号”远航写的《环球游记》中，我们读到：“回程途中我们遇到一群黑人姑娘在聚会，……我们笑着看了很久，还给了她们一些钱，这着实令她们欣喜一番，拿着钱尖声大笑起来，很远还能听到那愉悦的笑声。”

有趣的是，在班克斯在塔希提岛纵欲六十多年后，达尔文随着“小猎犬号”也来到了塔希提岛，岛上的土著女性同样引起了达尔文的注意，在《环球游记》中他写道：“我对这里妇女的外貌感到有些失望，然而她们却很爱美，把一朵白花或者红花戴在脑后的髮髻上……”接着他以居高临下的笔调描述了当地女性的几种发饰。

用今天的眼光来看，这些在别的民族土地上采集植物动物标本、测量地质水文数据等等的“科学考察”行为，有没有合法性问题？有没有侵犯主权的问题？这些行为得到当地人的同意了吗？当地人知道这些行为的性质和意义吗？他们有知情权吗？……这些问题，在今天的国际交往中，确实都是存在的。

也许有人会为这些帝国科学家辩解说：那时当地土著尚在未开化或半开化状态中，他们哪有“国家主权”的意识啊？他们也没有制止帝国科学家的考察活动啊？但是，这样的辩解是无法成立的。

姑不论当地土著当时究竟有没有试图制止帝国科学家的“科学考察”行为，现在早已不得而知，只要殖民者没有记录下来，我们通常就无法知道。况且殖民者有军舰有枪炮，土著就是想制止也无能为力。正如法拉所描述的：“在几个塔希提人被杀之后，一套行之有效的易货贸易体制建立了起来。”

即使土著因为无知而没有制止帝国科学家的“科学考察”行为，这事也很像一个成年人闯进别人的家，难道因为那家只有不懂事的小孩子，闯入者就可以随便打探那家的隐私、拿走那家的东西、甚至将那家的房屋土地据为己有吗？事实上，很多情况下殖民者就是这样干的。所以，所谓的“帝国科学”，其实是有着原罪的。

如果沿用上述比喻，现在的局面是，家家户户都不会只有不懂事的孩子了，所以任何外来者要想进行“科学探索”，他也得和这家主人达成共识，得到这家主人的允许才能够进行。即使这种共识的达成依赖于利益的交换，至少也不能单方面强加于人。

博物学在今日中国

博物学在今日中国之复兴，北京大学刘华杰教授提倡之功殊不可没。自刘教授大力提倡之后，各界人士纷纷跟进，仿佛昔日蔡锷在云南起兵反袁之“滇黔首义，薄海同钦，一檄遥传，景从恐后”光景，这当然是和博物学本身特点密切相关的。

无论在西方还是在中国，无论在过去还是在当下，为何博物学在它繁荣时尚的阶段，就会应者云集？深究起来，恐怕和博物学本身的特点有关。博物学没有复杂的理论结构，它的专业训练也相对容易，至少没有天文学、物理学那样的数理“门槛”，所以和一些数理学科相比，博物学可以有更多的自学成才者。这次编印的《西方博物学大系》，卷帙浩繁，蔚为大观，同样说明了这一点。

最后，还有一点明显的差别必须在此处强调指出：用刘华杰教授喜欢的术语来说，《西方博物学大系》所收入的百余种著作，绝大部分属于“一阶”性质的工作，即直接对博物学作出了贡献的著作。事实上，这也是它们被收入《西方博物学大系》的主要理由之一。而在中国国内目前已经相当热的博物学时尚潮流中，绝大部分已经出版的书籍，不是属于“二阶”性质（比如介绍西方的博物学成就），就是文学性的吟风咏月野草闲花。

要寻找中国当代学者在博物学方面的“一阶”著作，如果有之，以笔者之孤陋寡闻，唯有刘华杰教授的《檀岛花事——夏威夷植物日记》三卷，可以当之。这是刘教授在夏威夷群岛实地考察当地植物的成果，不仅属于直接对博物学作出贡献之作，而且至少在形式上将昔日“帝国科学”的逻辑反其道而用之，岂不快哉！

2018 年 6 月 5 日
于上海交通大学
科学史与科学文化研究院

《北美四足动物志》

约翰·詹姆斯·奥杜邦
（1785-1851）

1785年，约翰·詹姆斯·奥杜邦（John James Audubon）生于海地圣多明哥，是一个法国船长的私生子。他自幼跟随父亲亲近、观察自然，熟悉打猎和标本制作。十八岁时，他为逃避兵役而来到美国，替父亲看管费城郊外的产业，并与来自英格兰的富家移民小姐露西·贝克维尔成婚。

以自己的积蓄和父亲的产业为后盾，奥杜邦曾多方尝试商业投资与贸易经营，却屡屡不顺。与此同时，他被美国广袤壮丽的自然风光吸引，特别喜爱当地的各类鸟儿，对鸟类的观察研究与绘画技法也不断成长。1819年的经济危机彻底断送他经商的经济实力之后，奥杜邦开始全力投入绘制北美鸟类图鉴的事业，两个孩子则留在家中，由妻子露西抚养。

此后十多年间，奥杜邦的足迹踏遍北美各地，南抵佛罗里达，北至加拿大的拉布拉多，绘制了四百八十九种北美鸟类，旅行期间还发现了二十多个新种。1826年，他希望将这些作品结集出版，却遭到美国同行的排挤。在露西的建议下，他携带三百多幅作品乘船横渡大西洋，到利物浦寻求机会。他在英国大获成功，被誉为“美国的林中居民”。1839年，他出版了代表作《美洲鸟类图鉴》以及《鸟类志》，成为继富兰克林之后美国第二个英国皇家学会会员，并入选林奈学会。他不愧为鸟类及哺乳动物研究的大家，在世时也已对工业化、人类开拓将对自然界与生物造成的影响提出过警示，他的画像现在仍悬挂在白宫。

奥杜邦晚年最重要的作品是《北美四足动物志》，本书共三卷，翔实记录了北美地区多种重要胎生四足兽。和以往欧美学者仅靠购买剥制标本进行动物研究的方式不同，奥杜邦和巴赫曼亲身深入田野林间，亲自追踪观察动物习性和外观。与《美洲鸟类图鉴》一样，奥杜邦在设计插画时在科学严谨的基础上，以生动多变的场景增添画面的动感和趣味性，配以真实环境下的地貌、植物，并为读者详细描述相关动物的习性、分布、被人类捕猎的方式以及被捕杀的原因。

《北美四足动物志》的底稿创作于1846年至1847年间，由巴赫曼提供文本，奥杜邦绘制草稿。此时，他已罹患阿尔茨海默症，后期几近失智，所以书中大部分画作为其子约翰·伍德豪斯·奥杜邦协作完成。老奥杜邦生前，《北美四足动物志》仅出版了第一卷，他于1851年去世后，妻子露西协助雷夫·巴赫曼出版了后两卷。

露西·贝克维尔·奥杜邦
（1787-1874）

然而，若提到奥杜邦的毕生成就，就不能不提一直在他背后默默大力支持的妻子露西。

1787年，露西·贝克维尔生于英格兰一个富裕家庭。1801年，其父威廉·贝克维尔举家移民美国，在宾州费城近旁买下一片人称“福特沃野”的土地定居下来。两年后，约翰·詹姆斯·奥杜邦从海地逃兵役来到美国，邂逅十六岁的露西。奥杜邦家族的产业就在贝克维尔一家土地的隔壁，二人情投意合，露西也常与奥杜邦一起外出观察鸟类——这是后者的一大兴趣爱好。

1808年，二人正式结婚。如前所述，奥杜邦是美国历史上最伟大的博物学家和鸟类画家之一，早年却耗费二十年光阴从事商业投资，事实证明他完全不是这块料。他最有希望的一份产业，是与露西及亲戚合作投资的蒸汽磨谷机，但随着1819年经济危机到来，全部投资也泡了汤。此时，奥杜邦决定全身心投入观察、研究和绘制北美鸟类的事业，露西成为他的坚强后盾。

在一开始的七年中，奥杜邦一家穷困潦倒。虽然露西后来打趣说若自己会嫉妒谁，便会觉得每一只画中的鸟儿都是情敌，因为它们完全夺去了丈夫的心和思绪，但在奥杜邦最困难的时刻，却是她独立外出做家教贴补全部家用。露西受过良好的教育，体格强健，是家里的顶梁柱，奥杜邦的两个儿子都在博物学和出版界事业有成，也离不开她的教诲。不仅如此，奥杜邦在美国遭遇同行排挤，作品难以出版时，也是露西鼓励他去英国尝试，才得以成就其一世英名。1851年奥杜邦病逝后，也正是露西带领两个儿子，才令《北美四足动物志》得以问世。

1874年，露西·贝克维尔·奥杜邦去世。

THE

QUADRUPEDS

OF

NORTH AMERICA

BY

JOHN JAMES AUDUBON, F R. S., &c. &c.

AND

THE REV. JOHN BACHMAN, D. D., &c. &c.

VOL. I.

NEW-YORK:
PUBLISHED BY V. G. AUDUBON.
1851.

J. LUDWIG & CO., PRINTERS,
70 VESEY-STREET, N. Y.

INTRODUCTION.

In presenting the following pages to the public, the authors desire to say a few words explanatory of the subject on which they have written. The difficulties they have attempted to surmount, and the labour attending their investigations, have far exceeded their first anticipations.

Many of the "Quadrupeds of North America" were long since described by European authors, from stuffed specimens; and in every department of Natural History additions to the knowledge of the old writers have been making for years past; researches and investigations having been undertaken by scientific observers in all parts of the world, and many specimens accumulated in the Museums of Europe. Comparatively little, however, has of late been accomplished toward the proper elucidation of the animals which inhabit the fields, forests, fertile prairies, and mountainous regions of our widely-extended and diversified country.

The works of Harlan and of Godman were confined to the limited number of species known in their day. The valuable "Fauna Boreali Americana" of Richardson was principally devoted to the description of species which exist in the British Provinces, north of the United States; and the more recent work of Dr. Dekay professes to describe only the Quadrupeds of the State of New-York, although giving a catalogue of those noticed by authors as existing in other portions of North America.

Several American and European Zoologists have, however, at different times, given the results of their investigations in various scientific journals, thus making it important for us to examine numberless papers, published in different cities of Europe and America. We have, in all cases, sought to discover and give due credit to every one who has in this manner made known a new species; but as possibly some author may have published discoveries in a journal we have not seen, we must at once announce our conviction, that the task of procuring and reading all the zoological papers scattered through the pages of hundreds of periodicals, in many different languages, is beyond our power, and that no one can reasonably complain when we take the liberty of pronouncing for ourselves on new or doubtful species without hesitation, from the sources of knowledge to which we have access, and from our own judgment.

The geographical range which we have selected for our investigations is very extensive, comprising the British and Russian possessions in America, the whole of the United States and their territories, California, and that part of Mexico north of the tropic of Cancer; we having arrived at the conclusion, that in undertaking the natural history of a country, our researches should not be confined by the artificial boundaries of States—which may be frequently changed—but by those divisions the limits of which are fixed by nature, and where new *forms* mark the effects of a low latitude and warm climate. In this way America is divided into three parts:—North America, which includes all that country lying north of the tropics; Central or Tropical America, the countries within the tropics; and South America, all that country south of the tropic of Capricorn.

Within the tropical region peculiar forms are presented in every department of nature,—we need only instance the Monkey tribe among the animals, the Parrots among the birds, and the Palms among the plants.

A considerable portion of the country to which our attention has been directed, is at the present period an uncultivated and almost unexplored wild, roamed over by ferocious beasts and warlike tribes of Indians.

The objects of our search, Quadrupeds, are far less numerous than birds at all times, and are, moreover, generally nocturnal in their habits, and consequently obtained with far greater difficulty than the latter.

Although the *Genera* may be easily ascertained, by the forms and dental arrangements peculiar to each, many *species* so nearly approach each other in size, while they are so variable in colour, that it is exceedingly difficult to separate them, especially closely allied squirrels, hares, mice, shrews, &c., with positive certainty.

We have had our labours lightened, however, by many excellent friends and gentlemen in different portions of the country, who have, at great trouble to themselves, procured and sent us various animals—forwarded to us notes upon the habits of different species, procured works on the subject otherwise beyond our reach, and in many ways excited our warmest feelings of gratitude. Mr. J. K. TOWNSEND, of Philadelphia, allowed us to use the rare and valuable collection of Quadrupeds which he obtained during his laborious researches on the western prairies, the Rocky Mountains, and in Oregon, and furnished us with his notes on their habits and geographical distribution. SPENCER F. BAIRD, Esq., of Carlisle, Pennsylvania, aided us by carefully searching various libraries for notes and information in regard to species published in different journals, and also by obtaining animals from the wilder portions of his State, &c.; Dr. BARRITT, of Abbeville, S. C., prepared and mounted specimens of *Lepus aquaticus*, and several other species; Dr. THOMAS M. BREWER, of Boston, favoured us with specimens of a new species of shrew-mole (*Scalops Breweri*), and sundry arvicolæ; EDMUND RUFFIN, Esq., of Virginia, sent us several specimens of the rodentia inhabiting that State, and obliged us by communicating much information in regard to their geographical range; the late Dr. JOHN WRIGHT, of Troy, N. Y., furnished us valuable notes on the various species of quadrupeds found in the northern part of the State of New-York, and several specimens; Dr. WURDEMAN, of Charleston, supplied us with several specimens from Cuba, thereby enabling us to compare them with genera and species existing in America. To Professor LEWIS R. GIBBES, of the College of Charleston, we express our thanks, for several specimens of rare quadrupeds, and for his kindness in imparting to us much information and scientific knowledge.

Among others to whose zeal and friendship we are most indebted, we are proud to name: Dr. GEO. C. SHATTUCK and Dr. GEO. PARKMAN, of

Boston; J. Prescott Hall, Esq., James G. King, Esq., Major John Leconte, Mr. J. G. Bell, and Issachar Cozzens, of New-York; Hon. Daniel Wadsworth, of Hartford; W. O. Ayres, Esq., of Sag Harbour, Long Island; Edward Harris, Esq., of Moorestown, New Jersey; Dr. Samuel George Morton and Samuel Bispham, Esq., of Philadelphia; Wm. Case, Esq., Cleveland, Ohio; Ogden Hammond, Esq., of South Carolina; Gideon B. Smith, Esq., M.D., of Baltimore; Messrs. P. Chouteau, Jr. & Co., St. Louis; Sir George Simpson, of the Hudson's Bay Fur Company; John Martyn, Jr., Quebec; Mr. Fothergill, of Canada, &c., &c., &c.

For the sake of convenience and uniformity we have written in the plural number, although the facts stated, and the information collected, were obtained at different times by the authors in their individual capacities.

Without entering into details of the labours of each in this undertaking, it will be sufficient to add, that the history of the habits of our quadrupeds was obtained by both authors, either from personal observation or through the kindness of friends of science, on whose statements full reliance could be placed.

For the designation of species, and the letter-press of the present volume, the junior author is principally responsible.

In our Illustrations we have endeavoured (we hope not without success) to place before the public a series of plates, which are not only scientifically correct, but interesting to all, from the varied occupations, expressions, and attitudes, we have given to the different species, together with the appropriate accessories, such as trees, plants, landscapes, &c., with which the figures of the animals are relieved; and we have sought to describe those represented, so as not only to clear away the obscurity which had gathered over some species, but to make our readers acquainted with their habits, geographical distribution, and all that we could ascertain of interest about them, and the mode of hunting or destroying such as are pursued either to gratify the appetite, to furnish a rich fur or skin, or in order to get rid of dangerous or annoying neighbours.

QUADRUPEDS OF NORTH AMERICA.

GENUS LYNX.

DENTAL FORMULA.

$$\textit{Incisive}\ \frac{6}{6};\ \textit{Canine}\ \frac{1-1}{1-1};\ \textit{Cheek-Teeth}\ \frac{3-3}{3-3} = 28.$$

The teeth in animals of this genus, with the exception of there being one less on each side, in the upper jaw, do not differ from the dental arrangement of the genus Felis. The canine are very strong, there are but three molars on each side, above: The small false-molar, next to the canine, which exists in the larger species of *long-tailed* cats, such as the lion, tiger, panther, cougar, &c., as well as in the domestic or common cat, is wanting in the lynxes. There is one false-molar, or conical tooth on each side—one carnivorous, with three lobes and a tubercle or blunted heel, on the inner. The third cheek-tooth is rather small, and is placed transversely. In the lower jaw there are on each side, two false, compressed, simple molars, and one canine, which is bicusped.

The head is short, round, and arched; jaws short; tongue aculeated; ears short, erect, more or less tufted.

Fore-feet with five toes, hind-feet with only four; nails retractile. Tail shorter than the head, although nearly as long, in a few instances.

The species heretofore classed in the genus Felis have been so multiplied by the discoveries of late years in various parts of the world, that they have for some time demanded a careful examination, and the separation of such as present characters essentially different from the types of that genus, into other genera.

Some of the distinctive marks by which the genus Lynx is separated from the old genus FELIS, are the tufted ears and shorter bodies and tails of the lynxes, as well as the slight difference above mentioned in the dental arrangement of the two genera. In a note in the American Monthly Magazine, vol. i., p. 437, RAFINESQUE, in a few lines, proposed the genus LYNX, but gave no detailed characters, although he states that he had increased the species of this genus from four to fifteen ! in which supposition, alas, he was sadly mistaken.

Dr. DEKAY, in the "Natural History of New-York," a work published "By Authority" of the State, has adopted the genus LYNCUS, as established by GRAY.

We have not seen the work in which Mr. GRAY proposed this generic name, and are consequently unable to ascertain on what characters it was founded, and we prefer the more classical name of LYNX. The name Lynx was formerly applied to one of the species of this genus. It is derived from the Greek work λυγξ (*lugx*), a Lynx. Eight species of Lynx have been described ; one being found in Africa, two in Persia, one in Arabia, two in Europe, and two in North America.

LYNX RUFUS.—GULDENSTAED.

COMMON AMERICAN WILD CAT.—BAY LYNX.

PLATE I.—MALE.

L. Cauda capite paullo breviore, ad extremum supra nigra, apice subalbida ; auribus pagina posteriore maculo sub albido nigro marginato distinctis ; hyeme et auctumno rufo-fuscus ; vere et æstate cinereo-fuscus.

CHARACTERS.

Tail nearly as long as the head, extremity on the upper surface black, tipped with more or less white ; a whitish spot on the hinder part of the ear bordered with black ; general colour reddish-brown in autumn and winter, ashy brown in spring and summer ; soles naked.

SYNONYMES.

BAY LYNX, Pennant, Hist. Quadr., No. 171. Arctic Zool., vol. 1., p. 51.
FELIS RUFA, Guld. in Nov. Comm. Petross. xx., p. 499.
FELIS RUFA, Temm., Monog., &c., vol. 1., p. 141.
LYNX FASCIATUS, Rafin. in Amer. Month. Mag., 1817, p. 46.
LYNX MONTANUS, Idem, Ibid., pp. 46, 2.

Common American Wild-cat.

Male.

Lynx Floridanus, Idem, Ibid., pp. 4, 64.
Lynx Aureus, Idem, Ibid., pp. 46, 6.
Felis Carolinensis, Desm., Mamm., p. 231.
Felis Rufa, Godm., Amer. Nat. Hist., vol. iii., p. 239; Fig. in vol. I.

DESCRIPTION.

In size and form, this species bears some resemblance to small specimens of the female Canada Lynx, (*Lynx Canadensis*,) the larger feet and more tufted ears of the latter, however, as well as its grayer colour, will enable even an unpractised observer at a glance to distinguish the difference between the two species.

Head of moderate size, rounded; body rather slender; legs long; soles of feet naked; hind-feet webbed to within five-eighths of an inch of the claws; ears large, nearly triangular, erect, tipped with coarse hairs half an inch long, which drop out in summer; the inner surface thinly sprinkled with loose hairs, outer, thickly covered with short fur.

A ruff of elongated hairs surrounding the throat, more prominent in the male than female; tail short, slender, and slightly turned upwards. mammæ eight; four pectoral and four abdominal.

COLOUR.

The hind-head and back, yellowish-brown, with a dorsal line more or less distinct, of dark-brown, running from the shoulder to near the insertion of the tail. A few irregular longitudinal stripes on the back, of the same colour. The sides spotted with dark-brown, these spots being more distinct and in closer approximation in some specimens than in others.

Forehead obscurely striped with dark-brown. Over and beneath the eyes yellowish-white; whiskers nearly all white. Ears, outer surface, a triangular spot of dull white, dilated towards the outer margin, bordered with brownish-black; inner surface yellowish white. Under surface of body yellowish white, spotted with black; tail, above, barred with rufous and black, towards the extremity a broad band of black, tipped at the point and particularly in the centre with white; under surface of tail, light-gray, interspersed with small and irregular patches of black hairs.

Fore-feet on the upper surface, broadly, and towards the toes minutely, spotted with black on a light yellowish-brown ground; inner surface dull white, with two broad and several narrow bars of black; paws beneath, and hair between the soles, dark-brown. Hind-legs barred and spotted similarly to the fore-legs. Chin and throat dull white, with two black lines, commencing on a point on a line with the articulation of the lower jaw, where they form an acute angle, and thence diverge to the sides of the neck, and unite with the ruff, which is black, mixed with yellowish-brown and gray hairs.

The female is considerably smaller than the male, her body more slender, and her movements have a stronger resemblance, in their lightness and agility, to those of the common house-cat; the markings appear more distinct, and the rounded black spots on the back and sides, smaller and more numerous. There is in this species a considerable diversity in colour, as well as in size. In spring and early summer, before it has shed its winter coat, it is uniformly more rufous, and the black markings are less distinct, than after shedding its hair, and before the new hair is elongated in autumn to form the winter coat.

Our specimens obtained in summer and autumn, are of a light gray colour, with scarcely any mixture of rufous, and all the black markings are brighter and far more distinct than they are in those killed in the winter or spring months.

There are, however, at all seasons of the year, even in the same neighbourhood, strongly-marked varieties, and it is difficult to find two individuals precisely alike.

Some specimens are broadly marked with fulvous under the throat, whilst in others the throat as well as the chin are gray. In some the stripes on the back and spots along the sides are very distinctly seen, whilst in others they are scarcely visible, and the animal is grayish-brown above with a dark dorsal stripe. A specimen from the mountains of Pennsylvania presents this appearance strikingly, and is withal nearly destitute of the triangular marking under the throat, so that we hesitated for some time in referring it to this species. A specimen from Louisiana is of the same uniform colour above, but with more distinct linear markings on the face, and with coarse hair, not more than half the length of that of individuals from the Northern States. We obtained a specimen in Carolina, which in nearly every particular answers to the description of *Felis Carolinensis* of DESMAREST. If the various supposed new species of Wild Cat described by RAFINESQUE, HARLAN, DESMAREST, &c., are entitled to a place in our Fauna, on account of some peculiarity of colour, we have it in our power, from specimens before us, to increase the number to a considerable extent; but in doing so we think we should only swell the list of synonymes, and add to the confusion which already prevails in regard to some of the species belonging to this genus.

DIMENSIONS.

Adult Male.—[Fine Specimen.]

From point of nose to root of tail - - - -	30 inches.
Tail (vertebræ) - - - - - - - -	5 do.

Tail, to end of hair - - - - - - -	$5\frac{1}{2}$ inches.
From nose to end of skull - - - - - -	$4\frac{1}{2}$ do.
From nose, following the curvature of the head -	6 do.
Tufts on the ears - - - - - - -	$\frac{1}{2}$ do.
Breadth of ear - - - - - - - - -	$1\frac{5}{8}$ do.
Anterior length of ear - - - - - -	$1\frac{3}{4}$ do.
Length of neck - - - - - - -	4 do.

Weight 17lbs.

HABITS.

The general appearance of this species conveys the idea of a degree of ferocity, which cannot with propriety be considered as belonging to its character, although it will, when at bay, show its sharp teeth, and with outstretched claws and infuriated despair, repel the attacks of either man or dog, sputtering the while, and rolling its eyes like the common cat.

It is, however, generally cowardly when attacked, and always flies from its pursuers, if it can; and although some anecdotes have been related to us of the strength, daring, and fierceness of this animal, such as its having been known to kill at different times a sheep, a full-grown doe, attack a child in the woods, &c.; yet in all the instances that have come under our own notice, we have found it very timid, and always rather inclined to beat a retreat, than to make an attack on any animal larger than a hare or a young pig. In the American Turf Register, there is an interesting extract of a letter from Dr. Coleman, U. S. A., written at Fort Armstrong, Prairie du Chien, giving an account of a contest between an eagle and a Wild Cat. After a fierce struggle, in which the eagle was so badly wounded as to be unable to fly, the Cat, scratched and pierced in many places, and having had one eye entirely "gouged out" in the combat, was found lying dead.

In hunting at night for racoons and opossums, in which sport the negroes on the plantations of Carolina take great delight, a Cat is occasionally "treed" by the dogs; and the negroes, who seldom carry a gun, climb up the tree and shake him off as they would do a racoon, and although he fights desperately, he is generally killed by the dogs. During a botanical excursion through the swamps of the Edisto river, our attention was attracted by the barking of a small terrier at the foot of a sapling, (young tree.) On looking up, we observed a Wild Cat, about twenty feet from the ground, of at least three times the size of the dog, which he did not appear to be much afraid of. He seemed to have a greater dread of man, however, than of this diminutive specimen of the canine race, and leaped from the tree as we drew near.

The Wild Cat pursues his prey with both activity and cunning, sometimes bounding suddenly upon the object of his rapacity, sometimes with stealthy pace, approaching it in the darkness of night, seizing it with his strong retractile claws and sharp teeth, and bearing it off to his retreat in the forest.

The individual from which our figure was drawn had been caught in a steel-trap, and was brought to us alive. We kept it for several weeks; it was a fine male, although not the largest we have seen. Like most of the predacious animals, it grew fat in confinement, being regularly fed on the refuse parts of chickens and raw meat, as well as on the common brown rat.

The Bay Lynx (as this animal is sometimes called) is fond of swampy, retired situations, as well as the wooded sides of hills, and is still seen occasionally in that portion of the Alleghany mountains which traverses the States of Pennsylvania and New-York. It is abundant in the *Cane-brakes* (patches or thickets of the *Miegia Macrosperma*, of Michaux, which often extend for miles, and are almost impassable) bordering the lakes, rivers, and lagoons of Carolina, Louisiana, and other Southern and South Western States. This species also inhabits the mountains and the undulating or *rolling* country of the Southern States, and frequents the thickets that generally spring up on deserted cotton plantations, some of which are two or three miles long, and perhaps a mile wide, and afford, from the quantity of briars, shrubs, and young trees of various kinds which have overgrown them, excellent cover for many quadrupeds and birds. In these bramble-covered old fields, the " Cats " feed chiefly on the rabbits and rats that make their homes in their almost impenetrable and tangled recesses; and seldom does the cautious Wild Cat voluntarily leave so comfortable and secure a lurking place, except in the breeding season, or to follow in very sultry weather, the dry beds of streams or brooks, to pick up the cat-fish, &c., or cray-fish and frogs that remain in the deep holes of the creeks, during the drought of summer.

The Wild Cat not only makes great havoc among the chickens, turkeys, and ducks of the planter, but destroys many of the smaller quadrupeds, as well as partridges, and such other birds as he can surprise roosting on the ground. The hunters often run down the Wild Cat with packs of fox-hounds. When hard pressed by fast dogs, and in an open country, he ascends a tree with the agility of a squirrel, but the baying of the dogs calling his pursuers to the spot, the unerring rifle brings him to the ground, when, if not mortally wounded, he fights fiercely with the pack until killed. He will, however, when pursued by hunters with hounds, frequently elude both dogs and huntsmen, by an exercise of instinct, so closely bordering on reason, that we are bewildered in the attempt to separate it from the latter. No sooner does he become aware that the enemy is on his track, than, instead of taking a straight course for the deepest forest, he speeds to one of the largest old-fields overgrown with briery thickets, in the neighbourhood; and having reached this tangled maze, he runs in a variety of circles, crossing and re-crossing his path many times, and when he thinks the scent has been diffused sufficiently in different directions by this manœuvre, to puzzle both men and dogs, he creeps slyly forth, and makes for the woods, or for some well known swamp, and if he should be lucky enough to find a half-dried-up pond, or a part of the swamp, on which the clayey bottom is moist and sticky, he seems to know that the adhesive soil, covering his feet and legs, so far destroys the *scent*, that although the hounds may be in full cry on reaching such a place, and while crossing it, they will lose the track on the opposite side, and perhaps not regain it without some difficulty and delay.

At other times the "Cat," when chased by the dogs, gains some tract of "burnt wood," common especially in the pine lands of Carolina, where fallen and upright trees are alike blackened and scorched, by the fire that has run among them burning before it every blade of grass, every leaf and shrub, and destroying many of the largest trees in its furious course; and here, the charcoal and ashes on the ground, after he has traversed the burnt district a short distance, and made a few leaps along the trunk of a fallen tree, that has been charred in the conflagration, will generally put any hounds at fault. Should no such chance of safety be within his reach, he does not despair, but exerting his powers of flight to the utmost, increases his distance from the pursuing pack, and following as intricate and devious a path as possible, after many a weary mile has been run over, he reaches a long-fallen trunk of a tree, on which he may perchance at some previous time have baffled the hunters as he is now about to do. He leaps on to it, and hastily running to the farther end, doubles and returns to the point from which he gained the tree, and after running backward and forward repeatedly on the fallen trunk, he makes a sudden and vigorous spring, leaping as high up into a tree some feet distant, as he can; he then climbs to its highest forks, (branches,) and closely squatted, watches the movements of his pursuers. The dogs are soon at fault, for he has already led them through many a crooked path; the hunters are dispirited and weary, and perhaps the density of the woods, or the approach of night, favours him. The huntsmen call off their dogs from the fruitless search, and give up the chase; and shortly afterwards the escaped marauder descends leisurely to the earth, and wanders off in search of food, and to begin a new series of adventures.

In some parts of Carolina, Georgia, Mississippi, and Louisiana, the Wild Cat has at times become so great a nuisance as to have aroused the spirit of vengeance in the hearts of the planters, who are constant sufferers from his depredations. They have learned by experience, that one Cat will do as much mischief among the pigs and poultry as a dozen gray foxes. They are now determined to allow their hounds, which they had hitherto kept solely for the favourite amusement of deer hunting, and which had always been whipped-in from the trail of the Wild Cat, to pursue him, through thicket, briar patch, marsh, and morass, until he is caught or killed.

Arrangements for the Cat-hunt are made over night. Two oı three neighbours form the party, each one bringing with him all the hounds he can muster. We have seen thirty of the latter brought together on such occasions, some of which were not inferior to the best we have examined in England, indeed, great numbers of the finest fox-hounds are annually imported into Carolina.

At the earliest dawn, the party is summoned to the spot previously fixed on as the place of meeting. A horn is sounded, not low and with a single blast, as is usual in hunting the deer, lest the timid animal should be startled from its bed among the broom-grass (*Andropagon dissitiflorus*) and bound away out of the drive, beyond the reach of the hunter's double-barrel loaded with buckshot; but with a loud, long, and oft-repeated blast, wakening the echoes that rise from the rice-fields and marshes, and are reverberated from shore to shore of the winding sluggish river, until lost among the fogs and shadows of the distant forest.

An answering horn is heard half a mile off, and anon comes another response from a different quarter. The party is soon collected, they are mounted, not on the fleetest and best-blooded horses, but on the most sure-footed, (sometimes called "Old field Tackies,") which know how to avoid the stump-holes on the burnt grounds of the pine lands, which stand the fire of the gun, and which can not only go with tolerable speed, but are, to use a common expression, "tough as a pine knot." The hunters greet each other in the open-hearted manner characteristic of the Southern planter. Each pack of dogs is under the guidance of a coloured driver, whose business it is to control the hounds and encourage and aid them in the hunt. The drivers ride in most cases the fleetest horses on the ground, in order to be able, whilst on a deer hunt, to stop the dogs. These men, who are so important to the success of the chase, are possessed of a good deal of intelligence and shrewdness, are usually much petted, and regarding themselves as belonging to the aristocracy of the plantation, are apt to look down upon their fellow-servants as inferiors, and consider themselves privileged even to crack a joke with their masters. The drivers are ordered to stop the dogs if a deer should be started, a circumstance which often occurs, and which has saved the life of many a Cat, whose fate five minutes before this unlucky occurrence was believed to be sealed. Orders are given to destroy the Cat fairly, by running him down with the hounds, or if this cannot be done, then by shooting him if he ascends a tree or approaches within gun shot of the stand which the hunter has selected as the most likely place for him to pass near. The day is most auspicious—there is not a breath of wind to rustle the falling leaves, nor a cloud to throw its shadows over the wide joyous landscape. The dew-drops are sparkling on the few remaining leaves of the persimmon tree, and the asters and dog-fennel hang drooping beneath their load of moisture. The dogs are gambolling in circles around, and ever and anon, in spite of all restraint, the joyous note breaks forth—the whole pack is impatient for the chase, and the young dogs are almost frantic with excitement.

But we have not time for a farther description of the scene—whilst we are musing and gazing, the word is given, "Go!" and off start the hounds, each pack following its own driver to different parts of the old fields, or along the borders of the swamps and marshes. Much time, labour and patience are usually required, before the "Cat" can be found by the dogs: sometimes there is a sudden burst from one or the other of the packs, awakening expectation in the minds of the huntsmen, but the driver is not to be so easily deceived, as he has some dogs that never open at a rabbit, and the snap of the whip soon silences the riotous young babblers. Again there is a wild burst and an exulting shout, giving assurance that better game than a rabbit is on foot; and now is heard a distant shot, succeeded in a second of time by another, and for an instant all is still: the echoes come roaring up through the woods, and as they gradually subside, the crack of the whip is again heard stopping the dogs. The story is soon told: a deer had been started—the shot was too small—or the distance too great, or any other excuses (which are always at hand among hunters of fertile imagination) are made by the unsuccessful sportsman who fired, and the dogs are carried back to the "trail" of the Cat, that has been growing fresher and fresher for the last half hour. At length, "Trimbush," (and a good dog is he,) that has been working on the cold trail for some time, begins to give tongue, in a way that brings the other dogs to his aid. The drivers now advance to each other, encouraging their dogs; the trail becomes a drag; onward it goes through a broad marsh at the head of a rice-field. "He will soon be started now!" "He is up!" What a burst! you might have heard it two miles off—it comes in mingled sounds, roaring like thunder, from the muddy marsh and from the deep swamp. The barred owl, frightened from the monotony of his quiet life among the cypress trees, commences hooting in mockery as it were, of the wide-mouthed hounds. Here they come, sweeping through the resounding swamp like an equinoctial storm—the crackling of a reed, the shaking of a bush, a glimpse of some object that glided past like a shadow, is succeeded by the whole pack, rattling away among the vines and fallen timbers, and leaving a trail in the mud as if a pack of wolves in pursuit of a deer had hurried by. The Cat has gone past. It is now evident that he will not climb a tree. It is almost invariably the case, that where he can retreat to low swampy situations, or briar patches, he will not take a tree, but seeks to weary the dogs by making short windings among the almost impassable briar patches. He has now been twisting and turning half a dozen times in a thicket covering only three or four acres—let us go in and take our stand on the very trail where he last passed, and shoot him if we can. A shot is heard on the opposite edge of the thicket, and again all is still; but once more the pack is in full cry. Here he comes, almost brushing our legs as he dashes by and disappears in the bushes, before we can get sight of him and pull trigger. But we see that the dogs are every moment pressing him closer, that the marauder is showing evidences of fatigue and is nearly "done up." He begins to make narrower circles, there are restless flashes in his eye, his back is now curved upwards, his hair is bristled nervously forward, his tongue hangs out—we raise our gun as he is approaching, and scarcely ten yards off—a loud report—the smoke has hardly blown aside, ere we see him lifeless, almost at our very feet—had we waited three minutes longer, the hounds would have saved us the powder and shot!

One fine morning in autumn, when we had crossed the Ohio river at Henderson, in Kentucky, with the view of shooting some wild turkeys, geese, and perhaps a deer, we chanced to seat ourselves about fifty yards from a prostrate tree, and presently saw a Wild Cat leap on to it and go through the manœuvres we have described in a preceding page. He did not see us, and had scarcely reached one of the higher branches of a tall white-oak, after springing into it from the fallen tree, when we heard the dogs, which soon came up, with the hunters following not far behind. They asked, when they perceived us, whether we had seen the "Cat" that had given them the slip. Always willing to assist the hunter who has lost his game, and having no particular liking towards this species, we answered in the affirmative, and showed them the animal, closely squatted on a large branch some distance from the ground. One of the party immediately put his rifle to his shoulder and pulled the trigger: the Cat leaped from the branch into the air, and fell to the earth quite dead. Whilst residing in Louisiana some twenty years since, we chanced one afternoon to surprise one of these depredators. He had secured a hare, (commonly called rabbit,) and was so eagerly engaged in satisfying his hunger as not to observe us, until we were near the spot where he was partially concealed behind a rotten log. At sight of us, he squatted flat on the ground. As we looked at him, we heard a squirrel close by, and turned our head for an instant, but scarce had we glanced at the squirrel, when looking again for the Wild-Cat, he had disappeared, carrying the remains of the hare away with him.

About twenty miles from Charleston, South-Carolina, resides a worthy friend of ours, a gentleman well known for his skill in the sports of the field, his hospitality to both friends and strangers, and the excellent manner in which his plantation is managed. The plantation of Dr. Desel is, in short, the very place for one who likes the sight of several fine bucks hanging on the branches of an old Pecan-nut tree; while turkeys, geese, and poultry of other kinds, are seen in abundance in his well stocked poultry yards, affording certainty of good cheer to his visitors.

The Doctor's geese were nightly lodged near the house, in an enclosure which was rendered apparently safe, by a very high fence. As an additional security, several watch dogs were let loose about the premises; besides an excellent pack of hounds, which by an occasional bark or howl during the night, sounded a note of warning or alarm in case any marauder, whether biped or quadruped, approached.

Notwithstanding these precautions, a goose disappeared almost every night, and no trace of the ingress or egress of the robber could be discovered. Slow in attaching suspicion to his servants, the Dr. waited for time and watchfulness to solve the mystery. At length, the feathers, and other remains of his geese, were discovered in a marsh about a quarter of a mile from the house, and strong suspicions were fastened on the Wild-Cat; still, as he came at odd hours of the night, all attempts to catch or shoot him proved for a time unavailing.

One morning, however, he came about day-light, and having captured a good fat goose, was traced by the keen noses of the hounds. The chase was kept up for some time through the devious windings of the thickets, when his career of mischief was brought to a close by a shot from the gun of our friend the Doctor, who, in self-defence, became his executioner. Thus ended his career. In this respect he fared worse than he deserved, compared with those beings of a superior nature, who, not understanding that "*Honesty is the best policy*," outdo our Wild-Cat in his destructive habits, until the laws, so just and useful, when mildly, but always, enforced, put an effectual stop to their criminal proceedings.

The Wild-Cat is a great destroyer of eggs, and never finds a nest of grouse or partridge, wild turkey or other bird, without sucking every egg in it. Indeed, it will, if practicable, seize on both young and old birds of these and other species. Its "*penchant*" for a "*poulet au naturel*" has suggested the following method of capturing it in Georgia, as related to us by our friend Major Leconte, late of the United States Army.

A large and strong box-trap is constructed, and a chicken-cock (rooster), placed at the farthest end of it from the door, is tied by one leg, so that he cannot move. There is a stout wire partition about half way between the fowl and the door, which prevents the Cat when entering the trap, from seizing the bird. The trap is then set, so that when the animal enters, the open door closes behind him by a spring, (commonly the branch of some tree bent down for the purpose, and released by a trigger set at the entrance or just within the trap.) These traps are placed in different parts of the plantations, or in the woods, and the Wild-Cat is generally attracted by the crowing of the cock at early dawn of day.

Major Leconte has caught many of them by this artifice, on and about his plantations in the neighbourhood of Savannah, in Georgia; and this method of capturing the Wild-Cat is also quite common in South Carolina. Indeed, this species does not seem to possess the suspicion and cunning inherent in the fox, enabling the latter to avoid a trap of almost any kind. We have seen the Wild-Cat taken from the common log-traps set for racoons. We saw one in a cage, that had been caught in a common box-trap, baited with a dead partridge, and have heard intelligent domestics residing on the banks of the Santee river, state, that after setting their steel traps for otters, they frequently found the Wild-Cat caught in them instead.

When this animal discovers a flock of wild turkeys, he will generally follow them at a little distance for some time, and after having ascertained the direction in which they are proceeding, make a rapid detour, and concealing himself behind a fallen tree, or in the lower branches of some leafy maple, patiently wait in ambush until the birds approach, when he suddenly springs on one of them, if near enough, and with one bound secures it. We once, while resting on a log in the woods, on the banks of the Wabash river, perceived two wild turkey cocks at some distance below us, under the bank near the water, pluming and picking their feathers; on a sudden, one of them flew across the river, and the other we saw struggling in the grasp of a Wild-Cat, which almost instantly dragged it up the bank into the woods, and made off. On another occasion we observed an individual of this species, about nine miles from Charleston, in pursuit of a covey of partridges, (*Ortyx Virginiana*,)—so intent was the Cat upon its prey, that it passed within ten steps of us, as it was making a circle to get in advance and in the path of the birds,—its eyes were constantly fixed on the covey, and it stealthily concealed itself behind a log it expected the birds to pass. In a second attempt the marauder succeeded in capturing one of the partridges, when the rest in great affright flew and scattered in all directions.

An individual that was kept alive at Charleston, and afterwards for a short time at our house, in the city of New-York, showed its affinity to the domestic cat, by purring and mewing at times loud enough to be heard at some distance. At the former place its cry was several times mistaken for that of the common house-cat. In the woods, during the winter season, its loud catterwauling can be heard at the distance of a mile.

Although this species may perhaps be designated as nocturnal in its habits, it is, by no means, exclusively so, as is shown by the foregoing account. We have, in fact, in several instances, seen this Cat engaged in some predatory expedition in full sunshine, both in winter and summer.

It is not a very active swimmer, but is not averse to taking the water. We witnessed it on one occasion crossing the Santee river when not pursued, and at another time saw one swimming across some ponds to make its escape from the dogs. It has been observed, however, that when it has taken to the water during a hard chase, it soon after either ascends a tree or is caught by the hounds.

The domicile of the Wild-Cat is sometimes under an old log, covered with vines such as the *Smilax*, *Ziziphus volubilus*, *Rubus*, &c., but more commonly in a hollow tree. Sometimes it is found in an opening twenty or thirty feet high, but generally much nearer the ground, frequently in a cavity at the root, and sometimes in the hollow trunk of a fallen tree, where, after collecting a considerable quantity of long moss and dried leaves to make a comfortable lair, it produces from two to four young. These are brought forth in the latter end of March in Carolina; in the Northern States, however, the kittens appear later, as we have heard of an instance in Pennsylvania where two young were found on the 15th day of May, apparently not a week old. Our friend Dr. SAMUEL WILSON, of Charleston, a close observer of nature, has made the following note in our memorandum book: "April 15th, 1839, shot a female Wild-Cat as it started from its bed, out of which four young ones were taken; their eyes were not yet open." Our friend Dr. DESEL, whom we have already mentioned, saw three young ones taken out from the hollow of a tree which was thirty feet from the ground. On four occasions, we have had opportunities of counting the young, either in the nest or having been very recently taken from it. In every case there were three young ones. In one instance the nest was composed of long moss, (*Tillandsia usneoides,*) which seemed to have been part of an old, deserted, squirrel's nest.

We once made an attempt at domesticating one of the young of this species, which we obtained when only two weeks old. It was a most spiteful, growling, snappish little wretch, and showed no disposition to improve its habits and manners under our kind tuition. We placed it in a wooden box, from which it was constantly striving to gnaw its way out. It, one night, escaped into our library, where it made sad work among the books, (which gave us some valuable lessons on the philosophy of patience, we could not have so readily found among our folios,) and left the marks of its teeth on the mutilated window-sashes. Finally, we fastened it with a light chain, and had a small kennel built for it in the yard. Here it was constantly indulging its carnivorous propensities, and catching the young poultry, which it enticed within reach of its chain by leaving a portion of its food at the door of its house, into which it retreated until an opportunity offered to pounce on its unsuspecting prey. Thus it continued, growing, if possible, more wild and vicious every day, growling and spitting at every servant that approached it, until at last, an unlucky blow, as a punishment for its mischievous tricks, put an end to its life, and with it to one source of annoyance.

The Bay Lynx is generally in fine order, and often very fat. The meat is white, and has somewhat the appearance of veal. Although we omitted to taste it, we have seen it cooked, when it appeared savoury, and the persons who partook of it pronounced it delicious.

The muscular powers of this species are very great, and the fore-feet and legs are rather large in proportion to the body.

GEOGRAPHICAL DISTRIBUTION.

The geographical range of the Bay Lynx is very extensive, it being found to inhabit portions of the Continent from the tropics as far north as 60°. It abounds in Texas, Louisiana, Florida, Georgia, and both the Carolinas, and is found in all the States east of these, and likewise in New Brunswick, and Nova Scotia. We have seen it on the shores of the Upper Missouri more than a thousand miles above St. Louis. We examined one that had been taken a few hours before, by some hunters in Erie county, in the State of New-York, and have heard of its existing, although rather sparingly, in Upper Canada, where it has been occasionally captured.

GENERAL REMARKS.

We are not so fortunate as to possess any specimen from Oregon, or the regions west of the Rocky Mountains, to enable us to institute a close comparison, and therefore cannot be certain that the Cat described by LEWIS and CLARK, to which naturalists, without having seen it, have attached the name of *Felis fasciata*, or that the individual described by Dr. RICHARDSON, and referred by him to *Felis rufa*, are identical with the present species; yet as they do not present greater marks of difference than those observable in many other varieties of it, and as we have carefully examined several hundred specimens in the museums and private collections of Europe and America, and have, at this moment, upwards of twenty lying before us, that were obtained in various parts of the country, from Texas to Canada, our present conclusion is, that in the United States, *east and north of the Mississippi*, there are but two species of Lynx—the well known Canada Lynx, and the Bay Lynx—our present species, and that the varieties in colour, (especially in the latter animal,) have contributed to the formation of many imaginary species. Whatever may be the varieties, however, there are some markings in this species which are permanent, like the white ears and nose of the fox squirrel, (*Sc. Capistratus*,) and which serve to identify it through all the variations of sex, season, and latitude. All of them have naked soles, and the peculiar markings at the extremity of the slender tail, which terminates as abruptly as if it had been amputated. It may also be distinguished from any variety of the Canada Lynx, (*L. Canadensis*,) by a white patch behind the ear, which does not exist in the latter.

This peculiar mark is to be observed, however, in several species of the genus FELIS. We have noticed it in the jaguar, royal tiger, panther, ocelot, hunting-leopard, and other species.

GENUS ARCTOMYS, *Gmel., Cuv.*

DENTAL FORMULA.

$$\textit{Incisive } \frac{2}{2}; \quad \textit{Canine } \frac{0\text{—}0}{0\text{—}0}; \quad \textit{Molar } \frac{5\text{—}5}{4\text{—}4} = 22.$$

Incisors strong, narrow, and wedge-shaped, anterior surface rounded; molars, with the upper surface thick and heavy.

Head large, mouth small, and placed below; eyes large, ears short, paws strong; fore-feet with four toes and the rudiment of a thumb; hind-feet with five toes; nails strong, compressed; tail bushy; no cheek pouches.

The name *Arctomys*, is derived from two Greek words: αρκτος, (*arktos*,) a bear, and μυς, (*mus*,) a mouse.

There are, as far as we are informed, but eight known species of the genus as it is now defined, five on the Eastern Continent and three in North America.

ARCTOMYS MONAX.—LINN.

WOOD-CHUCK. MARYLAND MARMOT. GROUND-HOG.

PLATE II.—FEMALE AND YOUNG.

A. Supra fusco cinereus, subtus sub-rufus, capite, cauda, pedibusque fuscis, naso et buccis cinereis.

CHARACTERS.

Brownish-gray above; head, tail, and feet, dark-brown; nose and cheeks ashy-brown, under surface reddish.

SYNONYMES.

MUS MONAX, Linn., 12 ed., p. 81.
MARYLAND MARMOT, Penn., Arct. Zool., vol. i., p. 111.
MONAX, ou MARMOTTE DE CANADA, Buff., Supp. 111.
MARYLAND MARMOT, Godman, Nat. Hist. vol. ii., p. 100, figure.
MARYLAND MARMOT, Griffiths' Cuvier, vol. iii., p. 130, figure.

Nº 1. Plate II

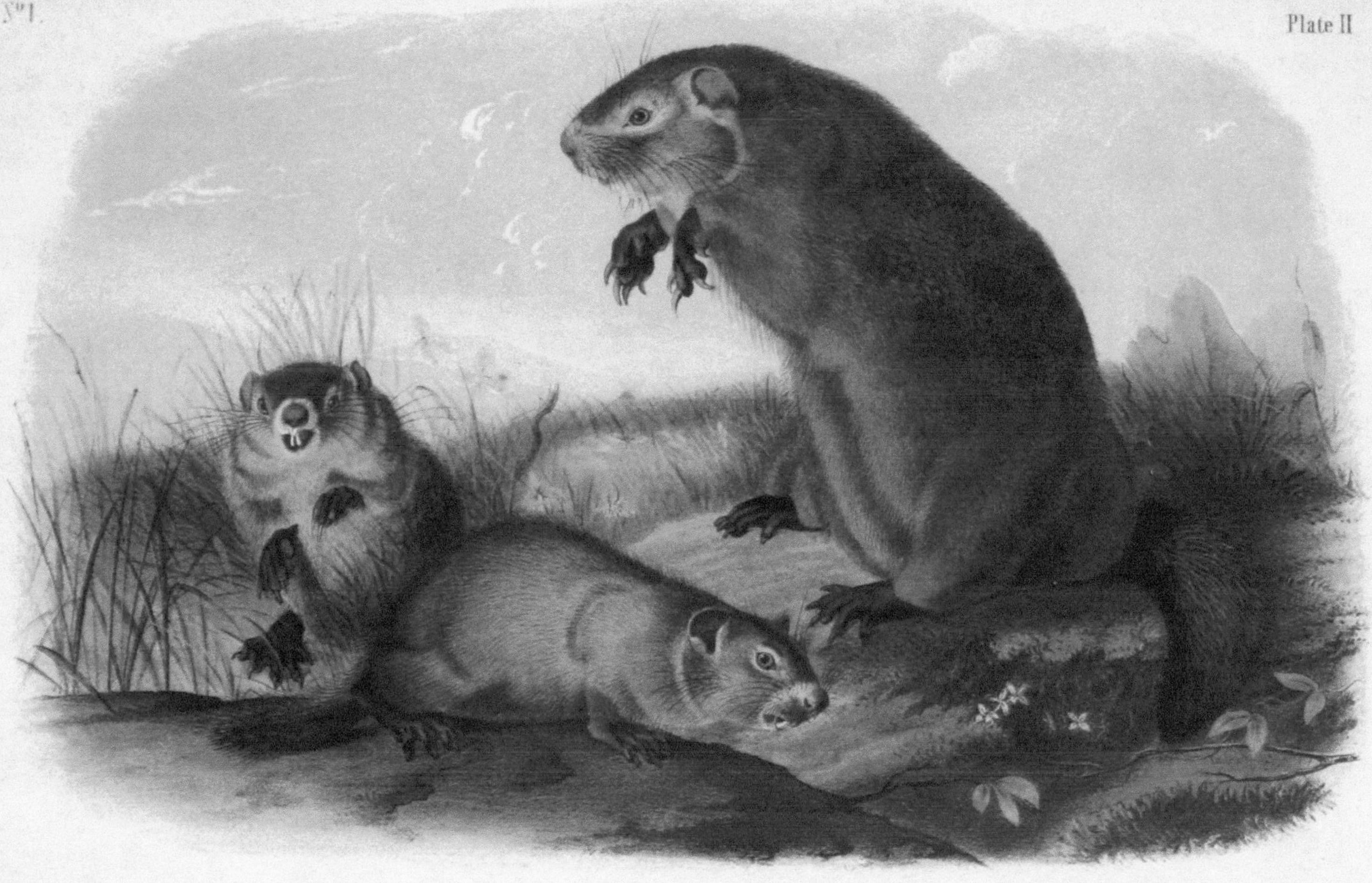

Maryland Marmot Woodchuck Groundhog

Drawn from Nature by J.J. Audubon F.R.S. F.L.S.

Old & Young.

Printed by Nagel & Weingaertner N.Y.

QUEBEC MARMOT, Pennant, Hist. Quad., 1st ed., No. 259.
MUS EMPETRA, Pallas, Glir., p. 75.
ARCTOMYS EMPETRA, Salt, Linn., Trans., vol. xiii., p. 24.
ARCTOMYS EMPETRA, Godman, Nat. Hist., vol. ii., p. 208.
ARCTOMYS MONAX, et ARCTOMYS EMPETRA, Sabine, Trans. Linnæan Soc., vol. xiii., pp. 582, 584.
ARCTOMYS EMPETRA, Richardson, Fauna Boreali Americana, p. 147, pl. 9.

DESCRIPTION.

The body is thick, and the legs are short, so that the belly nearly touches the ground. Head short and conical; ears short, rounded, and thinly clothed with hair on both surfaces; eyes moderate; whiskers numerous, extending to the ear; a membrane beneath the ears, on the posterior parts of the cheek, and a few setæ on the eye-brows; legs, short and muscular; fore-feet, with four toes, and the rudiment of a thumb, with a minute nail; hind-feet, with five toes. Toes long and well separated, palms naked, with tubercles at the roots of the toes. The middle toe longest—the first and third, which are nearly equal to each other, not much shorter; the extremity of the nail of the outer, extends only to the base of the nail of the adjoining toe; fore-claws moderately arched, obtuse and compressed; the soles of the hind-feet long, and naked to the heel; hind-feet semi-palmated; nails channelled near the ends. Tail bushy, partly distichous; body clothed with soft woolly fur, which is mixed with coarse long hairs.

COLOUR.

This species (like the foregoing one) is subject to many variations in the colour of its fur, which may account perhaps for its numerous synonymes. We will, however, describe the animal in its most common colouring.

The finer woolly fur is for two-thirds of its length from the roots upwards, of a dark ashy brown, with the extremities light yellowish-brown. The long hairs are dark brown for two-thirds of their length, tipped sometimes with reddish white, but generally with a silvery white. The general tint of the back is grizzly or hoary; cheeks, and around the mouth, light gray; whiskers black; head, nose, feet, nails and tail, dark brown; eyes black. The whole under surface, including the throat, breast, belly, and the fore and hind legs, reddish orange.

The specimens before us present several striking varieties of colour; among them is one from Lower Canada, coal-black with the exception of the nose and a patch under the chin, which are light gray; the fur is short, and very soft; and the tail less distichous than in other varieties of this species.

DIMENSIONS.

Adult Male.

From point of nose to root of tail - - -	$18\frac{3}{4}$ inches.
Tail (vertebræ) - - - - - - -	$3\frac{7}{8}$ do.
Tail, to end of hair - - - - - -	$5\frac{7}{8}$ do.
Ear, posteriorly - - - - - -	$\frac{3}{4}$ do.
Girth of body - - - - - - - -	17 do.
From fore to hind claw, when stretched - -	26 do.

We have found some difference in the length of the tail, in different individuals, it being, in some specimens, nearly seven inches long including the hair.

Weight 9lb. 11 oz.

HABITS.

In the Middle States many individuals of this species seem to prefer stony places, and often burrow close to or in a stone wall. When this is the case, it is very difficult to procure them, as they are secure from the attacks of dogs, and much labour would be necessary in removing the large stones, and digging up the earth in order to dislodge them.

From our own observations, we are obliged to contradict the following account given of the habits of this species. It has been said that "when about to make an inroad upon a clover field, all the marmots resident in the vicinity, quietly and cautiously steal towards the spot, being favoured in their march by their gray colour, which is not easily distinguished.

"While the main body are actively engaged in cropping the clover heads, and gorging their '*ample cheek-pouches*,' one or more individuals remain at some distance in the rear as sentinels. These watchmen sit erect, with their fore-paws held close to their breast, and their heads slightly inclined, to catch every sound which may move the air. Their extreme sensibility of ear enables them to distinguish the approach of an enemy long before he is sufficiently near to be dangerous, and the instant the sentinel takes alarm, he gives a clear shrill whistle, which immediately disperses the troop in every direction, and they speedily take refuge in their deepest caves. The time at which such incursions are made is generally about mid-day, when they are less liable to be interrupted than at any other period, either by human or brute enemies," (Godman, American Natural History, vol. ii., p. 102.)

We kept two of these animals alive for several weeks, feeding them on different grasses, potatoes, apples, and other fruits and vegetables. We found them to be very active at times, though fond of placing themselves in an erect posture, sitting on their rump, and letting their fore-legs and feet hang loosely down in the manner of our squirrels.

The old female, when approached, opened her mouth, showed her teeth, and made a rattling or clattering noise with the latter, evidently in anger. Neither the female nor the young appeared to become in any degree tame during the period we kept them. The former frequently emitted a shrill whistle-like noise, which is a note of alarm and anger, and may be heard when one is at a distance of about fifty yards from the animal. After we had made figures from those specimens, we examined their mouths, but did not find any pouches like those described by Dr. Godman, although there appeared to be a cavity, not larger than would admit a common green pea, and which was the only trace of any thing like a pouch in those we procured, and in all that have been observed by us.

When the Wood-Chuck is feeding, it keeps its erect position, inclining the head and fore-part of its body forward and sideways, so as to reach its food without extending the fore-legs and feet, which are drawn back under it; after getting a mouthful, it draws back its head again and brings its body to an upright posture by the muscular power of the hind-legs and feet. On being surprised or pursued, this species runs very fast for some eight or ten yards, and then frequently stops short and squats down close to the ground, watching to see if it has been observed; and will allow you to approach within a few feet, when it starts suddenly again, and again stops and squats down as before. Not unfrequently, under these circumstances, it hides its head beneath the dry leaves, or amid tufts of grass, to conceal itself from the pursuer. You may then generally capture or kill it with a stick. These animals bite severely, and defend themselves fiercely, and will, when unable to escape, turn and make battle with a dog of more than double their own size. Sometimes whilst they were lying down as if asleep, we have heard them make the clattering noise before spoken of, with their teeth; reminding us of a person's teeth chattering in an ague fit. When walking leisurely, they place their feet flat upon the ground at full length, arching the toes, however, as is the habit of squirrels. These Marmots sleep during the greater part of the day, stealing from their burrows early in the morning and towards evening. They climb trees or bushes awkwardly, and when they have found a comfortable situation in the sunshine, either on the branch of a tree, or on a bush, will remain there for hours. They clean their faces with the fore-feet, whilst sitting up on their hind-legs, like a squirrel, and frequently lick their fur in the manner of a cat, leaving the coat smoothed down by the tongue. The body of the Wood-Chuck is extremely flabby after being killed; its flesh is, however, tolerably good, although a little strong, and is frequently purchased by the humbler classes of people, who cook it like a roasting pig. Occasionally, and especially in autumn, it is exceedingly fat.

This species becomes torpid about the time the leaves have fallen from the trees in the autumn and the frosty air gives notice of the approach of winter, and remains burrowed in the earth until the grass has sprung up and the genial warmth of spring invites it to come forth.

We once observed one sunning itself at the mouth of its burrow, on the 23d of October, in the State of New-York; and in the same State, saw one killed by a dog on the first of March, when the winter's snow was yet lying in patches on the ground.

Where the nature of the country will admit of it, the Wood-Chucks select a projecting rock, in some fissure under which, they can dig their burrows. In other localities they dig them on the sides of hills, or in places where the surface of the ground is nearly level. These burrows or excavations are sometimes extended to the length of twenty or thirty feet from the opening; for the first three or four feet inclining obliquely downward, and the gallery being continued farther on, about on a level, or with a slight inclination upward to its termination, where there is a large round chamber, to which the occupants retire for rest and security, in which the female gives birth to her young, and where the family spend the winter in torpidity.

Concerning this latter most singular state of existence, we are gratified in being able to communicate the following facts, related to us by the Hon. DANIEL WADSWORTH, of Hartford, Connecticut. "I kept," said he to us, "a fine Wood-Chuck in captivity, in this house, for upwards of two years. It was brought to me by a country lad, and was then large, rather wild, and somewhat cross and mischievous; being placed in the kitchen, it soon found a retreat, in which it remained concealed the greater part of its time every day. During several nights it attempted to escape by gnawing the door and window sills; gradually it became more quiet, and suffered itself to be approached by the inmates of the kitchen, these being the cook, a fine dog, and a cat; so that ere many months had elapsed, it would lie on the floor near the fire, in company with the dog, and would take food from the hand of the cook. I now began to take a particular interest in its welfare, and had a large box made for its use, and filled with hay, to which it became habituated, and always retired when inclined to repose. Winter coming on, the box was placed in a warm corner, and the Wood-Chuck went into it, arranged its bed with care, and became torpid. Some six weeks having passed without its appearing, or having received any food, I had it taken out of the box, and brought into the parlour;—it was inanimate, and as round as a ball, its nose being buried as it were in the lower part of its abdomen, and covered by its tail; it was rolled over the carpet many times, but without effecting any apparent change in its lethargic condition; and being desirous to push the experiment as far as in my power, I laid it close to the fire, and having ordered my dog to lie down by it, placed the Wood-Chuck in the dog's lap. In about half an hour, my pet slowly unrolled itself, raised its nose from the carpet, looked around for a few minutes, and then slowly crawled away from the dog, moving about the room as if in search of its own bed! I took it up, and had it carried down stairs and placed again in its box, where it went to sleep, as soundly as ever, until spring made its appearance. That season advancing, and the trees showing their leaves, the Wood-Chuck became as brisk and gentle as could be desired, and was frequently brought into the parlour. The succeeding winter this animal evinced the same dispositions, and never appeared to suffer by its long sleep. An accident deprived me of my pet, for having been trodden on, it gradually became poor, refused food, and finally died extremely emaciated."

May we here be allowed to detain you, kind reader, for a few moments, whilst we reflect on this, one among thousands of instances of the all-wise dispensations of the Creator? Could any of the smaller species of quadrupeds, incapable, as many of them are, of migrating like the swift-winged inhabitants of the air to the sunny climes of the South, and equally unable to find any thing to subsist on among the dreary wastes of snow in the frost-bound lands of the North during winter, have a greater boon at the hands of Nature than this power of escaping the rigours and cold blasts of that season, and resting securely, in a sleep of insensibility, free from all cravings of hunger and all danger of perishing with cold, till the warm sun of spring once more calls them into life and activity? The Wood-Chuck and several other species of quadrupeds, whose organization in this respect differs so widely from general rules, may be said to have no winter in their year, but enjoy the delightful weather of spring, summer, and autumn, without caring for the approach of that season during which other animals often suffer from both cold and hunger.

"Whilst hunting one day, (said a good friend of ours, when we were last in Canada,) I came across a Wood-Chuck, called in Canada by the different names of Siffleur, Ground-Hog, and occasionally Marmot, with a litter of six or seven young ones by her side. I leaped from my horse, feeling confident that I could capture at least one or two of them, but I was mistaken; for the dam, which seemed to anticipate my evil designs, ran round and round the whole of her young 'chucks,' urging them towards a hole beneath a rock, with so much quickness—energy, I may call it—that ere I could lay hands on even one of her progeny, she had them all in the hole, into which she then pitched herself, and left me gazing in front of her well-secured retreat, thus baffling all my exertions!"

We have now and then observed this Marmot in the woods, leaning with its back against a tree and exposing its under parts to the rays of the hottest sun: on such occasions its head was reclining on its breast, the eyes were closed, the fore-legs hanging down, and it was apparently asleep, and presented a singular and somewhat ludicrous figure.

An intelligent naturalist has in his account of these animals, said that "their burrows contain large excavations in which they deposit stores of provisions." This assertion contradicts our own observation and experience. We are inclined to doubt whether storing up provisions at any or for any season of the year, can be a habit of this species. In the summer of 1814, in Rensselaer County, in the State of New-York, we marked a burrow which was the resort of a pair of Marmots. In the beginning of November the ground was slightly covered with snow, and the frost had penetrated to the depth of about half an inch. We now had excavations made in a line along the burrow or gallery of the Marmots; and at about twenty-five feet from the mouth of the hole, both of them were found lying close to each other in a nest of dried grass, which did not appear to have been any of it eaten or bitten by them. They were each rolled up, and looked somewhat like two misshapen balls of hair, and were perfectly dormant. We removed them to a hay stack, in which we made an excavation to save them from the cold. One of them did not survive the first severe weather of the winter, having, as we thought on examining them, been frozen to death. The other, the male, was now removed to a cellar, where he remained in a perfectly dormant state until the latter part of February, when he escaped before we were aware of his reanimation. We had handled him only two days previously, and could perceive no symptoms of returning vivacity. During the time he was in the cellar, there was certainly no necessity for a "store of provisions" for him, as the animal was perfectly torpid and motionless from the day he was caught, until, as just mentioned, he emerged from that state and made his escape.

In the month of May, or sometimes in June, the female brings forth her young, generally four or five in number. We have however on two occasions, counted seven, and on another eight, young in a litter. In about three weeks, they may be seen playing around the mouth of the burrow, where sitting on their hind-feet in the manner of the Kangaroo, they closely watch every intruder, retreating hastily into the hole at the first notes of alarm sounded by the mother.

The Wood-Chuck in some portions of our country exists in considerable numbers, although it is seldom found associating with any of its own species except while the young are still unable to provide for themselves, until which period they are generally taken care of by both parents.

When the young are a few months old they prepare for a separation, and dig a number of holes in the vicinity of their early domicile, some of which are only a few feet deep and are never occupied. These numerous burrows have given rise to the impression that this species lives in communities, which we think is not strictly the case.

GEOGRAPHICAL DISTRIBUTION.

We have found the Wood-Chuck in every State of the Union north-east of South Carolina, and throughout the Canadas, Nova Scotia, and New Brunswick. We have also a specimen from Hudson's Bay; but perhaps it is nowhere more plentiful than on the upper Missouri River, where we found its burrows dug in the loamy soil adjoining the shores, as well as in the adjacent woods. It is not found in the maritime districts either of North or South Carolina, but exists very sparingly in the mountainous regions of those States. We have also traced it along the eastern range of the Rocky Mountains as far south as Texas. A Marmot exists in California resembling the present species very nearly, but which will probably prove distinct from the latter, a point which time and a greater number of specimens must determine.

GENERAL REMARKS.

It will be observed that we have united *A. monax* with *A. empetra*, and have rejected the latter as a species. This must necessarily follow from the fact, that if there is but one species, the name *monax* having been first given, must be retained. SCHREBER appears to have committed the first error in describing from a young specimen of a variety of *A. monax* and erecting it into a new species. The old authors followed, and most of them being mere compilers, have constantly copied his errors. Mr. SABINE (Transactions Linn. Soc., vol. xiii., part 2, p. 584) described a specimen existing in the British Museum, as *A. empetra*, which we, after a careful examination, consider only a variety of *A. monax*. Mr. SABINE's description of the latter species is, as he informed us, compiled from various authors. Had he possessed a specimen, we think he would not have fallen into the common error. Dr. RICHARDSON, who appears not to have known the *A. monax*, also described it under the name of *A. empetra*, and gave a figure of it. We have, however, been unable to discover any specific differences between the specimens now before us and the one so accurately described and figured by him in the Fauna-boreali-Americana. We are, therefore, compelled to consider them all as identical.

The great varieties of colour to be observed in different specimens of this Marmot, together with the circumstance that no two of them are of the same size, have tended no doubt to confuse those who have described it. We have seen them of all colours, from black to brown, and from rufous to bluish-gray, although they are most frequently of the colour represented in the plate. We have received a specimen from an eminent British naturalist as *A. empetra*, obtained from Hudson's Bay, which does not differ from the present species, and which instead of being eleven inches in length, the size given to *A. empetra*, measures fifteen. As RICHARDSON'S species, moreover, was from seventeen to twenty inches in length, and as we compared his specimen (now in the museum of the Zoological Society of London) with several specimens of the Maryland Marmot, without observing the least specific difference between them, we consider it necessary to strike off the Canada Marmot, or *Arctomys empetra*, from the North American Fauna.

From the short and very unsatisfactory description, and the wretched figure of the Bahama Coney, contained in CATESBY, vol. ii., p. 79, plate 79, it is very difficult to decide either on the species or genus which he intended to describe. As however nearly all our writers on natural history have quoted his Bahama Coney as referring to the Maryland Marmot, we have carefully compared his descriptions and figure with this species, and have arrived at the conclusion that CATESBY described and figured one of the species of jutia, (*Capromys Fournieri*, Desm.,) and that his *Cuniculus Bahamiensis* has been therefore erroneously quoted as a synonyme of *A. monax*.

Nº 1.
Plate III.
Drawn on Stone by R. Trembly
Townsend's Rocky Mountain Hare
Male & Female
Drawn from Nature by J. J. Audubon F.R.S. F.L.S.
Printed by Nagel & Weingaertner N.Y.

GENUS LEPUS.—Linn.

DENTAL FORMULA.

Incisive $\frac{4}{2}$; *Canine* $\frac{0-0}{0-0}$; *Molar* $\frac{6-6}{5-5}$ = 28.

Upper incisors in pairs, two in front large and grooved, and two immediately behind, small; lower incisors square; molars, with flat crowns, and transverse laminæ of enamel. Interior of the mouth and soles of the feet furnished with hair; ears and eyes large; fore-feet with five toes; hind-feet with only four; hind-legs very long; tail short; mammæ, from six to ten.

The word Lepus is derived from the Latin, *lepus*, and Greek Eolic, λεπορις, (*leporis*,) a hare.

There are about thirty known species of this genus, of which rather the largest number (perhaps sixteen or seventeen species) exist in North and South America; while the remainder belong to the Eastern continent.

LEPUS TOWNSENDII.—Bach.

Townsend's Rocky Mountain Hare.

PLATE III.—Male and Female.

L. magnitudine, L. Americano par; auribus, cauda, cruribus tarsisque longissimis; supra diluti cinereus, infra albus.

CHARACTERS.

Size of the Northern hare, (L. Americanus:) ears, tail, legs, and tarsus, very long; colour above, light gray; beneath, white.

SYNONYMES.

Lepus Townsendii, Bach., Journal Acad. Nat. Sciences, Philadelphia, vol. viii., part 1, p. 90, pl. 2, (1839,) read Aug. 7, 1838.

DESCRIPTION

Body, long and slender; head, much arched; eyes large; ears, long; tail very long, (compared with others of the genus,) in proportion to the size of the animal; legs long and slender; tarsus very long. The whole conformation of this animal is indicative of great speed.

COLOUR.

Crown of the head, cheeks, neck, whole upper parts, and the front of the ears and legs, externally, gray; with a faint cream-coloured tinge. Hair, on back and sides, whitish, or silver gray, at the roots, followed by brownish-white, which is succeeded by black, subdued gradually to a faint yellowish-white, and finally tipped with black, interspersed with long silky hairs, some of which are black from their roots. On the chin throat, under surface, interior of legs, and the tail, (with the exception of a narrow dark line running longitudinally on the top,) the hair is pure white from the roots. Irides light hazel; around the eyes white; back part of the tips of the ears black; external two-thirds of the hinder part of the ears white, running down to the back part of the neck, and then blending with the colour of the upper surface; anterior third of the outer portion of the ear, the same gray colour as the back, fringed on the edge with long hairs, which are reddish fawn colour at the roots and white at the tips; interior of the ear very thinly covered with beautiful fine white hairs, being more thickly clothed near the edge, where it is grizzly-black and yellowish; edge, fringed with pure white, becoming yellowish toward the tip, and at the tip black. Moustaches for the most part white, black at the roots, a few hairs are pure white, others wholly black.

The specimen which was described and first published in the Transactions of the Academy of Natural Sciences of Philadelphia, was a female, procured by J. K. Townsend, Esq., on the Walla-Walla, one of the sources of the Columbia river.

Another specimen now in our possession, the dimensions of which are given below, is in summer pelage, having been obtained on the 9th June. There is scarcely a shade of difference in its general colour, although the points of many of the hairs are yellowish-white, instead of being tipped with black, as in the specimen obtained by Mr. Townsend. There is also a white spot on the forehead. The young is a miniature of the adult. We observe no other differences than that the colour is a little lighter, and the tail pure white.

DIMENSIONS.

Adult Male, (killed on the Upper Missouri river.)

From nose to root of tail - - - - - -	21½	inches.
Tail (vertebræ) - - - - - - - -	3⅛	do.
Do., to end of hair - - - - - - -	4¾	do.
Height of ear, posteriorly - - - - - -	5½	do.
Length of head in a direct line - - - - -	4⅝	do.
" " following the curvature - -	5¼	do.
" from heel to end of claw - - - - -	5⅝	do.

Weight, 6½ pounds.

Adult Female, (shot by Edward Harris, Esq., on the 27th July, 1843.)

From nose to root of tail - - - - - -	21	inches.
Tail (vertebræ) - - - - - - - -	3	do.
Do., to end of hair - - - - - - -	4½	do.
Height of ear, posteriorly - - - - - -	5¼	do.
Between the eyes - - - - - - -	2	do.
From nose to hind feet (stretched out) - - -	36	do.
Height from foot to shoulder - - - - -	13½	do.
Height to rump - - - - - - - -	14	do.

Young.

From nose to root of tail - - - - - -	12	inches.
Tail (vertebræ) - - - - - - - -	1¼	do.
Do., to end of hair - - - - - - -	2⅛	do.
Height of ear, posteriorly - - - - - -	2⅝	do.
Height from claw to shoulder - - -	7⅛	do.
Length of head in a direct line - - - - -	2¾	do.
" " following the curve - - -	3¾	do.
" from heel to end of claw - - - - -	3⅝	do.

HABITS.

We subjoin the following note, received from the original discoverer of this Hare, which contains some valuable information in regard to its habits :—"This species is common in the Rocky Mountains. I made particular inquiries both of the Indians and British traders, as to the changes it undergoes at different seasons, and they all agreed that it never was lighter coloured. We first saw it on the plains of the Blackfoot river, east of the mountains, and observed it in all similar situations during our route to the Columbia. When first seen, which was in July, it was lean and unsavory, having, like our common species, the larva of an insect imbedded in its neck; but when we arrived at Walla-Walla, in September, we found the Indians and the persons attached to the fort using it as a common article of food. Immediately after we arrived we were regaled with a dish of hares, and I thought I had never eaten anything more delicious. They are found in great numbers on the plains covered with wild wormwood, (*Artemesia.*) They are so exceedingly fleet that no ordinary dog can catch them. I have frequently surprised them in their forms and shot them as they leaped away, but I found it necessary to be very expeditious and to pull trigger at a particular instant, or the game was off among the wormwood and I never saw it again. The Indians kill them with arrows by approaching them stealthily as they lie concealed under the bushes, and in winter take them with nets. To do this, some one or two hundred Indians, men women and children, collect, and enclose a large space with a slight net about five feet wide, made of hemp; the net is kept in a vertical position by pointed sticks attached to it and driven into the ground. These sticks are placed about five or six feet apart, and at each one an Indian is stationed with a short club in his hand. After these arrangements are completed a large number of Indians enter the circle and beat the bushes in every direction. The frightened hares dart off towards the net, and in attempting to pass are knocked on the head and secured. Mr. PAMBRUN, the superintendent of Fort Walla-Walla, from whom I obtained this account, says that he has often participated in this sport with the Indians and has known several hundred to be thus taken in a day. When captured alive they do not scream like the common gray rabbit, (*L. Sylvaticus.*)" "This Hare inhabits the plains exclusively, and seems particularly fond of the vicinity of the aromatic wormwood. Immediately you leave these bushes in journeying towards the sea you lose sight of the Hare."

To the above account we added some farther information on our last visit to the far West. On the 8th June 1843 whilst our men were engaged in cutting wood and bringing it on board the steamer Omega, it being necessary in that wild region to stop and cut wood for fuel for the boat every day, one of the crew started a young Hare and after a short chase the poor thing squatted and was killed by a blow with a stick. It proved to be the young of *Lepus Townsendii*, was large enough to have left its dam, weighed rather more than one pound, and was a beautiful specimen. Its irides were pure amber colour and the eyes large, its hair was slightly curled This Hare was captured more than twelve hundred miles east of the Rocky Mountains. On the next day in the afternoon one of the negro fire-tenders being out with a rifle, shot two others, both old individuals; one of them was however cut in two by the ball and left on the spot. The hair, or fur, of this individual was slightly curled, as in the young one, especially along the back and sides, but shortly after the skins had been prepared this character disappeared. These specimens are now in our collection.

Pursuing our journey up the tortuous and rapid stream, we had not the good fortune to see any more of these beautiful animals until after our arrival at FORT UNION near the mouth of the Yellow Stone river, where we established ourselves for some time by the kind permission of the gentlemen connected with the fur trade.

On the 29th of July on our return from a buffalo-hunt, when we were some forty or fifty miles from the fort suddenly a fine hare leaped from the grass before us and stopped within twenty paces. Our friend, EDWARD HARRIS, Esq., was with us but his gun was loaded with ball and ours with large buck-shot intended for killing antelopes; we fired at it but missed: away it went, and ran around a hill, Mr. HARRIS followed, and its course being seen by Mr. BELL, who observed "Pussy" stealing carefully along with her ears low down trying to escape the quick eyes of her pursuers, the former gentleman came up to and shot her.

This species, like all others of the same family, is timid and fearful in the extreme. Its speed, we think, far surpasses that of the European hare, (*L. timidus.*)

If the *form* is indicative of character, this animal, from its slender body long hind legs and great length of tarsus must be the fleetest of the hares of the West.

These hares generally place or construct their forms under a thick willow bush, or if at a distance from the water-courses on the banks of which those trees grow, or when they are in the open prairie, they place them under the edge of some rock, or seek the shelter of a stone or large tuft of grass.

The Rocky Mountain Hare produces from four to six young in the year. As far as we have been able to ascertain it has but one litter. The young suck and follow the dam for about six weeks after which she turns them off and leaves them to provide for themselves. The flesh of this species resembles in flavour that of the European hare, but is white, nstead of dark-coloured, as is the case with the latter.

GEOGRAPHICAL DISTRIBUTION.

Although the entire geographical range of this species has not been well defined, yet it must be very considerable. It is found in great numbers, long ere the western traveller has passed the prairies, on the shores of the lower Missouri, and has a range of fifteen hundred miles east of the great Rocky Mountain Chain.

According to Mr. TOWNSEND it is common on the Rocky Mountains and exists in considerable numbers on the western side of that great chain; and if travellers have not confounded it with other species it extends southwardly as far as Upper California.

The period may arrive when civilization shall have drawn wealth and a large population into these regions. Then will in all probability this poor hare be hunted by greyhounds followed by gentlemen on horseback; and whilst the level plains of our vast prairies will afford both dogs and horsemen every opportunity of rapid pursuit, the great swiftness of this species will try their powers and test their speed to the utmost.

GENERAL REMARKS.

We have, since this species was first described had some misgivings in regard to its being entitled to the name by which we have designated it.

We had previously (Journ. Acad. Nat. Scien., vol. vii., part. 2, p. 349, and vol. viii., part 1, p. 80) described a species from the West, in its white winter colour, under the name of *L. campestris*. We had no other knowledge of its summer dress than that given us by LEWIS and CLARK. Being however informed by Mr. TOWNSEND, who possessed opportunities of seeing it in winter, that the present species never becomes white, we regarded it as distinct and bestowed on it the above name. We have been since assured by the residents of Missouri, that like the Northern hare, *Lepus Townsendii* assumes a white garb in winter, and it is therefore probable that the name will yet require to be changed to *L. campestris*. As, however, another hare exists on the prairies of the West, the specific characters of which have not yet been determined, we have concluded to leave it as it stands, supposing it possible that the white winter colour may belong to another species.

GENUS NEOTOMA.—Say et Ord.

DENTAL FORMULA.

$$Incisive\ \frac{2}{2};\ Canine\ \frac{0-0}{0-0};\ Molar\ \frac{3-3}{3-3} = 16.$$

Messrs. Say and Ord, who established this genus, having given an extended description of its teeth, &c., we shall present a portion of it in their own words.

"Molars, with profound radicles. *Superior jaw.*—Incisors even and slightly rounded on their anterior face: first molar with five triangles, one of which is anterior, two exterior, and two interior. Second molar with four triangles; one anterior, two on the exterior side, and a very small one on the interior side: third molar with four triangles; one anterior, two exterior, and a very minute one, interior.

"*Inferior jaw.*—Incisors even, pointed at top: first molar with four divisions or triangles, one anterior, a little irregular, then one exterior, one interior opposite, and one posterior: second molar, with four triangles anterior and posterior, nearly similar in form, an intermediate one opposite to the interior and exterior one: third molar with two triangles, and an additional small angle on the inner side of the anterior one. Tail hairy; fore-feet, four toed, with an armed rudiment of a fifth toe; hind-feet, five toed.

OBSERVATIONS.

The grinding surface of the molars differs somewhat from that of the molars of the genus Arvicola; but the large roots of the grinders constitute a character essentially different. The folds of enamel which make the sides of the crown, do not descend so low as to the edge of the alveolar processes; in consequence of this conformation, the worn down tooth of an old individual must exhibit insulated circles of enamel on the grinding surface.

Neotoma—Gr. νεος, (*neos*,) new; and τεμνω, (*temno*,) I cut or divide.

Two species of this genus have been described, both existing in North America.

NEOTOMA FLORIDANA.—Say et Ord.

Florida Rat.

PLATE IV.—Male, Female, and Young.

N. corpore robusto, plumbeo, quoad lineam dorsalem nigro mixto, facie et lateribus fusco-flavescentibus, infra albo; cauda corpore paullo curtiore, vellere molli.

CHARACTERS.

Body robust, lead colour, mixed with black on the dorsal line; face and sides ferruginous-yellow, beneath white, tail a little shorter than the body; fur soft.

SYNONYMES.

Mus Floridanus, Ord, Nouv. Bull. de la Société Philomatique, 1818.
Arvicola Floridanus, Harlan, Fauna Amer., p. 142.
" " Godman, Nat. Hist., vol. ii., p. 69.
Mus " Say, Long's Expedition, vol. i., p. 54.
Neotoma Floridana, Say et Ord, Journ. Acad. Nat. Sciences, Philadelphia, vol. iv., part. 2, p. 352, figure.
Neotoma Floridana, Griffiths, Animal Kingdom, vol. iii., p. 160, figure.

DESCRIPTION.

The form of our very common white-footed or field-mouse (*Mus leucopus*) may be regarded as a miniature of that of the present species; its body has an appearance of lightness and agility, bearing some resemblance to that of the squirrel; snout elongated; eyes large, resembling those of the common flying squirrel (*P. volucella*); ears large, prominent, thin, sub-ovate, clothed so thinly with fine hair as to appear naked; tail covered with soft hair; whiskers reaching to the ears; legs robust; toes annulate beneath; thumb, minute; in the palms of the fore-feet there are five tubercles, and in the soles of the hind-feet six, of which the three posterior are distant from each other; nails, concealed by hairs, which extend considerably beyond them; mammæ, two before, and four behind.

COLOUR.

The body and head are lead-colour, intermixed with yellowish and

Florida Rat.

Male Female & Young of different ages.

black hair; the black preuominating on the ridge of the back and head, forming an indistinct dorsal line of dark brown, gradually fading away into the brownish-yellow colour of the cheeks and sides; border of the abdomen and throat, buff; whiskers, white and black; feet, white; under surface of body, white, tinged with cream colour.

In a very young specimen, the colour is dark brown on the upper surface, and plumbeous beneath; differing so much from the adult, that the unpractised observer might easily be led to regard it as a new species.

DIMENSIONS.

Adult Male.

From nose to root of tail - - - - - -	8	inches.
Length of tail - - - - - - - -	$5\frac{1}{8}$	do.
From fore-claws to hind-claws, when stretched -	$13\frac{1}{4}$	do.
From nose to end of ears - - - - - -	$2\frac{1}{2}$	do.

Weight $7\frac{3}{4}$ ounces. Weight of an old Female, 8 ounces.

Young Male.

From nose to root of tail - - - - - -	$5\frac{1}{4}$	inches.
From fore-claws to hind-claws, when stretched -	$8\frac{1}{2}$	do.
From nose to end of ear - - - -	$2\frac{3}{8}$	do.
Length of tail - - - - - - - -	$4\frac{1}{4}$	do.

HABITS.

The specimens from which we drew the figures we have given on our plate, which represents this species in various ages and attitudes on the branch of a pine tree, were obtained in South Carolina, and were preserved alive for several weeks in cages having wire fronts. They made no attempt to gnaw their way out. On a previous occasion we preserved an old female with three young (which latter were born in the cage a few days after the mother had been captured) for nearly a year; by which time the young had attained the size of the adult. We fed them on corn, potatoes, rice, and bread, as well as apples and other fruit. They seemed very fond of corn flour, (Indian meal,) and for several months subsisted on the acorns of the live oak. (*Quercus virens.*)

They became very gentle, especially one of them which was in a separate cage. It was our custom at dark to release it from confinement, upon which it would run around the room in circles, mount the table we were in the habit of writing at, and always make efforts to open a particular drawer in which we kept some of its choicest food.

There are considerable differences in the habits of this species in various parts of the United States, and we hope the study of these peculiarities may interest our readers. In Florida they burrow under stones and the ruins of dilapidated buildings. In Georgia and South Carolina they prefer remaining in the woods. In some swampy situation in the vicinity of a sluggish stream, amid tangled vines interspersed with leaves and long moss, they gather a heap of dry sticks which they pile up into a conical shape, and which, with grasses, mud, and dead leaves, mixed in by the wind and rain, forms, as they proceed, a structure impervious to rain, and inaccessible to the wild-cat, racoon, or fox. At other times, their nest, composed of somewhat lighter materials, is placed in the fork (branch) of a tree.

About fifteen years ago, on a visit to the grave-yard of the church at Ebenezer, Georgia, we were struck with the appearance of several very large nests near the tops of some tall evergreen oaks (*Quercus aquaticus*) ; on disturbing the nests, we discovered them to be inhabited by a number of Florida rats of all sizes, some of which descended rapidly to the ground, whilst others escaped to the highest branches, where they were concealed among the leaves. These nests in certain situations are of enormous size. We have observed some of them on trees, at a height of from ten to twenty feet from the ground, where wild vines had made a tangled mass over head, which appeared to be larger than a cart wheel and contained a mass of leaves and sticks that would have more than filled a barrel.

Those specimens, however, which we procured on our journey up the Missouri river, were all caught in the hollows of trees which were cut down by the crew, as we proceeded, for fuel for our steamer. LEWIS and CLARK, in their memorable journey across the Rocky Mountains, found them nestling among clefts in the rocks, and also in hollow trees. In this region they appeared to be in the habit of feeding on the prickly pear or Indian fig, (*Cactus opuntia*,) the travellers having found large quantities of seeds and remnants of those plants in their nests. In the Floridas, Mr. BARTRAM also found this species. He says, "they are singular with respect to their ingenuity and great labour in the construction of their habitations, which are conical pyramids about three feet high, constructed with dry branches which they collect with great labour and perseverance and pile up without any apparent order; yet they are so interwoven with one another that it would take a bear or wild cat some time to pull one of these castles to pieces, and allow the animals sufficient time to secure a retreat with their young."

This is a very active rat, and in ascending trees, exhibits much of the agility of the squirrel, although we do not recollect having observed it leaping from branch to branch in the manner of that genus.

The Florida rat is, in Carolina, a very harmless species; the only depredation we have known it to commit, was an occasional inroad on the corn-fields when the grain was yet juicy and sweet. We have seen several whole ears of Indian corn taken from one of their nests, into which they had been dragged by these animals the previous night. They appear also to be very fond of the Chinquapin (*Castania pumila*), and we have sometimes observed around their nests traces of their having fed on frogs and cray-fish.

This species is nocturnal, or at least crepuscular, in its habits. In procuring specimens we were only successful when the traps had been set over night. Those we had in captivity scarcely ever left their dark chambers till after sunset, when they came forth from their dormitories and continued playful and active during a great part of the night. They were mild in their dispositions, and much less disposed to bite when pursued than the common and more mischievous Norway rat.

Whilst the young are small they cling to the teats of the mother, who runs about with them occasionally without much apparent inconvenience; and even when older, they still, when she is about to travel quickly, cling to her sides or her back. Thus on a visit from home, she may be said to carry her little family with her, and is always ready to defend them even at the risk of her life. We once heard a gratifying and affecting anecdote of the attachment to its young, manifested by one of this species, which we will here relate as an evidence that in some cases we may learn a valuable lesson from the instincts of the brute creation.

Our friend Gaillard Stoney Esq., sent us an old and a young Florida rat, obtained under the following circumstances. A terrier was seen in pursuit of a rat of this species, followed by two young about a third grown. He had already killed one of these, when the mother sprang forward and seized the other in her mouth, although only a few feet from her relentless enemy—hastened through a fence which for a moment protected her, and retreated into her burrow. They were dug out of the ground and sent to us alive. We observed that for many months the resting place of the young during the day was on the back of its mother.

From three to six are produced at a litter, by this species, which breeds generally twice a year; we have seen the young so frequently in March and August, that we are inclined to the belief that these are the periods of their reproduction. We have never heard them making any other noise than a faint squeak, somewhat resembling that of the brown rat. The very playful character of this species, its cleanly habits, its mild, prominent, and bright eyes, together with its fine form and easy susceptibility of domestication, would render it a far more interesting pet than many others that the caprice of man has from time to time induced him to select.

GEOGRAPHICAL DISTRIBUTION.

This species is very widely scattered through the country. It was brought from East Florida by Mr. ORD, in 1818, but not published until 1825. It was then supposed by him to be peculiar to Florida, and received its specific name from that circumstance. We had, however, obtained a number of specimens, both of this species and the cotton rat, (*Sigmodon hispidum,*) in 1816, in South Carolina, where they are very abundant. In Louisiana, Georgia, Alabama, Mississippi, Missouri, and the former States, it is a common species. Its numbers diminish greatly as we travel eastward. In North Carolina some specimens of it have been obtained. We observed a few nests among the valleys of the Virginia mountains; farther north we have not personally traced it, although we have somewhere heard it stated that one or two had been captured as far to the north as Maryland.

GENERAL REMARKS.

On a farther examination of BARTRAM'S work, which is also referred to by GODMAN (Nat. Hist., vol. ii., p. 21), we find his descriptions of the habits of this species very accurate; the first part of that article, however, quoted by Dr. GODMAN, is evidently incorrect. "The wood rat," says BARTRAM, "is a very curious animal; they are not half the size of the domestic rat, of a dark brown or black colour; thin tail, slender and shorter in proportion, and covered thinly with short hair." The error of BARTRAM, in describing one species, and applying to it the habits of another, seems to have escaped the observation of Dr. GODMAN. The cotton rat, or as it is generally called, wood rat (*Sigmodon hispidum*), answers this description of BARTRAM, in its size, colour, and tail; but it does not build "conical pyramids;" this is the work of a much larger and very different species—the Florida rat of this article.

The adoption of the genus NEOTOMA, when proposed by SAY and ORD, was met with considerable opposition by naturalists of that day, and some severe strictures were passed upon it by Drs. HARLAN and GODMAN. (See HARLAN, p. 143, GODMAN, vol. ii., p. 72.) They contended that the variations in the teeth that separated this species from *Mus* and *Arvicola*, were not sufficient to establish genuine distinctions.

More recently naturalists have, however, examined the subject calmly and considerately. It is certain that this genus cannot be arranged either under *Arvicola* or *Mus*, without enlarging the characters of one or the other of these genera. Another species, from the Rocky Mountains, has been discovered by Dr. Richardson, (*Neotoma Drummondii,*) and we feel pretty confident that the genus will be generally adopted.

GENUS SCIURUS.—Linn., Erxleb., Cuv., Geoff., Illiger.

Dental Formula.

$$Incisive\ \frac{2}{2};\quad Canine\ \frac{0-0}{0-0};\quad Molar\ \frac{4-4}{4-4}\ \text{or}\ \frac{5-5}{4-4} = 20\ \text{or}\ 22.$$

Body elongated; tail long and furnished with hairs; head large; ears erect; eyes projecting and brilliant; upper lip divided. Four toes before, with a tubercle covered by a blunt nail; five toes behind. The four grinders, on each side the mouth above and beneath, are variously tuberculated; a very small additional one in front, above, is in some species permanent, but in most cases drops out when the young have attained the age of from six to twelve weeks. Mammæ, eight; two pectoral, the others abdominal.

The squirrel is admirably adapted to a residence on trees, for which nature has designed it. Its fingers are long slender and deeply cleft, and its nails very acute and greatly compressed; it is enabled to leap from branch to branch and from tree to tree, clinging to the smallest twigs, and seldom missing its hold. When this happens to be the case, it has an instinctive habit of grasping in its descent at the first object which may present itself; or if about to fall to the earth, it spreads itself out in the manner of the flying squirrel, and thus by presenting a greater resistance to the air is enabled to reach the ground without injury and recover itself so instantaneously, that it often escapes the teeth of the dog that watches its descent and stands ready to seize upon it at the moment of its fall. It immediately ascends a neighbouring tree, emitting very frequently a querulous bark, which is either a note of fear or of triumph.

Although the squirrel moves with considerable activity on the ground, it rather runs than leaps; on trees, however, its activity and agility are surprising, and it is generally able to escape from its enemies and conceal itself in a few moments, either among the thick foliage, in its nest, or in a hollow tree. The squirrel usually conveys its food to the mouth by the fore-paws. Nuts, and seeds of all kinds, are held by it between the rudimental thumbs and the inner portions of the palms. When disturbed or alarmed, it either drops the nut and makes a rapid retreat, or seizes it with the incisors, and carries it to its hole or nest.

All American species of this genus, as far as we have been able to become acquainted with their habits, build their nests either in the fork of a tree, or on some secure portion of its branches. The nest is hemispherical in shape, and is composed of sticks, leaves, the bark of trees, and various kinds of mosses and lichens. In the vicinity of these nests, however, they have a still more secure retreat in some hollow tree, to which they retire in cold or in very wet weather, and where their first litter of young is generally produced.

Several species of squirrels collect and hide away food during the abundant season of autumn, to serve as a winter store. This hoard is composed of various kinds of walnuts and hickory nuts, chesnuts, chinquepins, acorns, corn, &c., which may be found in their vicinity. The species, however, that inhabit the Southern portions of the United States, where the ground is seldom covered with snow, and where they can always derive a precarious support from the seeds, insects, and worms, which they scratch up among the leaves, &c., are less provident in this respect; and of all our species, the chickaree, or Hudson's Bay squirrel, (*Sc. Hudsonius*,) is by far the most industrious, and lays up the greatest quantity of food.

In the spring the squirrels shed their hair, which is replaced by a thinner and less furry coat; during summer their tails are narrower and less feathery than in autumn, when they either receive an entirely new coat, or a very great accession of fur; at this season also, the outer surfaces of the ears are more thickly and prominently clothed with fur than in the spring and summer.

Squirrels are notorious depredators on the Indian-corn fields of the farmer, in some portions of our country, consuming great quantities of this grain, and by tearing off the husks exposing an immense number of the unripe ears to the mouldering influence of the dew and rain.

The usual note emitted by this genus is a kind of tremulous, querulous bark, not very unlike the quacking of a duck. Although all our larger squirrels have shades of difference in their notes which will enable the practised ear to designate the species even before they are seen, yet this difference cannot easily be described by words. Their bark seems to be the repetition of a syllable five or six times, quack–quack–quack–quack–qua—commencing low, gradually raising to a higher pitch, and ending with a drawl on the last letter in the syllable. The notes, however, of the smaller Hudson's Bay squirrel and its kindred species existing on the Rocky Mountains, differ considerably from those of the larger squirrels; they are sharper, more rapidly uttered, and of longer continuance; seeming intermediate between the bark of the latter and the chipping calls of the ground-squirrels, (TAMIAS.) The barking of the squirrel may be heard occasionally in the forests during all hours of the day, but is uttered most frequently in the morning and afternoon. Any sudden noise in the woods, or the distant report of a gun, is almost certain, during fine weather, to be succeeded by the barking of the squirrel. This is either a note of playfulness or of love. Whilst barking it seats itself for a few moments on a branch of a tree, elevates its tail over its back towards the head, and bending the point backwards continues to jerk its body and elevate and depress the tail at the repetition of each successive note. Like the mocking bird and the nightingale, however, the squirrel, very soon after he begins to sing, (for to his own ear, at least, his voice must be musical,) also commences skipping and dancing; he leaps playfully from bough to bough, sometimes pursuing a rival or his mate for a few moments, and then reiterating with renewed vigour his querulous and monotonous notes.

One of the most common habits of the squirrel is that of dodging around the tree when approached, and keeping on the opposite side so as to completely baffle the hunter who is alone. Hence it is almost essential to the sportsman's success that he should be accompanied by a second person, who, by walking slowly round the tree on which the squirrel has been seen beating the bushes and making a good deal of noise, causes him to move to the side where the gunner is silently stationed waiting for a view of him to fire. When a squirrel is seated on a branch and fancies himself undiscovered, should some one approach he immediately depresses his tail, and extending it along the branch behind him, presses his body so closely to the bark that he frequently escapes the most practised eye. Notwithstanding the agility of these animals, man is not their only nor even their most formidable enemy. The owl makes a frequent meal of those species which continue to seek their food late in the evening and early in the morning. Several species of hawks, especially the red-tailed (*Buteo borealis*), and the red-shouldered (*Buteo lineatus*), pounce upon them by day. The black snake, rattle snake, and other species of snakes, can secure them; and the ermine, the fox, and the wild cat, are incessantly exerting their sagacity in lessening their numbers.

The generic name Sciurus is derived from the Latin *sciurus*, a squirrel, and from the Greek σκιουρος (*skiouros*), from σκια (*skia*), a shade, and ουρα (*oura*), a tail.

There are between sixty and seventy species of this genus known to authors; about twenty well determined species exist in North America.

Drawn on Stone by R Trembly

Richardson's Columbian Squirrel,

Male & Female.

Drawn from Nature by J.J. Audubon F.R.S, F.L.S.

Printed by Nagel & Weingærtner N.Y.

SCIURUS RICHARDSONII.—BACH.

RICHARDSON'S COLUMBIAN SQUIRREL.

PLATE V.—MALE AND FEMALE.

S. cauda corpore breviore, apice nigro; supra griseus, subtus sub-albidus, S. Hudsonico minor.

CHARACTERS.

Smaller than Sciurus Hudsonius; tail shorter than the body; rusty gray above, whitish beneath; extremity of the tail black.

SYNONYMES.

BROWN SQUIRREL, Lewis and Clarke, vol. iii., p. 37.
SCIURUS HUDSONIUS, var. B. Richardson, Fauna Boreali Americana, p. 190.
SCIURUS RICHARDSONII, Bachman, Proceedings Zool. Soc., London, 1838, (read Aug. 14, 1838.)
SCIURUS RICHARDSONII, Bach., Mag. Nat. Hist., London, new series, 1839, p. 113.
" " Bach., Silliman's Journal.

DESCRIPTION.

The upper incisors are small and of a light yellow colour; the lower are very thin and slender, and nearly white. The first or deciduous molar, as in all the smaller species of pine squirrel that we have examined, is wanting.

The body of this diminutive species is short, and does not present that appearance of lightness and agility which distinguishes the *Sciurus Hudsonius*. Head less elongated, forehead more arched, and nose a little more blunt, than in that species. Ears short; feet of moderate size; the third toe on the fore-feet but slightly longer than the second; claws, compressed, arched, and acute; tail shorter than the body. Thumb nail broad, flat, and blunt.

COLOUR.

Fur on the back, dark plumbeous from the roots, tipped with rusty brown and black, giving it a rusty gray appearance. It is less rufous than *Sciurus Hudsonius*, and lighter coloured than *Sciurus Douglassii.*

Feet, on their upper surface rufous; on the shoulders, forehead, ears, and along the thighs, there is a slight tinge of the same colour. Whiskers, (which are a little longer than the head,) black. The whole of the under surface, as well as a line around the eyes and a small patch above the nostrils, bluish-gray. The tail for about one-half its length presents on the upper surface a dark rufous appearance, many of the hairs being nearly black, pointed with light rufous. At the extremity of the tail and along it for about an inch and three-quarters, the hairs are black, a few of them slightly tipped with rufous. Hind-feet, from the heel to the palms thickly clothed with short adpressed light-coloured hairs; palms naked. The sides are marked by a line of black, commencing at the shoulder and terminating abruptly on the flanks; this line is about two inches in length, and four lines wide.

DIMENSIONS.

Length of head and body - - - - - -	$6\frac{1}{4}$	inches.
Tail (vertebræ) - - - - - - -	$3\frac{3}{4}$	do.
Do., including fur - - - - - - - -	5	do.
Height of ear posteriorly - - - - - -	$\frac{3}{8}$	do.
Do., including fur - - - - - -	$\frac{5}{8}$	do.
Palm and middle fore-claw - - -	$1\frac{3}{8}$	do.
Sole and middle hind-claw - - - -	$1\frac{7}{8}$	do.

HABITS.

The only knowledge we have obtained of the habits of this species, is contained in a note from Mr. TOWNSEND, who obtained the specimen from which the above description was taken. He remarks: "It is evidently a distinct species. Its habits are very different from the *Sciurus Hudsonius.* It frequents the pine trees in the high ranges of the Rocky Mountains west of the Great Chain, feeding upon the seeds contained in the cones. These seeds are large and white, and contain a good deal of nutriment. The Indians eat a great quantity of them, and esteem them good.

"The note of this squirrel is a loud jarring chatter, very different from the noise of *Sciurus Hudsonius.* It is not at all shy, frequently coming down to the foot of the tree to reconnoitre the passenger, and scolding at him vociferously. It is, I think, a scarce species."

GEOGRAPHICAL DISTRIBUTION.

LEWIS and CLARK speak of the "Brown Squirrel" as inhabiting the banks of the Columbia river. Our specimen is labelled, Rocky Moun tains, Aug. 12, 1834. From Mr. Townsend's account, it exists on the mountains a little west of the highest ridge. It will be found no doubt to have an extensive range along those elevated regions.

In the Russian possessions to the Northward, it is replaced by the Downy Squirrel, (*Sc. lanuginosus*,) and in the South, near the Californian Mountains, within the Territories of the United States, by another small species.

GENERAL REMARKS.

The first account we have of this species is from Lewis and Clark, who deposited a specimen in the Philadelphia Museum, where it still exists. We have compared this specimen with that brought by Mr. Townsend, and find them identical. The description by Lewis and Clark (vol. iii., p. 37) is very creditable to the close observation and accuracy of those early explorers of the untrodden snows of the Rocky Mountains and the valleys beyond, to Oregon.

"The small brown Squirrel," they say, "is a beautiful little animal, about the size and form of the red squirrel (*Sc. Hudsonius*) of the Atlantic States and Western lakes. The tail is as long as the body and neck, and formed like that of the red squirrel; the eyes are black; the whiskers long and black, but not abundant; the back, sides, head, neck, and outer parts of the legs, are of a reddish brown; the throat, breast, belly, and inner parts of the legs, are of a pale red; the tail is a mixture of black and fox-coloured red, in which the black predominates in the middle, and the red on the edges and extremity. The hair of the body is almost half an inch long, and so fine and soft that it has the appearance of fur. The hair of the tail is coarser and double in length. This animal subsists chiefly on the seeds of various species of pine and is always found in the pine country."

Dr. Richardson, who had not seen a specimen, copied in his excellent work, (*Fauna Boreali Americana*, p. 19,) the description of Lewis and Clark, from which he supposed this species to be a mere variety of the *Sc. Hudsonius*. We had subsequently an opportunity of submitting a specimen to his inspection, when he immediately became convinced it was a different species.

The difference between these two species can indeed be detected at a glance by comparing specimens of each together. The present species, in addition to its being a fourth smaller,—about the size of our little chipping squirrel (*Tamias Lysteri*)—has less of the reddish brown on the upper surface, and may always be distinguished from the other by the blackness of its tail at the extremity.

GENUS VULPES.—Cuv.

DENTAL FORMULA.

$$\textit{Incisive}\ \frac{6}{6};\ \textit{Canine}\ \frac{1-1}{1-1};\ \textit{Molar}\ \frac{6-6}{7-7} = 42.$$

Muzzle pointed; pupils of the eyes forming a vertical fissure; upper incisors less curved than in the Genus Canis. Tail long, bushy, and cylindrical.

Animals of this genus generally are smaller, and the number of species known greater, than among the wolves; they diffuse a fœtid odour, dig burrows, and attack none but the weaker quadrupeds or birds, &c.

The characters of this genus differ so slightly from those of the genus Canis, that we were induced to pause before removing it from the subgenus in which it had so long remained. As a general rule, we are obliged to admit that a large fox is a wolf, and a small wolf may be termed a fox. So inconveniently large, however, is the list of species in the old genus Canis, that it is, we think, advisable to separate into distinct groups, such species as possess any characters different from the true Wolves.

Foxes, although occasionally seen abroad during the day, are nocturnal in their habits, and their character is marked by timidity, suspicion and cunning. Nearly the whole day is passed by the Fox in concealment, either in his burrow under ground, in the fissures of the rocks, or in the middle of some large fallen-tree-top, or thick pile of brush-wood, where he is well hidden from any passing enemy.

During the obscurity of late twilight, or in the darkness of night, he sallies forth in search of food; the acuteness of his organs of sight, of smell, and of hearing, enabling him in the most murky atmosphere to trace and follow the footsteps of small quadrupeds or birds, and pounce upon the hare seated in her form, or the partridge, grouse, or turkey on their nests.

Various species of squirrels, field-rats, and moles, afford him a rich repast. He often causes great devastation in the poultry yard; seizes on the goose whilst grazing along the banks of the stream, or carries off the lamb from the side of its mother.

The cautious and wary character of the Fox, renders it exceedingly

Drawn on Stone by R. Trembly

American Cross Fox.

Drawn from Nature by J. J. Audubon F.R.S. F.L.S.

Printed by Nagel & Weingærtner N.Y.

difficult to take him in a trap of any kind. He eludes the snares laid for him, and generally discovers and avoids the steel-trap, however carefully covered with brush-wood or grasses.

In the Northern States, such as Pennsylvania and New-York, and in New England, the rutting season of the Fox commences in the month of February. During this period he issues a succession of rapid yells, like the quick and sharp barking of a small dog. Gestation continues from 60 to 65 days. The cubs are from 5 to 9 in number, and like young puppies, are born with hair and are blind at birth. They leave their burrows generally when three or four months old, and in all predatory expeditions each individual goes singly, and plunders on his own account, and for his own especial benefit.

The Generic name is derived from the Latin word *vulpes*, a Fox.

There are about twelve well-known species belonging to this genus —four of which exist in North America.

VULPES FULVUS.—Desm: *var. Decussatus.*—Pennant.

American Cross Fox.

PLATE VI.—Male.

V. cruce nigra supra humeros, subtus linea longitudinali nigra, auribus pedibusque nigris.

CHARACTERS.

A cross on the neck and shoulders, and a longitudinal stripe on the under surface, black; ears and feet black.

SYNONYMES.

Renard Barré, Tsinantontongue, Sagard Theodat., Canada, p 745.
European Cross Fox, var. B., Cross Fox, Pennant, Arct., Zool., vol. i., p. 46.
Canis Decussatus, Geoff., Coll. du Mus.
Canis Fulvus, Sabine, Franklin's Journal, p. 656.
" " var. B., (decussatus) Rich., Fauna Boreali Americana, p. 93.

DESCRIPTION.

Form, agrees in every particular with that of the common red fox, (*V. fulvus.*) Fur, rather thick and long, but not thicker or more elongated than in many specimens of the red fox that we have examined. Soles of the feet densely clothed with short woolly hair, so that the callous spots at the roots of the nails are scarcely visible. A black longitudinal stripe, more or less distinct, on the under surface.

COLOUR.

Front of the head, and back, dark gray; the hairs being black at the roots, yellowish white near the ends, and but slightly tipped with black; so that the light colour of the under part of each hair showing through, gives the surface a gray tint; with these hairs a few others are mixed that are black throughout their whole length.

The soft fur beneath these long hairs is of a brownish black. Inner surface of ears, and sides of the neck from the chin to the shoulders, pale reddish yellow; sides, behind the shoulders towards the top of the back, slightly ferruginous; under surface, to the thighs, haunches, and under part of the root of tail, pale ferruginous. Fur underneath the long hair, yellowish. Tail dark brown; fur beneath, reddish yellow; the long hairs, yellowish at base, broadly tipped with black; at the extremity of the tail a small tuft of white hair. Nose, outer surface of ear, chin, throat, and chest, black. A line along the under surface for half its length, and broadest at its termination, black; a few white hairs intermixed, but not a sufficient number to alter the general colour. The yellowish tint on each side of the neck and behind the shoulders, is divided by a longitudinal dark brown band on the back, crossed at right angles by another running over the shoulders and extending over the fore-legs, forming a cross. There is another cross, yet more distinctly marked, upon the chest; a black stripe, extending downward from the throat towards the belly, being intersected by another black line, which reaches over the chest from the inside of one fore-leg to the other. Hence, the name of this animal does not originate in its ill-nature, or by reason of its having any peculiarly savage propensity, as might be presumed, but from the singular markings we have just described.

DIMENSIONS.

Adult Male.

From nose to root of tail - - - -	24¾	inches.
Tail, (vertebræ) - - - - - - -	12½	do.
Tail, to end of hair - - - - - - -	16	do.
From nose to end of ear - - - - - -	8	do.
" " to eyes - - - - -	2½	do.

Weight, 14 pounds.

HABITS.

In our youth we had opportunities whilst residing in the northern part of the State of New York, of acquiring some knowledge of the habits of the fox and many other animals, which then were abundant around us.

Within a few miles dwelt several neighbours who vied with each other in destroying foxes and other predacious animals, and who kept a strict account of the number they captured or killed each season. As trappers, most of our neighbours were rather unsuccessful—the wary foxes, especially, seemed very soon, as our western hunters would say, to be "up to trap." Shooting them by star-light from behind a hay-stack in the fields, when they had for some time been baited and the snow covered the ground so that food was eagerly sought after by them, answered pretty well at first, but after a few had been shot at, the whole tribe of foxes —red, gray, cross, and black—appeared to be aware that safety was no longer to be expected in the vicinity of hay-stacks, and they all gave the latter a wide berth.

With the assistance of dogs, pick-axes, and spades, our friends were far more successful, and we think might have been considered adepts. We were invited to join them, which we did on a few occasions, but finding that our ideas of sport did not accord precisely with theirs, we gradually withdrew from this club of primitive fox-hunters. Each of these sportsmen was guided by his own "rules and regulations" in the "chase;" the horse was not brought into the field, nor do we remember any scarlet coats. Each hunter proceeded in the direction that to him seemed best—what he killed he kept—and he always took the shortest possible method he could devise, to obtain the fox's skin. He seldom carried a gun, but in lieu of it, on his shoulder was a pick-axe and a spade and in his pocket a tinder box and steel.

A half-hound, being a stronger and swifter dog than the thorough bred, accompanied him, the true foxhound being too slow and too noisy for his purpose; we remember one of these half-bred dogs which was of great size and extraordinary fleetness; it was said to have a cross of the greyhound.

In the fresh-fallen and deep snows of mid-winter, the hunters were most successful. During these severe snow storms, the ruffed grouse, (*Tetrao umbellus,*) called in our Eastern States the partridge, is often snowed up and covered over; or sometimes plunges from on wing into the soft snow, where it remains concealed for a day or two. The fox occasionally surprises these birds, and as he is usually stimulated at this inclement season by the gnawings of hunger, he is compelled to seek for food by day as well as by night; his fresh tracks may be seen in the fields, along the fences, and on the skirts of the farm-yard, as well as in the deep forest.

Nothing is easier than to track the Fox under these favourable circumstances, and the trail having been discovered, it is followed up, until Reynard is started. Now the chase begins: the half-hound yells out, in tones far removed from the mellow notes of the thorough-bred dog, but equally inspiriting perhaps, through the clear frosty air, as the solitary hunter eagerly follows as fast as his limited powers of locomotion will admit. At intervals of three or four minutes, the sharp cry of the dog resounds, the Fox has no time to double and shuffle, the dog is at his heels almost, and speed, speed, is his only hope for life. Now the shrill baying of the hound becomes irregular; we may fancy he is at the throat of his victim; the hunter is far in the rear, toiling along the track which marks the course so well contested, but occasionally the voice of his dog, softened by the distance, is borne on the wind to his ear. For a mile or two the Fox keeps ahead of his pursuer, but the latter has the longest legs, and the snow impedes him less than it does poor Reynard; every bound and plunge into the snow diminishes the distance between the fox and his relentless foe. Onward they rush through field, fence, brushwood, and open forest, the snow flying from bush and briar as they dart through the copse or speed across the newly-cleared field. But this desperate race cannot last longer. The fox must gain his burrow, or some cavernous rock, or he dies. Alas! he has been lured too far away from his customary haunts and from his secure retreat, in search of prey; he is unable to reach his home; the dog is even now within a foot of his brush. One more desperate leap, and with a sudden snappish growl he turns upon his pursuer and endeavours to defend himself with his sharp teeth. For a moment he resists the dog, but is almost instantly overcome. He is not killed, however, in the first onset; both dog and fox are so fatigued that they now sit on their haunches facing each other, resting, panting, their tongues hanging out, and the foam from their lips dropping on the snow. After fiercely eyeing each other for a while, both become impatient—the former to seize his prey, and the latter to escape. At the first leap of the fox, the dog is upon him; with renewed vigour he seizes him by the throat, and does not loose his hold until the snow is stained with his blood, and he lies rumpled, draggled, with blood-shot eye and frothy open mouth, a mangled carcass on the ground.

The hunter soon comes up: he has made several *short cuts*, guided by the baying of his hound; and striking the deep trail in the snow again, at a point much nearer to the scene of the death-struggle, he hurries toward the place where the last cry was heard, and pushes forward in a half run until he meets his dog, which on hearing his master approach generally advances towards him and leads the way to the place where he has achieved his victory.

We will now have another hunt, and pursue a Fox that is within reach of his burrow when we let loose our dog upon him. We will suppose him "started;" with loud shouts we encourage our half-hound; he dashes away on the Fox's track, whilst the latter, with every muscle strained to the utmost is shortening the distance between himself and his stronghold; increasing his speed with his renewed hopes of safety, he gains the entrance to his retreat, and throws himself headlong into it rejoicing at his escape. Whilst yet panting for breath, he hears his foe barking at the entrance of his burrow, and flatters himself he is now beyond a peradventure safe. But perhaps we do injustice to his sagacity; he may have taken refuge in his hole well aware of the possibility of his being attacked there—yet what better could he do? However this may be, he has escaped one enemy by means of a swift pair of heels, and has only to dread the skill, perseverance and invention of the hunter, who in time comes up, rigged out pretty much as we have already described him, with spade, pick-axe, flint and steel.

On arriving at the spot where the Fox has been (in select phrase) "holed," the sportsman surveys the place, and if it is on level ground where he can use the spade, throws off his coat, and prepares for his work with a determination to have "that" fox, and no mistake! He now cuts a long slender stick, which he inserts in the hole to ascertain in what direction he shall dig the first pit. The edge or mouth of the burrow is generally elevated a little above the adjacent surface of the ground by the earth which the Fox has brought from within; and this slight embankment serves to keep out the rain water, that might otherwise flow in from the vicinity in stormy weather.

The burrow at first inclines downward for four or five feet at an angle of about twenty-five degrees; it then inclines upward a little, which is an additional security against inundations, and is continued at a depth of about three or four feet from the surface, until it reaches a point where it is divided into two or three galleries.

This dividing point the hunter discovers after sinking three or four pits—it is generally twenty or thirty feet from the entrance of the burrow. The excavation is now made larger and the earth and rubbish thrown out, the dog is placed in the hole thus laid open, and his aid is sought to ascertain into which branch of the gallery the Fox has retreated. There are seldom any tortuous windings beyond the spot whence the galleries diverge—the Fox is not far off. The stick is again inserted, and either reaches him, and the hunter is made aware of his whereabouts by his snapping at it and growling, which calls forth a yelp of fierce anxiety from the dog ; or, as frequently happens, the Fox is heard digging for life, and making no contemptible progress through the earth. Should no rocks or large roots interfere, he is easily unearthed, and caught by the dog.

It however very frequently occurs, that the den of the Fox is situated on the mountain side ; and that its winding galleries run beneath the enormous roots of some stately pine or oak ; or it may be amongst huge masses of broken rock, in some fissure of too great depth to be sounded, and too contracted to be entered by man or dog. What is then to be done? Should a "dead-fall" be set at the mouth of the hole, the Fox will (unless the ground be frozen too hard) dig another opening, and not go out by the old place of egress ; place a steel-trap before it, and he will spring it without being caught. He will remain for days in his retreat, without once exposing himself to the danger of having a dog snapping at his nose, or a load of duck-shot whistling round his ears. Our hunter, however, is not much worried with such reflections as we have just made ; he has already gathered an armful or two of dry wood, and perhaps some resinous knots, or bits of the bark of the pine-tree ; he cuts up a portion into small pieces, pulls out his tinder-box, flint, and steel, and in a few moments a smart fire is lighted within the burrow ; more wood is thrown on, the mass pushed further down the hole, and as soon as it begins to roar and blaze freely, the mouth is stopped with brush-wood covered with a few spadefuls of earth, and the den is speedily exhausted of pure air, and filled with smoke and noxious gases.

There is no escape for the Fox—an enemy worse than the dog or the gun is destroying him ; he dies a protracted, painful death by suffocation! In about an hour the entrance is uncovered, large volumes of smoke issue into the pure air, and when the hunter's eye can pierce through the dense smoky darkness of the interior, he may perhaps discern the poor Fox extended lifeless in the burrow, and may reach him with a stick. If not quite dead, the Fox is at least exhausted and insensible ; this is sometimes the case, and the animal is then knocked on the head.

The number of Foxes taken by our neighbours, in the primitive mode of hunting them we have attempted to describe, was, as nearly as we can now recollect, about sixty every winter, or an average of nearly twenty killed by each hunter. After one or two seasons, the number of Foxes in that part of the country was sensibly diminished, although the settlements had not increased materially and the neighbourhood was at that time very wild.

At this time Pennant's Marten (*Mustela Canadensis*) was not very scarce in Rensselaer county, and we had three different specimens brought to us to examine.

These, the people called Black Foxes. They were obtained by cutting down hollow trees in which they were concealed, and to which their tracks on the snow directed the hunters.

We cannot now find any note in regard to the number of Cross Foxes taken, as compared to the Red, Gray, and Black Foxes; about one-fourth of the whole number captured, however, were Gray Foxes, and we recollect but a single one that was perfectly black with the exception of a white tip at the end of its tail, like the specimen figured in our work.

On examining several packages of Fox skins at Montreal, we saw about four specimens only of the Cross Fox, and three of the Black Fox, in some three hundred skins. We were informed during our recent visit to the upper Missouri country, that from fifty to one hundred skins of the Cross Fox were annually procured by the American Fur Company from the hunters and Indians.

The specimen from which our drawing was made, was caught in a steel-trap by one of its fore-feet, not far from the falls of Niagara, and was purchased by J. W. Audubon of the proprietor of the "Museum" kept there to gratify the curiosity of the travellers who visit the great Cataract.

Dr. Richardson (Fauna Boreali Americana, p. 93) adheres to the opinion of the Indians, who regard the Cross Fox of the fur traders as a mere variety of the Red Fox. He says, "I found on inquiry that the gradations of colour between characteristic specimens of the Cross and Red Fox are so small, that the hunters are often in doubt with respect to the proper denomination of a skin; and I was frequently told, "This is not a Cross Fox yet, but it is becoming so." It is worthy of remark, moreover, that the European Fox (*Vulpes vulgaris*) is subject to similar varieties, and that the "*Canis crucigera* of Gesner differs from the latter animal in the same way that the American Cross Fox does from the red one."

We have had several opportunities of examining *C. crucigera* in the museums of Europe, and regard it as a variety of the common European Fox, but it differs in many particulars from any variety of the American Red Fox that we have seen.

The Cross Fox is generally regarded as being more wary and swift of foot than the Red Fox; with regard to its greater swiftness, we doubt the fact. We witnessed a trial of speed between the mongrel greyhound already referred to in this article, and a Red Fox, in the morning, and another between the same dog and a Cross Fox, about noon on the same day. The former was taken after an hour's hard run in the snow, and the latter in half that time, which we accounted for from the fact that the Cross Fox was considerably the fattest, and from this circumstance became tired out very soon. We purchased from a country lad a specimen of the Cross Fox in the flesh, which he told us he had caught with a common cur dog, in the snow, which was then a foot in depth.

In regard to the cunning of this variety there may be some truth in the general opinion, but this can be accounted for on natural principles; the skin is considered very valuable, and the animal is always regarded as a curiosity; hence the hunters make every endeavour to obtain one when seen, and it would not be surprising if a constant succession of attempts to capture it together with the instinctive desire for self-preservation possessed by all animals, should sharpen its wits and render it more cautious and wild than those species that are less frequently molested. We remember an instance of this kind which we will here relate.

A Cross Fox, nearly black, was frequently seen in a particular cover. We offered what was in those days considered a high premium for the animal in the flesh. The fox was accordingly chased and shot at by the farmers' boys in the neighbourhood. The autumn and winter passed away, nay, a whole year, and still the fox was going at large. It was at last regarded by some of the more credulous as possessing a charmed life, and it was thought that nothing but a silver ball could kill it. In the spring, we induced one of our servants to dig for the young Foxes that had been seen at the burrow which was known to be frequented by the Cross Fox. With an immense deal of labour and fatigue the young were dug out from the side of a hill; there were seven. Unfortunately we were obliged to leave home and did not return until after they had been given away and were distributed about the neighbourhood.

Three were said to have been black, the rest were red. The blackest of the young whelps was retained for us, and we frequently saw at the house of a neighbour, another of the litter that was red, and differed in no respect from the Common Red Fox. The older our little pet became, the less it grew like the Black, and the more like the Cross Fox. It was, very much to our regret, killed by a dog when about six months old, and as far as we can now recollect, was nearly of the colour of the specimen figured in our work.

The following autumn, we determined to try our hand at procuring the enchanted fox which was the parent of these young varieties, as it could always be started in the same vicinity. We obtained a pair of fine fox-hounds and gave chase. The dogs were young, and proved no match for the fox, which generally took a straight direction through several cleared fields for five or six miles, after which it began winding and twisting among the hills, where the hounds on two occasions lost the scent and returned home.

On a third hunt, we took our stand near the corner of an old field, at a spot we had twice observed it to pass. It came at last, swinging its brush from side to side, and running with great rapidity, three-quarters of a mile ahead of the dogs, which were yet out of hearing.—A good aim removed the mysterious charm: we killed it with squirrel-shot, without the aid of a silver bullet. It was nearly jet-black, with the tip of the tail white. This fox was the female which had produced the young of the previous spring that we have just spoken of; and as some of them, as we have already said, were Cross Foxes and others Red Foxes, this has settled the question in our minds, that both the Cross Fox and the Black Fox are mere varieties of the Red.

J. W. Audubon brought the specimen he obtained at Niagara, alive to New-York, where it was kept for six or seven weeks. It fed on meat of various kinds: it was easily exasperated, having been much teased on its way from the Falls. It usually laid down in the box in which it was confined, with its head toward the front and its bright eyes constantly looking upward and forward at all intruders. Sometimes during the night it would bark like a dog, and frequently during the day its movements corresponded with those of the latter animal. It could not bear the sun-light shining into its prison, and continued shy and snappish to the last.

The fur of the Cross Fox was formerly in great demand; a single skin sometimes selling for twenty-five dollars; at present, however, it is said not to be worth more than about three times the price of that of the Red Fox.

GEOGRAPHICAL DISTRIBUTION.

This variety seems to originate only in cold climates; hence we have not heard of it in the southern parts of the States of New-York and Pennsylvania, nor farther to the South. In the northern portions of the State of New-York, in New Hampshire, Maine, and in Canada, it is occasionally met with, in locations where the Red Fox is common. It also exists in Nova Scotia and Labrador. There is a Cross Fox on the Rocky Mountains, but we are not satisfied that it will eventually prove to be this variety.

GENERAL REMARKS.

The animal referred to by Sagard Theodat in his History of Canada, under the name of Renard Barré, Tsinantontongue, was evidently this variety. PENNANT probably also referred to it, (vol. i., p. 46,) although he blended it with the European *V. Crucigera* of GESNER, and the ***Korsraef*** of the Swedes. GEOFF (Collect. du Mus.) described and named it as a true species. DESMAREST (Mamm., p. 203, 308) and CUVIER (Dict. des Sc. Nat., vol. viii., p. 566) adopted his views. It is given under this name by SABINE (Franklin's Journ., p. 656.) HARLAN (Fauna, p. 88) published it as a distinct species, on the authority and in the words of DESMAREST. GODMAN, who gave the Black or Silver Fox (*A. argentatus*) as a true species, seemed doubtful whether the Cross Fox might not prove a "mule between the Black and Red Fox." RICHARDSON, under the name of the American Cross Fox, finally described it as a mere variety of the Red Fox.

We possess a hunter's skin, which we obtained whilst on the Upper Missouri, that differs greatly from the one we have described, in its size, markings, and the texture of its fur. The body, from point of nose to root of tail, is 33 inches long; tail to end of fur 18½; the skin is probably stretched beyond the natural size of the animal; but the tail, which is very large in circumference, is, we think, of its proper dimensions. The hair is long, being on the neck, sides, and tail, five inches in length; the under fur, which is peculiarly soft, is three inches long. There is scarcely a vestige of the yellowish-brown of our other specimen on the whole body; but the corresponding parts are gray. The tail is irregularly clouded and banded, the tip for three inches white. The colour of the remaining portions of the body does not differ very widely from the specimen we have described. The ears, nose, and paws of this specimen (as in most hunters' skins) are wanting. It is not impossible that this may be a variety of a *larger species* of Red Fox, referred to by LEWIS and CLARK, as existing on both sides of the Rocky Mountains.

Drawn on Stone by R. Trembly.

Carolina Grey Squirrel.

Male & Female.

Drawn from Nature by J. J. Audubon F.R.S.F.L.S. Printed by Nagel & Weingærtner, N.Y.

SCIURUS CAROLINENSIS.—GMEL.

CAROLINA GRAY SQUIRREL.

PLATE VII.—MALE AND FEMALE.

S. griseus supra, subtus albus, colorem haud mutaris, S. migratorii, minor. Cauda corpore breviore, S. migratorii angustiore.

CHARACTERS.

Smaller than the Northern Gray Squirrel, (Sciurus Migratorius,) tail narrower than in that species, and shorter than the body; above, rusty gray; beneath, white; does not vary in colour.

SYNONYMES.

ECUREUIL GRIS DE LA CAROLINE, Bosc., vol. ii., p. 96, pl. 29.
SCIURUS CAROLINENSIS, Bach., Monog., Proceedings Zool. Soc., London, August, 1838, Mag., Nat. Hist., 1839, p. 113.

DESCRIPTION.

This species, which has been many years known, and frequently described, has been always considered by authors as identical with the Gray Squirrel of the Northern States (*Sciurus migratorius*). There are, however, so many marked differences in size, colour and habit, that any student of nature can easily perceive the distinction between these two allied species.

Head shorter, and space between the ears proportionately broader than between those of the Northern Gray Squirrel; nose sharper than in that animal. Small anterior molar in the upper jaw permanent, (as we have invariably found it to exist in all the specimens we have examined;) it is considerably larger than in *S. migratorius*, and all our specimens which give indications of the individual having been more than a year old when killed, instead of having a small, thread-like, single tooth, as in the latter species, have a distinct double tooth with a double crown. The other molars are not much unlike those of *S. migratorius* in form, but are shorter and smaller,—the upper incisors being nearly a third shorter.

Body, shorter and less elegant in shape, and not indicating the quickness and vivacity by which *S. migratorius* is eminently distinguished.

The ears, which are nearly triangular, are so slightly clothed with hair on their interior surfaces, that they may be said to be nearly naked; externally they are sparsely clothed with short woolly hair, which however does not extend as far beyond the margins as in other species. Nails shorter and less crooked; tail shorter, and without the broad distichous appearance of that of the Northern Gray Squirrel.

COLOUR.

Teeth, light orange; nails, brown, lightest at the extremities; whiskers, black; on the nose and cheeks, and around the eyes, a slight tinge of rufous gray.

Fur on the back, for three-fourths of its length, dark plumbeous, succeeded by a slight indication of black, edged with yellowish-brown in some of the hairs, giving it on the surface a dark grayish-yellow tint. In a few specimens there is an obscure shade of light brown along the sides, where the yellowish tint predominates, and a tinge of this colour is observable on the upper surface of the fore-legs, above the knees. Feet, light gray; tail, for three-fourths of its length from the root yellowish-brown; the remainder black, edged with white; throat inner surface of the legs and belly, white.

This species does not run into varieties, as do the Northern Gray Squirrel and the Black Squirrel; the specimens received from Alabama, Florida and Louisiana, scarcely present a shade of difference from those existing in South Carolina, which we have just described.

DIMENSIONS.

Length of head and body - - - - - -	$9\frac{1}{2}$	inches.
" tail (vertebræ) - - - - - -	$7\frac{1}{3}$	do.
" " to end of hair - - - - -	$9\frac{1}{2}$	do.
Height of ear - - - - - - - - -	$\frac{1}{2}$	do.
Palm to end of middle claws - - - - -	$1\frac{1}{4}$	do.
Heel to end of middle nail - - - - - -	$2\frac{1}{2}$	do.
Length of fur on the back - - - - - -	$\frac{1}{2}$	do.
Breadth of tail (with hair extended) - - -	3	do.

HABITS.

This species differs as much in its habits from the Northern Gray Squirrel as it does in form and colour. From an intimate acquaintance with the habits of the latter, we are particularly impressed with the peculiarities of the present species. Its bark has not the depth of tone of that of the Northern species, and is more shrill and querulous. Instead of mounting high on the tree when alarmed, which the latter always does, the *Sc. Carolinensis* generally plays round the trunk, and on the side opposite to the observer, at a height of some twenty or thirty feet, often concealing itself beneath the Spanish moss (*Tillandsia Usneoides*) which hangs about the tree. When a person who has alarmed one of these Squirrels remains quiet for a few moments, it descends a few feet and seats itself on the first convenient branch, in order the better to observe his movements.

It is, however, capable of climbing to the extremity of the branches and leaping from tree to tree with great agility, but is less wild than the Northern species, and is almost as easily approached as the chickaree, (*Sc. Hudsonius.*) One who is desirous of obtaining a specimen, has only to take a seat for half an hour in any of the swamps of Carolina and he will be surprised at the immense number of these squirrels that may be seen running along the logs or leaping among the surrounding trees. A great many are killed, and their flesh is both juicy and tender.

The Carolina Gray Squirrel is sometimes seen on high grounds among the oak and hickory trees, although its usual haunts are low swampy places or trees overhanging streams or growing near the margin of some river. In deep cypress swamps covered in many places with several feet of water during the whole year, it takes up its constant residence, moving among the entwined branches of the dense forest with great facility. Its hole in such situations may sometimes be found in the trunk of a decayed cypress. On the large tupelo trees, (*Nyssa aquatica,*) which are found in the swamps, many nests of this species, composed principally of Spanish moss and leaves, are every where to be seen. In these nests, or in some woodpecker's hole, they produce their young. These are five or six in number, and are brought forth in March; it is well ascertained also that the female litters a second time in the season, probably about mid-summer.

This species has one peculiarity which we have not observed in any other. It is in some degree nocturnal, or at least crepuscular, in its habits. In riding along by-paths through the woods, long after sunset, we are often startled by the barking of this little Squirrel, as it scratches among the leaves, or leaps from tree to tree, scattering over the earth the seeds of the maple, &c., which are shaken off from the uppermost branches as it passes over them.

This species is seldom, if ever, seen in company with the Fox Squirrel,

(*Sc. Capistratus,*) or even found in the same neighbourhood; this arises probably not so much from any antipathy to each other, as from the fact that very different localities are congenial to the peculiar habits of each.

We have observed the Carolina Gray Squirrel on several occasions by moonlight, as actively engaged as the Flying Squirrel usually is in the evening; and this propensity to prolong its search after food or its playful gambols until the light of day is succeeded by the moon's pale gleams, causes it frequently to fall a prey to the Virginian owl, or the barred owl; which last especially is very abundant in the swamps of Carolina, where, gliding on noiseless pinions between the leafy branches, it seizes the luckless Squirrel ere it is aware of its danger, or can make the slightest attempt to escape. The gray fox and the wild cat often surprise this and other species by stratagem or stealth. We have beheld the prowling lynx concealed in a heap of brushwood near an old log, or near the foot of a tree frequented by the Squirrel he hopes to capture. For hours together will he lie thus in ambush, and should the unsuspicious creature pass within a few feet of him, he pounces on it with a sudden spring, and rarely fails to secure it.

Several species of snakes, the rattle-snake, (*Crotalus durassus,*) black snake, (*Coluber constrictor,*) and the chicken snake, (*Coluber quadrivittatus,*) for instance, have been found on being killed, to have a Squirrel in their stomach; and the fact that Squirrels, birds, &c., although possessing great activity and agility, constitute a portion of the food of these reptiles, being well established, the manner in which the sluggish serpent catches animals so far exceeding him in speed, and some of them endowed with the power of rising from the earth and skimming away with a few flaps of their wings, has been the subject of much speculation. Some persons have attributed a mysterious power, more especially to the rattle-snake and black snake—we mean the power of *fascinating*, or as it is commonly called, *charming*.

This supposed faculty of the serpent has, however, not been accounted for. The basilisk of the ancients killed by a look; the eye of the rattle-snake is supposed so to paralyze and at the same time attract its intended prey, that the animal slowly approaches, going through an infinite variety of motions, alternately advancing and retreating, until it finally falls powerless into the open jaws of its devourer.

As long as we are able to explain by natural deductions the very singular manœuvres of birds and squirrels when "fascinated" by a snake, it would be absurd to imagine that anything mysterious or supernatural is connected with the subject; and we consider that there are many ways of accounting for all the appearances described on these oc : si ·· s. Fear and surprise cause an instinctive horror when we find ourselves unexpectedly within a foot or two of a rattle-snake; the shrill, startling noise proceeding from the rattles of its tail as it vibrates rapidly, and its hideous aspect, no doubt produce a much greater effect on birds and small quadrupeds. It is said that the distant roar of the African lion causes the oxen to tremble and stand paralyzed in the fields; and HUMBOLDT relates that in the forests of South America the mingled cries of monkeys and other animals resound through the whole night, but as soon as the roar of the Jaguar, the American tiger, is heard, terror seizes on all the other animals, and their voices are suddenly hushed. Birds and quadrupeds are very curious, also, and this feeling prompts them to draw near to strange objects. "Tolling" wild ducks and loons, as it is called, by waving a red handkerchief or a small flag or by causing a little dog to bound backward and forward on a beach, has long been successfully practised by sportsmen on the Chesapeake Bay and elsewhere.

The Indians attract the reindeer, the antelope, and other animals, until they are within bow-shot, by waving a stick to which a piece of red cloth is attached, or by throwing themselves on their backs and kicking their heels up in the air. If any strange object is thrown into the poultry-yard, such as a stuffed specimen of a quadruped or a bird, &c., all the fowls will crowd near it, and scrutinize it for a long time. Every body almost may have observed at some time or other dozens of birds collected around a common cat in a shrubbery, a tortoise, or particularly a snake. The Squirrel is remarkable for its fondness for "sights," and will sometimes come down from the highest branch of a tree to within three feet of the ground, to take a view of a small scarlet snake, (*Rhinostoma coccinea*,) not much larger than a pipe-stem, and which, having no poisonous fangs, could scarcely master a grasshopper. This might be regarded by believers in the fascinating powers of snakes as a decided case in favour of their theories, but they would find it somewhat difficult to explain the following circumstances which happened to ourselves. After observing a Squirrel come down to inspect one of the beautiful little snakes we have just been speaking of, the reptile being a rare species was captured and secured in our carriage box. After we had driven off, we recollected that in our anxiety to secure the snake we had left our box of botanical specimens at the place where we had first seen the latter, and on returning for it, we once more saw the Squirrel darting backward and forward, and skipping round the root of the tree, eyeing with equal curiosity the article we had left behind; and we could not help making the reflection that if the little snake had "charmed" the Squirrel, the same "fascinating" influence was exercised by our tin box!

Quadrupeds and birds have certain antipathies: they are capable of experiencing many of the feelings that appertain to mankind; they are susceptible of passion, are sometimes spiteful and revengeful, and are wise enough to know their "natural enemies" without a formal introduction. The blue jay, brown thrush, white-eyed fly-catcher, and other little birds, are often to be heard scolding and fluttering about a thicket in which some animal is concealed; and on going to examine into the cause of their unwonted excitement, you will probably see a wild cat or fox spring forth from the covert. Every one familiar with the habits of our feathered tribes must have seen at times the owl or buzzard chased by the smallest birds, which unite on such occasions for the purpose of driving off a common enemy; in these cases the birds sometimes approach too near, and are seized by the owl. We once observed some night-hawks (*Chordeiles Virginianus*) darting round a tree upon which an owl was perched. Whilst looking on, we perceived the owl make a sudden movement and found that he had caught one of them in his sharp claws, and notwithstanding the cries and menaces of the others he instantly devoured it.

Birds dart in the same manner at snakes, and no doubt are often caught by passing too near—shall we therefore conclude that they are fascinated?

One of the most powerful "attractions" which remain to be considered, is the love of offspring. This feeling, which is so deeply rooted in the system of nature as to be a rule almost without an exception, is manifested strongly by birds and quadrupeds; and snakes are among the most to be dreaded destroyers of eggs and young birds and of the young of small species of viviparous animals; is it not likely therefore that many of the (supposed) cases of fascination that are related, may be referred to the intrepidity of the animals or birds, manifested in trying to defend their young or drive away their enemy from their vicinity? In our work, the "Birds of America," we represented a mocking-bird's nest attacked by a rattle-snake, and the nest of a red thrush invaded by a black snake; these two plates each exhibit several birds assisting the pair whose nest has been robbed by the snake, and also show the mocking-bird and thrush courageously advancing to the jaws even of their enemy. These pictures were drawn after the actual occurrence before our eyes of the scenes which we endeavoured to represent in them; and supposing a person but little acquainted with natural history to have seen the birds, as we did, he might readily have fancied that some of them at least were fascinated, as he could not probably have been near enough to mark the angry expression of their eyes, and see their well concealed nest.

Our readers will, we trust, excuse us for detaining them yet a little longer on this subject, as we have more to say of the habits of the rattle-snake in connexion with the subject we are upon.

This snake, the most venomous known in North America, subsists wholly on animal food; it digests its food slowly, and is able to exist without any sustenance for months, or even years, in confinement; during this time it often increases in size, and the number of its rattles is augmented. In its natural state it feeds on rabbits, squirrels, rats, birds, or any other small animals that may come in its way. It captures its prey by lying in wait for it, and we have heard of an instance in which one of these snakes remained coiled up for two days before the mouth of the burrow of the Florida rat, (*Neotoma Floridana*,) and on its being killed it was found to have swallowed one of these quadrupeds.

As far as we have been able to ascertain, it always strikes its intended prey with its fangs, and thus kills it before swallowing it. The bite is sudden, and although the victim may run a few yards after it is struck, the serpent easily finds it when dead. Generally the common species of rattle-snake refuses all food when in a cage, but occasionally one is found that does not refuse to eat whilst in captivity. When a rat is turned loose in a cage with one of these snakes, it does not immediately kill it, but often leaves it unmolested for days and weeks together. When, however, the reptile, prompted either by irritation or hunger, designs to kill the animal, it lies in wait for it, cat-like, or gently crawls up to it and suddenly gives the mortal blow, after which, it very slowly and deliberately turns it over into a proper position and finally swallows it.

We have seen a rattle-snake in a very large cage using every means within its power and exerting its cunning for a whole month, before it could succeed in capturing a brown thrush that was imprisoned with it. At night the bird roosted beyond the reach of the snake, and during the day-time it was too cautious in its movements, and too agile, snatching up its food at intervals, and flying instantly back to its perch, to be struck by the unwieldy serpent. We now added a mouse to the number of the inmates of the cage; the affrighted animal retreated to a corner, where the snake, slowly crawling up to it, with a sudden blow darted his fangs into and killed it; soon after which he swallowed it. About a week after this adventure, the snake again resumed his attempts to capture the thrush, and pursued it all round the cage.

This experiment offered a fair opportunity for the rattle-snake to exert its powers of fascination, had it possessed any; but as it did not exhibit them, we do not hesitate to say that it was entirely destitute of any faculty of the kind.

After some hours' fruitless manœuvring, the snake coiled itself up near the cup of water from which the bird drank. For two days the thrush avoided the water; on the third, having become very thirsty, it showed a constant desire to approach the cup; the snake waited for it to come within reach, and in the course of the day struck at it two or three times; the bird darted out of its way, however, and was not killed until the next day.

If, notwithstanding these facts, it is argued, that the mysterious and inexplicable power of *fascination* is possessed by the snake, because birds have been seen to approach it, and with open wings and plaintive voice seemed to wait upon its appetite, we must be prepared to admit that the same faculty is possessed by other animals. On a certain day, we saw a mocking-bird exhibiting every appearance, usually, according to descriptions, witnessed when birds are under the influence of fascination. It approached a hog which was occupied in munching something at the foot of a small cedar. The bird fluttered before the grunter with open wings, uttered a low and plaintive note, alighted on his back, and finally began to peck at his snout. On examining into the cause of these strange proceedings, we ascertained that the mocking-bird had a nest in the tree, from which several of her younglings had fallen, which the hog was eating! Our friend, the late Dr. WRIGHT, of Troy, informed us that he witnessed a nearly similar scene betwen a cat-bird and a dog which had disturbed her brood, on which occasion the cat-bird went through many of the movements generally ascribed to the effect of fascination.

GEOGRAPHICAL DISTRIBUTION.

We have received a specimen of this Squirrel which was procured in the market at New Orleans, where it is said to be exceedingly rare. We have not traced it farther to the South. It is the most abundant species in Florida, Georgia and South Carolina. We have seen it in the swamps of North Carolina, but have no positive evidence that it extends farther to the northward than that State. We have obtained it in Alabama, and in Mississippi we are told it is found in the swamps. Nothing has been heard of it west of the Mississippi river.

GENERAL REMARKS.

This species was first described by GMELIN, and afterwards noticed and figured by Bosc. The descriptions in Harlan, Godman, and all other authors who have described this species under the name of *Sciurus Carolinensis*, refer to the *Northern* Gray Squirrel. We believe we were the first to observe and point out the distinctive characters which separate the present species from *S. migratorius*, the Gray Squirrel of the North.

GENUS TAMIAS.—Illiger.

$$Incisive\ \frac{2}{2};\ Canine\ \frac{0}{0};\ Molar\ \frac{5\text{—}5}{4\text{—}4} = 22.$$

Upper incisors, smooth; lower ones, compressed and sharp; molars, with short, tuberculous crowns.

Nose, pointed; lip, cloven; ears, round, short, not tufted or fringed; cheek-pouches, ample.

Tail, shorter than the body, hairy, sub-distichous, somewhat tapering. Mammæ, exposed; feet, distinct, ambulatory; fore-feet, four toed, with a minute blunt nail in place of a thumb; hind-feet five toed; claws, hooked.

This genus differs from Sciurus in several important particulars. The various species that have been discovered have all the same characteristics, and strongly resemble each other in form, in their peculiar markings and in their habits. In shape they differ from the true squirrels and approach to the spermophiles; they have a sharp convex nose adapted to digging in the earth; they have longer heads, and their ears are placed farther back than those of squirrels; they have a more slender body and shorter extremities. Their ears are rounded, without any tufts on the borders or behind them. They have cheek-pouches, of which all squirrels are destitute; their tails are roundish, narrow, seldom turned up, and only sub-distichous.

The species belonging to this genus are of small size, and are all longitudinally striped on the back and sides.

Their notes are very peculiar; they emit a chipping clucking sound differing very widely from the quacking chattering cry of the squirrels.

They do not mount trees unless driven to them from necessity, but dig burrows, and spend their nights and the season of winter under ground.

They are, however, more closely related to the squirrels than to the spermophiles. The third toe from the inner side is slightly the longest, as in the former; whilst in the latter, the second is longest, as in the marmots. The genus Tamias is therefore nearly allied to the squirrels, whilst the spermophiles approach the marmots.

Authentic species of the genus Sciurus are already very numerous, and as we have now a number of species, to which constant additions are making by the explorers of our Western regions, which by their cheek-pouches, their markings, and habits, can be advantageously separated

N°2 Plate VIII

Chipping Squirrel, Hackee

Drawn from Nature by J.J. Audubon F.R.S, F.L.S. Printed by Nagel & Weingærtner N.Y.

from that genus, no doubt naturalists will arrange them in the genus TAMIAS.

When this genus was first established by ILLIGER, but a single species was satisfactorily known, and naturalists were unwilling to separate it from the squirrels, to which it bears so strong an affinity; but we are now, however, acquainted with six species, and doubt not that a few more years of investigation will add considerably to this number. We have consequently adopted the genus TAMIAS of that author.

The word Tamias is derived from the Greek ταμιας, (*tamias,*) a keeper of stores—in reference to its cheek-pouches.

One species of this genus exists in the Northern portions of the Eastern continent; four in North, and one in South, America. We also possess an undescribed species, the habitat of which is at present unknown to us.

TAMIAS LISTERI.—RAY.

CHIPPING SQUIRREL, HACKEE, &c.

PLATE VIII.—MALE, FEMALE, AND YOUNG (First Autumn).

T. dorso fusco-cinereo, striis quinque nigris, et duobus luteo-albis longitudinalibus ornato; fronte et natibus fusco-luteis; ventre albo.

CHARACTERS.

Brownish gray on the back; forehead and buttocks brownish orange; five longitudinal black stripes and two yellowish white ones on the back; under surface white.

SYNONYMES.

ECUREUIL SUISSE, Sagard Theodat, Canada, p. 746, A. D. 1636.
GROUND SQUIRREL, Lawson's Carolina, p. 124.
" " Catesby, Carol. vol. ii., p. 75.
EDWARDS, vol. iv., p. 181. Kalm, vol. i., p. 322.
SCIURUS LYSTERI, Ray, Synops. Quad., p. 216, A. D. 1693.
LE SUISSE, Charlevoix, Nouv. Fr., vol. v., p. 196.
STRIPED DORMOUSE, Pennant, Arc. Zool., 4 vols., vol. i., p. 126.
SCIURUS CAROLINENSIS, Brisson, Reg. Anim., p. 155, A. D. 1756.
ECUREUIL SUISSE, (Desm. Enc. Mamm.,) Nota, p. 339, Esp., 547.

SCIURUS STRIATUS, Harlan, Fauna, p. 183.
" " Godman, Nat. Hist., vol. ii., p. 142.
SCIURUS (TAMIAS) LYSTERI, Rich., F. B. A., p. 181, plate 15.
" " " Doughty's Cabinet Nat. Hist., vol. i., p. 169, pl. 15.
SCIURUS STRIATUS, DeKay, Nat. Hist. of N. Y., part 1, p. 62, pl. 16, fig. 1.

DESCRIPTION.

Body, rather slender; forehead, arched; head, tapering from the ears to the nose, which is covered with short hairs; nostrils, opening downwards, margins and septum naked; whiskers, shorter than the head. A few bristles on the cheeks and above the eye-brows; eyes, of moderate size; ears, ovate, rounded, erect, covered with short hair on both surfaces, not tufted, the hair on those parts simply covering the margins. Cheek-pouches, of tolerable size, extending on the sides of the neck to a little below the ear, opening into the mouth between the incisors and molars. Fore-feet, with four slender, compressed, slightly-curved claws, and the rudiment of a thumb, covered with a short blunt nail; hind-feet, long and slender, with five toes, the middle toe being a little the longest. Tail, rather short and slender, nearly cylindrical above, dilated on the sides, not bushy, sub-distichous. Hair on the whole body short and smooth, but not very fine.

COLOUR.

A small black spot above the nose; forehead, yellowish brown; above and beneath the eyelids, white; whiskers and eyelashes, black; a dark brown streak running from the sides of the face through the eye and reaching the ear; a yellowish brown stripe extending from near the nose, running under the eye to behind the ear, deepening into chesnut-brown immediately below the eye, where the stripe is considerably dilated.

Anterior portion of the back, hoary gray, this colour being formed by a mixture of gray and black hairs. Colour of the rump, extending to a little beyond the root of the tail, hips, and exterior surface of the thighs, reddish fawn, a few black hairs sprinkled among the rest, not sufficiently numerous to give a darker shade to those parts. A dark dorsal line commencing back of the head is dilated on the middle of the back, and runs to a point within an inch of the root of the tail; this line is brownish on the shoulder, but deepens into black in its progress downwards.

On each flank there is a broad yellowish-white line, running from the shoulder to the thighs, bordered on each side with black. The species may be characterized by its having five black and two white stripes on a gray ground. The flanks, sides, and upper surface of feet and ears, are reddish-gray; whole under surface white, with no line of demarcation between the colours of the back and belly. Tail, brown at its root, afterwards grayish-black, the hair being clouded and in some places banded with black; underneath, reddish-brown, with a border of black, edged with light gray.

There are some varieties observable among specimens procured in different States of the Union. We have noted it, like the Virginian deer, becoming smaller in size at it was found farther to the South. In Maine and New Hampshire it is larger than in the mountains of Carolina and Louisiana, and the tints of those seen at the North were lighter than the colouring of the Southern specimens we have examined. We possess an albino, sent to us alive, snow-white, with red eyes; and also another specimen jet-black. We have, however, found no intermediate varieties, and in general we may remark that the species of this genus are not as prone to variations in colour as those of the true Squirrels.

DIMENSIONS.

	Inches.	Lines.
Length of head and body - - - - -	6	3
" head - - - - - - - -	1	6
" tail (vertebræ) - - - - - -	3	7
" tail, including fur - - - -	4	7
Height of ear - - - - - - - -	0	4
Breadth of ear - - - - - - - -	0	3½

HABITS.

The Chipping Squirrel, as this little animal is usually called, or Ground Squirrel, as it is named almost as frequently, is probably, with the exception of the common flying squirrel, (*Pteromys volucella*,) one of the most interesting of our small quadrupeds. It is found in most parts of the United States, and being beautifully marked in its colouring, is known to every body. From its lively and busy habits, one might consider it among the quadrupeds as occupying the place of the *wren* among the feathered tribes. Like the latter, the Ground Squirrel, full of vivacity, plays with the utmost grace and agility among the broken rocks or uprooted stumps of trees about the farm or wood pasture; its clucking resembles the chip, chip, chip, of a young chicken, and although not musical, like the song of the little winter wren, excites agreeable thoughts as it comes on the air. We fancy we see one of these sprightly Chipping Squirrel as he runs before us with the speed of a bird, skimming along a log o fence, his chops distended by the nuts he has gathered in the woods; he makes no pause till he reaches the entrance of his subterranean retreat and store-house. Now he stands upright, and his chattering cry is heard, but at the first step we make towards him, he disappears. Stone after stone we remove from the aperture leading to his deep and circuitous burrow; but in vain is all our labour—with our hatchets we cut the tangled roots, and as we follow the animal, patiently digging into his innermost retreat, we hear his angry, querulous tones. We get within a few inches of him now, and can already see his large dark eyes; but at this moment out he rushes, and ere we can "grab" him, has passed us, and finds security in some other hiding place, of which there are always plenty at hand that he is well accustomed to fly to; and we willingly leave him unmolested, to congratulate himself on his escape.

The Chipping Squirrel makes his burrow generally near the roots of trees, in the centre of a decayed stump, along fences or old walls, or in some bank, near the woods whence he obtains the greater portion of his food.

Some of these retreats have two or three openings at a little distance from each other. It rarely happens that this animal is caught by digging out its burrow. When hard pressed and closely pursued it will betake itself to a tree, the trunk of which it ascends for a little distance with considerable rapidity, occasionally concealing itself behind a large branch, but generally stopping within twelve or fifteen feet of the ground, where it often clings with its body so closely pressed to the trunk that it is difficult to detect it; and it remains so immovable that it appears like a piece of bark or some excrescence, till the enemy has retired from the vicinity, when it once more descends, and by its renewed clucking seems to chuckle over its escape.

We are doubtful whether this species can at any time be perfectly tamed. We have preserved it in cages from time to time, and generally found it wild and sullen. Those we had, however, were not young when captured.

At a subsequent period we obtained in the State of New-York five or six young ones almost half grown. We removed them to Carolina, where they were kept during winter and spring. They were somewhat more gentle than those we had formerly possessed, occasionally took a filbert or a ground-nut from the fingers, but never became tame enough to be handled with safety, as they on more than one occasion were disposed to test the sharpness of their teeth on our hand.

The skin which covered the vertebræ of their tails was so brittle that nearly all of them soon had mutilated them. They appeared to have some aversion to playing in a wheel, which is so favourite an amusement of the true squirrels. During the whole winter they only left their nest to carry into it the rice, nuts, Indian corn, &c., placed in their cage as food.

Late in the following spring, having carried on our experiments as far as we cared to pursue them, we released our pets, which were ocoasionally seen in the vicinity for several months afterward, when they disappeared.

We were once informed of a strange carnivorous propensity in this species. A lady in the vicinity of Boston said to us, "We had in our garden a nest of young robins, (*Turdus migratorius*,) and one afternoon as I was walking in the garden, I happened to pass very close to the tree on which this nest was placed; my attention was attracted by a noise which I thought proceeded from it, and on looking up I saw a Ground Squirrel tearing at the nest, and actually devouring one of the young ones. I called to the gardener, who came accompanied by a dog, and shook the tree violently, when the animal fell to the earth, and was in an instant secured by the dog." We do not conceive that the unnatural propensity in the individual here referred to, is indicative of the genuine habit of this species, but think that it may be regarded as an exception to a general rule, and referred to a morbid depravity of taste sometimes to be observed in other genera, leading an individual to feed upon that which the rest of the species would loathe and reject. Thus we have known a horse which preferred a string of fish to a mess of oats; and mocking-birds, in confinement, kill and devour jays, black-birds, or sparrows.

We saw and caught a specimen of this beautiful TAMIAS in Louisiana, that had no less than sixteen chinquapin nuts (*Castanea pumila*) stowed away in its cheek-pouches. We have a specimen now lying before us, sent from Pennsylvania in alcohol, which contains at least one and a half table-spoonfuls of Bush trefoil (*Hedysarum cannabinum*) in its widely-distended sacks. We have represented one of our figures in the plate with its pouches thus filled out.

This species is to a certain extent gregarious in its habits. We had marked one of its burrows in autumn which we conceived well adapted to our purpose, which was to dig it out. It was in the woods on a sandy piece of ground and the earth was strewed with leaves to the depth of eight inches, which we believed would prevent the frost from penetrating to any considerable depth. We had the place opened in January, when the ground was covered with snow about five inches deep. The entrance of the burrow had been closed from within. We followed the course of the small winding gallery with considerable difficulty. The hole descended at first almost perpendicularly for about three feet. It then continued with one or two windings, rising a little nearer the surface until it had advanced about eight feet, when we came to a large nest made of oak leaves and dried grasses. Here lay, snugly covered, three Chipping Squirrels. Another was subsequently dug from one of the small lateral galleries, to which it had evidently retreated to avoid us. They were not dormant, and seemed ready to bite when taken in the hand; but they were not very active, and appeared somewhat sluggish and benumbed, which we conjectured was owing to their being exposed to sudden cold from our having opened their burrow.

There was about a gill of wheat and buckwheat in the nest; but in the galleries we afterwards dug out, we obtained about a quart of the beaked hazel nuts, (*Corylus rostrata,*) nearly a peck of acorns, some grains of Indian corn, about two quarts of buckwheat, and a very small quantity of grass seeds. The late Dr. John Wright, of Troy, in an interesting communication on the habits of several of our quadrupeds, informs us, in reference to this species, that "It is a most provident little creature, continuing to add to its winter store, if food is abundant, until driven in by the severity of the frost. Indeed, it seems not to know when it has enough, if we may judge by the surplus left in the spring, being sometimes a peck of corn or nuts for a single Squirrel. Some years ago I watched one of these animals whilst laying up its winter store. As there were no nuts to be found near, I furnished a supply. After scattering some hickory nuts on the ground near the burrow, the work of carrying in was immediately commenced. It soon became aware that I was a friend, and approached almost to my feet for my gifts. It would take a nut from its paws, and dexterously bite off the sharp point from each end, and then pass it to its cheek-pouch, using its paws to shove it in, then one would be placed on the opposite side, then again one along with the first, and finally, having taken one between its front teeth, it would go into the burrow. After remaining there for five or ten minutes it would reappear for another load. This was repeated in my presence a great number of times, the animal always carrying four nuts at a time, and always biting off the asperities."

We perceive from hence that the Chipping Squirrels retire to winter quarters in small families in the early part of November, sooner or later according to the coldness or mildness of the season, after providing a store of food in their subterranean winter residence. When the snows are melted from the earth in early spring, they leave the retreat to which they had resorted during the first severe frosts in autumn. We have seen them sunning themselves on a stump during warm days about the last of February, when the snows were still on the earth here and there in patches a foot deep; we remarked, however, that they remained only for half an hour, when they again retreated to their burrows.

The young are produced in May, to the number of four or five at a birth, and we have sometimes supposed from the circumstance of seeing a young brood in August, that they breed twice a year.

The Chipping Squirrel does but little injury to the farmer. It seldom disturbs the grain before it is ripe, and is scarcely more than a gleaner of the fields, coming in for a small pittance when the harvest is nearly gathered. It prefers wheat to rye, seems fond of buckwheat, but gives the preference to nuts, cherry-stones, the seeds of the red gum, or pepperidge, (*Nyssa Multiflora,*) and those of several annual plants and grasses.

This species is easily captured. It enters almost any kind of trap without suspicion. We have seen a beautiful muff and tippet made of a host of little skins of this TAMIAS ingeniously joined together so as to give the appearance of a regular series of stripes around the muff, and longitudinally along the sides of the tippet. The animals had in most cases been captured in rat-traps.

There is, besides, a simple, rustic, but effectual mode of hunting the Ground Squirrel, to which we are tempted to devote a paragraph.

Man has his hours of recreation, and so has the school-boy; while the former is fond of the chase, and keeps his horses, dogs and guns, the latter when released from school gets up a little hunt agreeable to his own taste and limited resources. The boys have not yet been allowed to carry fire-arms, and have been obliged to adhere to the command of a careful mother—"don't meddle with that gun, Billy, it may go off and kill you." But the Chip Muck can be hunted without a gun, and Saturday, the glorious weekly return of their freedom and independence from the crabbed schoolmaster and the puzzling spelling-book, is selected for the important event.

There are some very pleasing reminiscences associated with these little sports of boyhood. The lads, hurried by delightful anticipations, usually meet half an hour before the time appointed. They come with their "shining morning faces" full of glee and talking of their expected success. In lieu of fire-arms they each carry a stick about eight feet long. They go along the old-fashioned worm-fences that skirt the woods,—a crop of wheat or of buckwheat has just been gathered, and the little Hackee is busily engaged in collecting its winter store.

In every direction its lively chirrup is heard, with answering calls from adjacent parts of the woods, and here and there you may observe one mounted on the top of a fence-stake, and chipping away as it were in exultation at his elevated seat. One of the tiny huntsmen now places his pole on a fence rail, the second or third from the bottom, along which the Ground Squirrel is expected to pass; a few yards behind him is another youngster, ready with his stick on another rail, in case the Chip Muck escapes the first enemy. One of the juveniles now makes a circuit, gets behind the little Hackee and gives a blow on the fence to drive him toward the others, who are eagerly expecting him. The unsuspecting little creature, with a sweep of his half-erected tail, quickly descends from the top of the fence along a stake, and betaking himself to some of the lower rails, makes a rapid retreat. If no stone-heaps or burrows are at hand, he runs along the winding fence, and as he is passing the place where the young sportsmen are lying in wait, they brush the stick along the rail with the celerity of thought, hitting the little creature on the nose, and knocking him off. "He is ours," is the exulting shout, and the whole party now hurry to the spot. Perhaps the little animal is not dead, only stunned, and is carried home to be made a pet. He is put into a calabash, a stocking, or a small bag prepared for the occasion by some fond little sister, who whilst sewing it for her brother half longed to enjoy the romp and the sport herself. Reader, don't smile at this group of juvenile sportsmen; older and bigger "boys" are often engaged in amusements not more rational, and not half so innocent.

Several species of hawks are successful in capturing the Chipping Squirrel. It furnishes also many a meal for the hungry fox, the wild cat, and the mink; but it possesses an enemy in the common weasel or ermine, (*mustela erminea*) more formidable than all the rest combined. This bloodthirsty little animal pursues it into its dwelling, and following it to the farthest extremity, strikes his teeth into its skull, and like a cruel savage of the wilderness, does not satiate his thirst for blood until he has destroyed every inhabitant of the burrow, old and young, although he seldom devours one fifth of the animals so wantonly killed. We once observed one pursue a Chipping Squirrel into its burrow. After an interval of ten minutes it reappeared, licking its mouth, and stroking its fur with its head by the aid of its long neck. We watched it as it pursued its way through a buckwheat field, in which many roots and stumps were yet remaining, evidently in quest of additional victims. On the following day we were impelled by curiosity to open the burrow we had seen it enter. There we found an old female ground squirrel and five young, half-grown, lying dead, with the marks of the weasel's teeth in their skulls.

GEOGRAPHICAL DISTRIBUTION.

The Chipping Squirrel has a pretty wide geographical range. It is common on the northern shores of Lakes Huron and Superior; and has been traced as far as the fiftieth degree of north latitude. In the Eastern, Northern, and Middle States, it is quite abundant; it exists along the whole of the Alleghany range, and is found in the mountainous portions of South Carolina, Georgia, and Alabama. In the alluvial districts of Carolina and Georgia it disappears. We have never found it nearer the seaboard of South Carolina than at Columbia, one hundred and ten miles from Charleston, where it is very rare. It is found in Tennessee and throughout Louisiana.

GENERAL REMARKS.

We have at the head of this article endeavoured to preserve TAMIAS as a valuable genus distinct from SCIURUS. We hope we have offered such reasons as will induce naturalists to separate this interesting and increasing little group, mostly of American species, from the squirrels, to which they bear about the same affinity as do the marmot squirrels (SPERMOPHILUS) to the true marmots (ARCTOMYS). We will now inquire whether the present species (*Tamias Lysteri*) is a foreigner from Siberia, naturalized in our Western world; or whether it is one of the aborigines of our country, as much entitled to a name as the grisly bear or the cougar.

Two of our American naturalists, HARLAN and GODMAN, supposed that it was the Asiatic species, the *S. striatus* of KLEIN, PALLAS, SCHREBER, and other authors; Dr. RICHARDSON (1829) believed that the descriptions given of *Sciurus striatus* did not exactly correspond with American specimens, and as he had no opportunity of instituting a comparison, he adopted the specific name of RAY, *Sciurus* (TAMIAS) *Lysteri*, for our species; and quoted what PALLAS had written in regard to the habits of the Asiatic animal, as applying to those of our little Chipping Squirrel. Very recently (1842) Dr. DEKAY, in the work on American quadrupeds, published by order of the State of New-York, has again referred it to *S. striatus* of LINNÆUS, and endeavoured to prove the identity of the two species from European writers. We suspect he had no opportunity of making a comparison from actual specimens.

Reasoning from analogy in regard to the species of birds or quadrupeds found to be identical on both continents, we should be compelled to admit that if our species is the *S. striatus* of Asia, it presents a solitary exception to a long-established general rule. That many species of water-birds, such as geese, ducks, gulls, auks, and guillemots, which during the long days of summer crowd toward the polar regions to engage in the duties and pleasures of reproduction, should be found on both continents, cannot be a matter of surprise; and that the ptarmigan, the white snow-bird, Lapland long-spur, &c., which resort annually to them, should at that season take wing and stray to either continent, is so probable a case, that we might think it strange if it were otherwise. Neither need we regard it as singular if a few quadrupeds, with peculiar constitutions and habits suited to the polar regions, should be inhabitants of the northern portions of both continents. Thus the polar bear, which delights in snow and ice, and which is indifferent as to whether it is on the land or on an iceberg at sea; the reindeer, which exists only in cold regions, and which by alternately swimming and walking can make its way over the icy waters in winter, and over rivers and arms of the sea in summer, and which migrates for thousands of miles; the beaver, which is found all over our continent, on the banks of the Mackenzie river leading into the polar sea in latitude 68°, and in the Russian settlements near Behring's Straits; the ermine, which riots in the snow-drifts, and has been found as far to the north as man has ever travelled; and the common wolf, which is a cosmopolite, exhibits itself in all colours, and strays from the tropics to the north pole; may be found on both continents without surprising us: but if this little land-animal, the Chipping Squirrel, which is unable to swim, and retires to the earth in cold weather, should be found both in Asia and America, it would oppose all our past experience in regard to American quadrupeds, and be the only exception to a long and universally admitted theory. The highest northern range in which this species has ever been seen is above Lake Huron, as far as latitude 50°; from thence there is a distance of more than 90° of longitude and 18° of latitude before we reach its Asiatic range, and in its migrations either way it would have to cross Behring's Straits, and traverse regions which even in summer are covered with snow and ice. From the above facts and from our knowledge of the adaptation of various animals for extensive migrations, we must conclude that this species cannot possibly exist on both continents, even admitting the correctness of the supposition that these continents had in some former age been united.

Dr. Richardson says, (p. 181,) "I am not aware that the identity of the species on the two continents has been established by actual comparison." In this he was quite correct. At the period when his valuable work on American quadrupeds was published, nearly all the figures and many of the descriptions of *Tamias striatus* of the Eastern continent were taken from American specimens of *Tamias Lysteri;* and the authors supposing them to be identical, were not sufficiently cautious to note this important fact.

In 1838 we carried to Europe, American specimens of nearly all those species which had their congeners on the Eastern continent. We were surprised at finding no specimen of the *T. striatus* in the museums of either England or France. At Berlin, however, an excellent opportunity was afforded us for instituting a comparison. Through the kindness of Dr. Lichtenstein, superintendent of the museum, we were permitted to open the cases, examine several specimens in a fine state of preservation, and compare them with our American species, which we placed beside them. The differences, at first sight were so striking that we could only account for their ever having been considered identical, from the fact that the descriptions of the old authors were so loose and unsatisfactory that many minute but important characteristics had not been noted. The following memorandum was made by us on the occasion:—"The *Tamias striatus* differs so widely from our American Chipping Squirrel or Hackee, that it is unnecessary to be *very* minute in making the comparison. The two species can always be distinguished from each other by one remarkable characteristic, which I have observed running through all the specimens. The stripes on the Asiatic (*T. striatus*) running over the back extend to the root of the tail; whilst those on the American (*T. Lysteri*) do not reach so far by a full inch. There are many other differences which may as well be noticed. *T. striatus* is a little the largest, the stripes on the back are situated nearer each other, and are broader than in the other species; the stripes on each side of the back are nearly black instead of yellowish-brown; on each side of the black stripe on the centre of the back of *Tamias Lysteri*, there is a broad space of reddish-gray. In *T. striatus* this part of the animal is yellowish; being an alternate stripe of black and yellowish-white. The tail of the latter is black towards the extremity, and tipped with white; its tail and ears also are larger than those of *T. Lysteri:* in short, these two species differ as widely from each other as *Tamias Lysteri* differs from the four-lined ground squirrel of Say, (*T. quadrivittatus.*)

GENUS SPERMOPHILUS. F. CUVIER.

DENTAL FORMULA.

$$Incisive\ \frac{2}{2};\ Canine\ \frac{0-0}{0-0};\ Molar\ \frac{5-5}{4-4} = 22.$$

The dentition of the Spermophiles differs from that of the true marmots in the following particulars. The first longitudinal eminence (colline) is nearly obliterated, and the curve (talon) which unites the second to the third is prolonged much more internally, which makes the molars of the Spermophiles more narrow transversely than longitudinally, as compared with those of the marmots. The teeth of the souslik (*Spermophilus citillus*) were examined by F. CUVIER, and considered as typical of this genus.

Nose, convex; ears, generally short; cheek-pouches.

Body, rather short; mammæ, pectoral and abdominal, from eight to twelve.

Feet, of moderate length, adapted for walking on the ground; nails, less in size than those of the marmots, less hooked than those of the squirrels; on the fore-feet, four toes, with the rudiment of a thumb, protected by a blunt nail; second toe from the thumb longest, as in the marmots, and not the third, as in the squirrels; hind-feet, with five toes.

Tail, generally rather short, and always shorter than the body; in several of the species capable of a slightly distichous arrangement.

The species belonging to this genus differ from the true marmots, not only in their teeth, as shown above, but also in several other striking particulars. They have cheek-pouches, of which the marmots are destitute. They are by no means clumsy, and in form are rather slender, and possess a degree of lightness and agility approaching the activity of the squirrels.

With the genus TAMIAS they assimilate so closely, that some of the species present intermediate characters, and authors may well differ as to which genus they ought to be referred to. Thus *Tamias quadrivittatus*, and *Spermophilus lateralis*, seem to form a connecting link between these two genera. It is to be recollected, however, that analogous cases exist, not only among the mammalia, but in every class of animals, and more especially in birds.

No. 2
Plate IX
Drawn on Stone by R. Trembly
Parry's Marmot Squirrel
Drawn from Nature by J.J. Audubon F.R.S. F.L.S.
Printed by Nagel & Weingaertner N.Y.

In referring again to the dentition of these allied genera, we may remark that the anterior molar of the upper jaw, which is deciduous and falls out at an early period in most species of true squirrels, remains permanently in all species of the genus TAMIAS and is smaller than in the Spermophiles. These genera differ also in the form and length of their claws. The long nails of the latter, the second claw, moreover, being longest, places them near the marmots; while the shorter, weaker, and more arched nails of the ground squirrels, in which the third claw, besides, is the longest, approximates them more nearly to the true squirrels.

The clucking notes of the chipping squirrels are replaced in the marmot-squirrels by the shrill whistling or chattering sounds emitted by the marmots.

The generic appellation Spermophilus, is derived from the Greek words σπερμα (*sperma*), a seed, and φιλος (*philos*), a lover.

There are now twelve species of this genus known as existing in North America, and three in Europe, and a few are set down as belonging to Asia and Africa. Some of the latter may, however, after more careful examination, be found to belong to the genus ARCTOMYS.

SPERMOPHILUS PARRYI.—RICHARDSON.

PARRY'S MARMOT-SQUIRREL.—PARRY'S SPERMOPHILE.

PLATE IX.—MALE.

S. flavo-cinereus, supra albo variegatus, genis, lateribus, ventre, pedibusque flavis; fronte aureo, pilis ex flavo et nigro; ad radices flavis, apice nigris.

CHARACTERS.

General colour, yellowish-gray; upper parts, mottled with white; cheeks, sides, under parts of the body, and feet, yellow; fore-part of the head, deep rich yellow; the hairs varied with yellow and black; at the roots chiefly deep yellow, and at the points principally black

SYNONYMES.

GROUND-SQUIRREL, Hearne's Journey, pp. 141 and 386.
QUEBEC MARMOT, Forster, Phil. Trans., vol. lxii. p. 378.
ARCTOMYS ALPINA, Parry, Second Voyage, p. 61, narrative.
ARCTOMYS PARRYI, Richardson, Parry's Second Voyage, App., p. 316.
ARCTOMYS (SPERMOPHILUS) PARRYI, Rich., Fauna Boreali Americana, p. 158, **pl. 10.**
SEEK-SEEK, ESQUIMAUX,—THOE-THIAY ROCK-BADGER, CHIPEWYANS, Rich.

DESCRIPTION.

This marmot-squirrel, although far from being as thick and heavy as the Maryland marmot, is not nearly so light and graceful as most of the other species of this genus, especially *Sp. Douglassii;* and in form resembles the marmots more than it does the ground squirrels. The forehead is arched, the nose rather short, thick, and closely covered with short hair; ears, short, triangular, and situated above the auditory opening; eyes, .prominent, and of moderate size; a few rather slender hairs over the eyes; along the cheeks are whiskers, arranged in five rows. Cheek-pouches, of medium dimensions, and opening into the mouth immediately behind the molars.

Legs and feet rather short and stout; toes well separated; nails long; feet covered with short hairs; palms of the fore-feet naked; soles of hind-feet for half an inch next the heel clothed with hair, the remainder naked. Tail, rather flat, rounded at base, hairs becoming longer towards the extremity; sub-distichous. The under fur on every part of the body, soft, glossy, and of a silky appearance.

COLOUR.

Hairs of the back, black at the roots, annulated above with black, nearer the tips yellowish-white or white; extreme tips black.

The longest hairs black; the under, black at the base, then whitish, and shaded into brown at the points. The whole upper surface is irregularly and thickly spotted with white; the spots confluent, especially over the shoulders; on the belly the under-fur is abundant, very soft and silky; grayish-black at the base, and yellowish-white at the tips; the visible portion of the longer hairs, deep yellow on the sides of the body, and paler yellow on the belly. Feet, yellow; hairs on the toes a pale yellow; claws blackish-brown; the hinder half of the tarsus covered beneath with brownish hairs; upper surface of the head, as far back as the eyes, of a deep rich yellow; around the eyes whitish; cheeks yellow; chin, throat, and sides of the muzzle, yellowish-white; tail, at base, coloured like the body; in the middle, the hairs are yellowish, with two rings or bars of black at the tips. The hairs on the under surface of the tail are chiefly of a rusty or brownish-red colour; moustaches black.

DIMENSIONS.

	Inches.	Lines.
From nose to root of tail - - - - -	11	6
Tail (vertebræ) - - - - - - - -	4	6
Tail, to end of hair - - - - - -	6	0
From heel to end of claw - - - - -	2	3
From ear to point of nose - - - - -	2	0
Height of ear - - - - - - - -	0	2½

HABITS.

The only account we have of this handsome spermophile is that given by its talented discoverer, who says of it,—

"It is found generally in stony districts, but seems to delight chiefly in sandy hillocks amongst rocks, where burrows, inhabited by different individuals, may be often observed crowded together. One of the society is generally observed sitting erect on the summit of the hillocks, whilst the others are feeding in the neigbourhood. Upon the approach of danger, he gives the alarm, and they instantly betake themselves to their holes, remaining chattering, however, at the entrance until the advance of the enemy obliges them to retire to the bottom. When their retreat is cut off, they become much terrified, and seeking shelter in the first crevice that offers, they not unfrequently succeed only in hiding the head and fore-part of the body, whilst the projecting tail is, as usual with them when under the influence of terror, spread out flat on the rock. Their cry in this season of distress strongly resembles the loud alarm of the Hudson's Bay squirrel, and is not very unlike the sound of a watchman's rattle. The Esquimaux name of this animal, *Seek-Seek*, is an attempt to express this sound. According to HEARNE, they are easily tamed, and are very cleanly and playful in a domestic state. They never come abroad during the winter. Their food appears to be entirely vegetable; their pouches being generally observed to be filled, according to the season, with tender shoots of herbaceous plants, berries of the Alpine arbutus, and of other trailing shrubs, or the seeds of bents, grasses, and leguminous plants. They produce about seven young at a time."

Captain Ross mentions that some of the dresses of the Esquimaux at Repulse Bay, were made of the skins of this species; these people also informed him that it was very abundant in that inhospitable region.

GEOGRAPHICAL DISTRIBUTION.

According to Dr. RICHARDSON, "this spermophile inhabits the barren grounds skirting the sea-coast, from Churchill, in Hudson's Bay, round by Melville's Peninsula, and the whole northern extremity of the Continent to Behring's Straits, where specimens precisely similar were procured by Captain BEECHEY. It abounds in the neighbourhood of Fort Enterprise, near the southern verge of the barren grounds in latitude 65°, and is also plentiful on Cape Parry, one of the most northern parts of the continent."

GENERAL REMARKS.

Our description of this rare animal was drawn up from a specimen deposited by Dr. RICHARDSON in the museum of the Zoological Society of London, which was said to have been the identical skin from which his description was taken.

We possess another specimen, presented to us by Dr. RICHARDSON, which is a little longer in the body and shorter in the tail than the one we have just spoken of; the body being 12½ inches in length, and the tail (vertebræ) 3½ inches, including fur 5 inches. The forehead and buttocks of this specimen are reddish-brown.

No. 2. Plate X

Common American Shrew Mole

Male & Female.

Drawn from Nature by J. J. Audubon F.R.S.F.L.S

Printed by Nagel & Weingaertner N.Y.

GENUS SCALOPS.—Cuvier.

DENTAL FORMULA.

Incisive $\frac{2}{4}$; *Molar* $\frac{3-3}{3-3}$; *False-Molars* $\frac{6-6}{3-3}$ = 36.

or

Incisive $\frac{2}{4}$; *Molar* $\frac{6-6}{6-6}$; *False-Molars* $\frac{4-4}{3-3}$ = 44.

Head, long, terminated by an extended, cartilaginous, flexible, and pointed muzzle; eyes and ears, concealed by the hair, and very minute. Hind-feet, short and slender, with five toes and delicate hooked nails; fore-feet (or hands) broad; claws, long and flat, fitted for excavating the earth.

The name Scalops is derived from the Greek σκαλλω, (*skallo*,) and from the Latin *scalpo*, I scrape.

The various species included in this genus, which approaches very closely to the genus Talpa, of Europe, (European mole,) are, we believe, confined to North America. There are, so far as we have been informed, only five species known at the present time.

SCALOPS AQUATICUS.—Linn.

Common American Shrew Mole.

PLATE X.—Male and Female.

S. magnitudine Talpæ Europeæ similis; corpore cylindrato, lanugine sericea, argenteo-cinereo induto

CHARACTERS.

Size of the European mole, (Talpa;) body, cylindrical; fur, velvety; colour, silvery-grayish-brown.

SYNONYMES.

SOREX AQUATICUS, Linn. Syst. Nat., 12th ed. corrected, vol. i., p. 74.
TALPA FUSCA, Pennant, Brit. Zool., Quadrupeds, 314.
SCALOPS CANADENSIS, Desm., Mam., p. 115.
SCALOPE DE CANADA, Cuv., Règne Animal, p. 134.
SHREW MOLE, Godman, Nat. Hist. vol. i., p. 84, pl. 5, fig. 3.
SCALOPS CANADENSIS, Harlan, Fauna, p. 32 Young.
" PENNSYLVANICA, Harlan, Fauna, p., 33. Adult.
" CANADENSIS, Emmons, Report on Quads. of Mass., p. 15.
" AQUATICUS, Bachman, Observations on the Genus Scalops, Boston Jour Nat. Hist., vol. iv., No. 1., p. 28, 1842.
" AQUATICUS, Dekay, Nat. Hist. of the State of New-York, p. 15.

DESCRIPTION.

Adult:—Teeth 36, corresponding with the first dental formula of this genus, given on the preceding page; incisors of moderate size, rounded on their front surface and flattened posteriorly. Immediately behind the incisors, two minute teeth on each side, crowded together—succeeded by four large false-molars, of a cylindrical shape, and pointed; the fourth smallest, the fifth a little larger and slightly lobed, and the sixth, which is the largest, more conspicuously lobed; followed by three true molars, each furnished with three sharp tubercles.

In the lower or inferior jaw, sixteen teeth; the two posterior incisors very small, succeeded on each side by another much larger, pointed, and extending forward; three false-molars which succeed these are pointed, and the third and largest slightly lobed; three true molars composed of two parallel prisms, terminated each by three points, and "presenting one of their angles on the outer side, and one of their faces on the internal surface; the two first of equal size, the other somewhat smaller." Part of the above description is in the words of Dr. GODMAN, from his very correct and interesting article on the Shrew Mole, (vol. i., p. 82,) which corresponds exactly with the results of our own investigations of the teeth of this animal, made at various times, during a period of several years.

Young.—We have found in specimens less than a year old, that the two small thread-like teeth inserted behind the incisors in the upper jaw were entirely wanting, as also the fourth lateral incisor on each side, leaving vacant spaces between them, and presenting the appearance ascribed to them by Baron CUVIER and by DESMAREST; the last mentioned teeth are first developed, the former appearing when the animal is full grown and all the edentate spaces between the molars are filled up.

Body, thick and cylindrical: neck, short, so that the head appears almost as if attached directly to the shoulders; snout, naked, cartilaginous, and very flexible, extending five lines beyond the incisors; the under surface projects a little beyond the nostrils, which are oblong and open on the upper surface near each other; mouth, large, and when open resembling somewhat (although in miniature) that of the hog; eyes, concealed by the fur, apparently covered by an integument, and so minute that they can with great difficulty be found. The orifice in the skin in which the eye is placed is not of larger diameter than would admit a bristle. No external ear; there is, however, a very small circular aperture leading to the ear, about three quarters of an inch behind the eye. The fore-arms are concealed by the skin and the palms only are visible, they are broad, and might be thought not unlike hands; they are thinly clothed with hair, and bordered with stiff hairs; the fingers are united at the base of the claws; nails, large, slightly curved, nearly convex above, and flattened on the inner surface; hind-feet, small and slender, naked on the under surface, and apparently above, although a close inspection shows the upper surface to be covered with fine short hairs; nails, small, a little arched, and compressed; tail, short, round, appears naked, but is very sparingly clothed with short adpressed hairs. On the inside of the thighs, near the tail, is a gland about half an inch long, from which a disagreeable musky odour issues, which makes the animal offensive to delicate olfactories. All our other shrew moles possess similar glands, and we have perceived the musky smell still remaining strong in skins that had been prepared and stuffed several weeks.

COLOUR.

Snout and palms, in the living animal, pinkish flesh-colour; chin, feet, and tail, dull white; hair on the body, about five lines in length, very soft, smooth, and lustrous; for three-fourths of its length, plumbeous; tips light-brown, giving the surface of the hair, above, a dark-brown colour, which varies in different lights, sometimes exhibiting black, silver-gray, or purple, reflections.

There are many variations in the colouring of different individuals of this species, but none of them permanent: we possess some specimens which are nearly black, and others of a light cream-colour; we also have a specimen, the tail of which is clothed with short hairs, with a considerable tuft at the extremity. From these and similar differences in various other animals, it is not surprising that authors have described in their works many as new, which, on being closely examined afterwards, prove to be mere accidental varieties of some well-known species.

DIMENSIONS.

Adult male.	Inches.	Lines.
From nose to root of tail	5	8
Tail	0	8
Breadth of palm	0	5
A specimen from Carolina.		
From nose to root of tail	4	7
Tail	0	9
Breadth of palm	0	6

HABITS.

Whilst almost every farmer or gardener throughout the Northern and Eastern States is well acquainted with this curious animal, as far as the mere observation of its meandering course through his fields and meadows, his beds of green peas or other vegetables, is concerned, but few have arrived at proper conclusions in regard to the habits of the Shrew Mole; and it is generally caught and killed whenever practicable; the common idea being, that the Mole feeds on the roots of tender plants, grasses, &c.; while the fact that the animal devours great quantities of earth-worms, slugs, and grubs, all hurtful to the fruit trees, to the grasses, and the peas and other vegetables, seems to be unknown, or overlooked.

In justice to the farmer and gardener, however, we must say, that the course taken occasionally by this species, directly along a row of tender plants, throwing them out of the earth, as it does, or zig-zag across a valuable bed or beautiful lawn, is rather provoking, and we have ourselves caused traps to be set for moles, being greatly annoyed by their digging long galleries under the grass on our sloping banks, which during a heavy shower soon filled with water, and presently increased to large gutters, or deep holes, requiring repairs forthwith. At such times also, a Mole-track through loose soil where there is any descent, will be found by the gardener, perchance, to have become a miniature ravine some twenty or thirty yards in length, and a few (anticipated) bushels of carrots are destroyed. In neglected or sandy soils, one of these gutters becomes deep and wide in a short time, and we may perhaps not err in hazarding the opinion that some of the unsightly ravines which run almost through large estates, occasionally might be traced to no higher origin than the wandering of an unlucky mole!

We kept one of this species alive for some days, feeding it altogether upon earth-worms, but we soon found it difficult to procure a sufficient supply; forty or fifty worms of moderate size did not appear too much for its seemingly insatiable appetite. At the expiration of four days, another of this species which we had in confinement would not touch any vegetable substances, although the cage was filled with clods covered with fine clover, pieces of sweet apples, bread, &c.

We were much interested in observing, that no matter how soiled its coat might have become in the cage, it would resume its beauty and glossiness after the mole had passed and re-passed through the earth eight or ten times, which it always accomplished in a few minutes. We frequently remarked with surprise the great strength of this animal, which enabled it to lift the lid or top of a box in which it was kept, although it was large and heavy; the box-top was not however fastened down. Seating ourselves quietly in the room, after putting back the mole into the box, the animal supposing itself no longer watched, very soon raised its body against the side of the box, which was partly filled with earth, and presently its snout was protruded through the small space between the box and the cover; and after a few efforts the creature got his fore-feet on to the edge of the box, raised itself over the latter, and fell upon a table on which we had placed the box. It immediately ran to the edge of the table, and thence tumbled on to the floor; this, however, did not at all incommode it, for it made off to a dark corner of the room at once, and remained there until again replaced in its prison.

When this Mole was fed on earth-worms, (*Lumbricus terrenus,*) as we have just related, we heard the worms crushed in the strong jaws of the animal, with a noise somewhat like the grating of broken glass, which was probably caused by its strong teeth gnashing on the sand or grit contained in the bodies of the worms. These were placed singly on the ground near the animal, which after smelling around for a moment turned about in every direction with the greatest activity, until he felt a worm, when he seized it between the outer surface of his hands or fore-paws, and pushed it into his mouth with a continually repeated forward movement of the paws, cramming it downward until all was in his jaws.

Small-sized earth-worms were despatched in a very short time; the animal never failing to begin with the anterior end of the worm, and apparently cutting it as he eat, into small pieces, until the whole was devoured. On the contrary, when the earth-worm was of a large size, the Mole seemed to find some difficulty in managing it, and munched the worm sideways, moving it from one side of its mouth to the other. On these occasions the gritting of its teeth, which we have already spoken of, can be heard at the distance of several feet.

We afterwards put the Mole into a large wire rat-trap, and to our surprise saw him insert his fore-paws or hands between the wires, and force them apart sufficiently to give him room to pass out through them at once, and this without any great apparent effort. It is this extraordinary muscular power in the fore-paws and arms, that enables the Shrew Moles to traverse the galleries they excavate with so much rapidity, in doing which they turn the backs of their palms or hands toward each other, push them forward as far as the end of their snout, and then open and bring them round backward, in the manner of a person moving his hands and arms when swimming. When running along on the surface of the ground, they extend the fore-legs as far forward as they will reach, turning the backs of the hands or paws (as just mentioned) towards each other, and placing them edge-wise, instead of flat on the earth as might be supposed, and in this manner they run briskly and without any awkward movement, crossing beaten-roads or paved walks, and sometimes running swiftly twenty or thirty feet before they can get into the ground.

The Shrew Mole varies somewhat in its habits, according to our observations: for while a solitary individual will occasionally for some weeks occupy and root up a large plot of grass or a considerable portion of a garden, and on his being caught in a trap, the place will remain free from fresh Mole-tracks for a long period, proving that all the mischief was the work of a single Mole, at other times we have caught several out of one gallery on the same day; and while excavating a root-house, the lower part of which was rock, four of these animals came during the night through one gallery and tumbled down into the pit, where, the rock preventing their digging a way out, they were found in the morning. No others ever came through that gallery while the cellar was in progress, and those thus caught may probably have been one family.

Although generally known to run through the same galleries often, so much so that the most common method of capturing them is to set a trap anywhere in one of these tracks to intercept them when again passing through it, we have known a trap to remain set in a fresh track for eleven days before the animal passed that way, when it was caught; and we are of opinion that many of their tracks are only passed through once, as this animal is known to travel from one field or wood to another, and probably the only galleries they regularly traverse are those adjacent to the spot they have selected for rearing their young. In relation to this subject, Dr. Godman says—

"It is remarkable how unwilling they are to relinquish a long frequented burrow; I have frequently broken down or torn off the surface of the same burrow for several days in succession, but would always find it repaired at the next visit. This was especially the case with one individual whose nest I discovered, which was always repaired within a short time, as often as destroyed. It was an oval cavity, about five or seven inches in length by three in breadth, and was placed at about eight inches from the surface in a stiff clay. The entrance to it sloped obliquely downwards from the gallery about two inches from the surface; three times I entirely exposed this cell, by cutting out the whole superincumbent clay with a knife, and three times a similar one was made a little beyond the situation of the former, the excavation having been continued from its back part. I paid a visit to the same spot two months after capturing its occupant, and breaking up the cell, all the injuries were found to be repaired, and another excavated within a few inches of the old one. Most probably numerous individuals, composing a whole family, reside together in these extensive galleries. In the winter they burrow closer to the streams, where the ground is not so deeply frozen."

This species whilst beneath the earth's surface seems to search for food with the same activity and untiring perseverance that are observable in animals that seek for their provender above ground. It works through the earth not only in a straight-forward direction, but loosens it to the right and left, beneath and above, so that no worm or insect can escape it. When in contact with any one of the objects of which it has been in search, it seizes it with remarkable quickness both with its fore-feet and its sharp teeth, drawing itself immediately backward with its prize, upon which it begins to prey at once. The Shrew Mole passes through loose soil with nearly the same ease and speed that it displays in running, or "scrabbling" along above ground. It moves backward almost as rapidly as it goes forward. The nose is often seen protruded above the surface of the ground.

The snout of this species, although apparently delicate, is most powerfully muscular, as well as flexible; the animal can turn it to the right or left, upward or downward, and at times inserts it in its mouth, as if for the purpose of cleansing it, and then suddenly withdraws it with a kind of smack of its lips; this habit we observed three times in the course of a few minutes. The Shrew Mole is exceedingly tenacious of life; it cannot easily be put to death, either by heavy pressure or strangling, and a severe blow on the head seems to be the quickest mode of despatching it.

Although this species, as we have seen, feeds principally on worms, grubs, &c., we have the authority of our friend Ogden Hammond, Esq., for the following example either of a most singular perversity of taste, or of habits hitherto totally unknown as appertaining to animals of this genus, and meriting a farther inquiry. While at his estate near Throg's Neck, on Long Island Sound, his son, who is an intelligent young lad, and fond of Natural History, observed in company with an old servant of the family, a Shrew Mole in the act of swallowing, or devouring, a common toad—this was accomplished by the Mole, and he was then killed, being unable to escape after such a meal, and was taken to the house, when Mr. HAMMOND saw and examined the animal, with the toad partially protruding from its throat. This gentleman also related to us some time ago, that he once witnessed an engagement between two Moles, that happened to encounter each other in one of the *noon-day* excursions this species is so much in the habit of making. The combatants sidled up to one another like two little pigs, and each tried to root the other over, in attempting which their efforts so much resembled the manner of two boars fighting, that the whole affair was supremely ridiculous to the beholder, although no doubt to either of the bold warriors the consequences of an overthrow would have been very serious ; for the conqueror would vent his rage upon the fallen hero, and punish him severely with his sharp teeth. We have no doubt these conflicts generally take place in the love season, and are caused by rivalry, and that some "fair Mole" probably rewards the victor. When approached, the Moles attempted to escape, but were both shot on the spot, thus falling victims to their own passions ; and if we would read aright, affording us an instructive lesson, either as individuals, or in a national point of view.

The Shrew Moles are able to work their way so rapidly, that in soft or loamy soil it is almost impossible for the most active man to overtake and turn them out with a spade, unless he can see the spot where they are working by the movement of the earth, in which case they can be thrown out easily by sticking the spade in front of them or at one side of their gallery, and with a quick movement tossing them on to the surface.

They have been known to make a fresh track after rain, during one night, several hundred yards in length ; oftentimes they proceed for a considerable distance in nearly a straight or direct line, then suddenly begin to excavate around and across a small space of not more than a few feet in diameter, until you could hardly place your foot on a spot within this subterranean labyrinth without sinking through into their track ; at this time they are most probably in pursuit of worms, or other food, which may be there imbedded.

Although cold weather appears to us to put a stop to the movements of the Mole, we do not feel by any means certain that such is the case ; and very probably the hardness of the ground when frozen, and the depth at which the Mole is then obliged to seek his food, may be a sufficient reason for our seeing no traces of this busy creature's movements during cold winter weather. We have, however, often perceived their tracks after a day or two of warm weather in January, and have repeatedly observed them about during a thaw, after the first autumnal frosts had occurred. In Carolina there are not many weeks in a winter in which we are not able to find here and there traces of the activity of the Mole. We admit, however, that even in this comparatively mild climate, they appear to be far less active in winter than at other seasons.

From the foregoing facts we are inclined to think the Mole does not become torpid at any time; and in corroboration of this idea, we find that the animal is not at any season found in high Northern latitudes. Dr. RICHARDSON thinks "the absence of the Shrew Mole from these countries is owing to the fact that the earth-worm on which the Scalops, like the common Mole, principally feeds, is unknown in the Hudson's Bay countries."

The idea commonly entertained by uninformed persons, that Moles have no eyes, is an error; although our own experience confirms the opinion of others, that they appear to possess the power of seeing only in a very limited degree. We must not forget, however, that a wise Providence has adapted their organs of vision to the subterraneous life they lead. Shut out from the light of the sun by a law of nature requiring them to search for food beneath the earth's surface, these animals would find a large pair of eyes one of the greatest of evils, inasmuch as they would be constantly liable to be filled with sand; thus causing inflammation, blindness, and eventually death.

It is not, however, beyond the reach of possibility, nor contrary to the economy of Nature, to suppose that during the night, when this species is seen occasionally above ground, or when engaged in running or fighting, or for purposes we have not yet discovered, this animal may have the power of expanding its minute orbs, and drawing back the hair that entirely conceals its eyes. This, however, is a mere conjecture, which we have thrown out for the consideration of those who are fond of investigating Nature in her minutest operations.

The inquiry has often been made, if the Shrew Mole does not feed upon the grains or roots of the corn, peas, potatoes, &c., planted in rows or in hills, why is it that this pest so ingeniously and so mischievously follows the rows, and as effectually destroys the young plants as if it had consumed them? We answer, it is not the spirit of mischief by which the Mole is actuated; it is the law of self-preservation. In the rows where these seeds have been sown, or these vegetables planted, the ground has been manured; this, and the consequent moisture around the roots of the plants, attracts worms and other insects that are invariably found in rich moist earth. To the accusations made against the Shrew Mole as a destroyer of potatoes, and other vegetables, he might often with great truth plead an alibi. LECONTE's pine mouse, (*Arvicola pinetorum,*) is usually the author of the mischief, whilst all the blame is thrown upon the innocent Shrew Mole. We are, moreover, inclined to think that whilst the earth-worm is the general, it is by no means the only food of the latter, and we had an opportunity of discovering to our cost, that when in captivity, this species relishes other fare. We preserved one in a cage in Carolina, during a winter, for the purpose of ascertaining on what kind of food it was sustained, and whether it became dormant. It at no time touched grains or vegetables; the lower part of the cage was filled with a foot of moist earth, in which we occasionally placed a pint of earth-worms. It devoured pieces of beef, and for a week was engaged in demolishing a dead pigeon. Until the middle of January we found it every day actively running through the earth in search of worms. Suddenly, however, it seemed to have gone to winter quarters, as we could see no more traces of its customary burrowing. We now carefully searched for it in the box, to ascertain its appearance in a dormant state. But the little creature had forced itself through the wooden bars, and was gone. We examined every part of the room without success, and finally supposed it had escaped through the door. The cage of the Mole had been set on a box, full of earth, in which the chrysolides of some sixty or seventy species of rare butterflies, moths, and sphinges, had been carefully deposited. In this box we a few days afterwards heard a noise, and on looking, discovered our little fugitive. On searching for our choice insects we found not one left; they had all been devoured by the Shrew Mole. This greatly disappointed us, and put an end to all our hopes of reading the following spring a better lesson on entomology than ever could have been taught us—either by FABRICIUS, SPENCE, or KIRBY.

We had an opportunity on two different occasions of examining the nests and young of the Shrew Mole. The nests were about eight inches below the surface, the excavation was rather large and contained a quantity of oak leaves on the outer surface, lined with soft dried leaves of the crab-grass, (*Digitaria sanguinalis.*) There were galleries leading to this nest, in two or three directions. The young numbered in one case, five, and in another, nine.

Our kind friend, J. S. HAINES, Esq, of Germantown, near Philadelphia, informed us that he once kept several Shrew Moles in confinement for the purpose of investigating their habits, and that having been neglected for a few days, the strongest of them killed and ate up the others; they also devoured raw meat, especially beef, with great avidity.

GEOGRAPHICAL DISTRIBUTION.

The Shrew Mole is found inhabiting various parts of the country from Canada to Kentucky, in considerable numbers, and is abundant in Carolina, Georgia, Louisiana and Florida. It is, according to RICHARDSON, unknown in Labrador, the Hudson's Bay Territories, and probably North of Latitude 50°. We did not see any of them in our trip up the Missouri river, and there are none to be found on the dry prairies of the regions immediately east of the great Rocky Mountain chain. The figures in our plate were drawn from specimens procured near the City of New-York. We mention this locality because the colours differ a little from others that we have seen, and that have been described.

GENERAL REMARKS.

In restoring to this animal the specific name of its first describer, we have adhered to a rule, from which, to prevent the repetition of synonymes we should never depart unless under very peculiar circumstances. The name "*Aquaticus*," certainly does not apply to the habits of this species, as although it is fond of the vicinity of moist ground where the earthworm is most abundant, yet it is nowise aquatic. The name of DESMAREST, however, viz., "*Canadensis*," is equally objectionable, as it is far more common in the Southern portion of the United States than in Canada.

Some differences of opinion are observable in the works of authors in regard to the number of teeth which characterize this species.

Although the genus was, until recently, composed of but a single acknowledged species (*Scalops Canadensis* of DESM.), its systematic arrangement has caused great perplexity among Naturalists. LINNÆUS placed it among the Shrews (SOREX), and PENNANT among the Moles (TALPA), Baron CUVIER finally established for it a new genus (SCALOPS), in which it now remains. The specimen, however, which he made the type of the genus, contained but thirty teeth. The upper jaw had but three lateral incisors or false-molars on each side; leaving considerable intermediate spaces between the incisors and true molars. In this dental arrangement he was followed by DESMAREST, Dr. HARLAN, GRIFFITH, and nearly all the Naturalists of that period. Subsequently, however, FREDERICK CUVIER gave a correct description of the teeth, which he found amounted to thirty-six. Dr. HARLAN finding a skeleton from the vicinity of Philadelphia, which in its dental arrangement corresponded generally with the characters given by FRED. CUVIER, considered it a new species, and described it under the name of *Sc. Pennsylvanica* (see Fauna Americana, p. 33).

Dr. RICHARDSON described a specimen which was obtained on the Columbia river (F. B. A., p. 9), which contained forty-four teeth, very differently arranged. This animal he referred to our common Shrew Mole, supposing that the difference in the dentition, as observed by different authors, was owing to their having examined and described specimens of different ages.

In 1840, Professor EMMONS (Report on the Quadrupeds of Massachusetts) characterizes the genus as having 44 teeth. In 1842, Dr. DEKAY (Nat. History of the State of New-York, p. 15) has very erroneously given as a character, its having from 34 to 46 teeth, and states that he had once seen the skull of one of this species containing 44 teeth.

In an article in the Boston Journal (vol. iv., No. i., p. 26, 1842), we endeavoured to explain and correct the contradictory views of former authors, and we feel confident we have it in our power to account for the skull seen by Dr. DEKAY containing forty-four teeth.

The specimens examined by Baron CUVIER, DESMAREST and Dr. HARLAN, each containing but 30 teeth, were evidently young animals, with their dentition incomplete. One half of the specimens now lying before us present the same deficiency in the number of teeth; they also exhibit the edentate spaces between the incisors and grinders remarked by those authors. We have, in deciding this point, compared more than fifty specimens together. Those on the other hand that were examined by F. CUVIER and Dr. GODMAN, and the skeleton of Dr. HARLAN's *Scalops Pennsylvanica*, containing 36 teeth, were adults of the same species. Dr. RICHARDSON's specimen was a new species (*Scalops Townsendii*), having 44 teeth, (see Journ. Acad. Nat. Sc., Philadelphia, vol. viii., p. 58). With regard to the skull seen by Dr. DEKAY, we have no doubt of its having belonged to *Scalops Brewerii* (see Bost. Journ. Nat. Hist., vol. iv., p. 32), which has 44 teeth, and is not uncommon in the State of New-York, as we obtained four specimens from our friend, the late Dr. WRIGHT, who procured them in the vicinity of Troy.

No. 3. Plate XI

Northern Hare – (Old & Young)

Summer pelage.

Drawn from Nature by J. J. Audubon F.R.S. F.L.S.

Drawn on Stone by R. Trembly

Printed by Nagel & Weingaertner N.Y.

LEPUS AMERICANUS.—Erxleben.

Northern Hare.

PLATE XI.—Fig. 1, Male; Fig. 2, Young Female. Summer Pelage.
PLATE XII.—Winter Pelage.

L. hyeme albus; pilis tricoloribus, apice albis, ad radices cœruleis, medio fulvis; æstate, supra rufo-fuscus, infra albus, auribus capite paullo brevioribus; L. Sylvatica paullo robustior. L. Glacialis minor.

CHARACTERS.

Size, larger than the gray rabbit (Lepus Sylvaticus), less than the Polar hare; (L. Glacialis). Colour in summer, reddish-brown above, white beneath; in winter, white; roots of the hairs, blue; nearer the surface, fawn-colour, and the tips, white; ears, a little shorter than the head.

SYNONYMES.

Lievre (Quenton Malisia), Sagard Theodat, Canada, p. 747. 1636.
Swedish Hare, Kalm's Travels in North America, vol. ii., p. 45. 1749.
American Hare, Philos. Trans., London, vol. lxii., pp. 11, 376. 1772.
Lepus Americanus, Erxleben, Syst. regni Animalis, p. 330. 1777.
" Nanus, Schreber, vol. ii., p. 881, pl. 234, fig.
" Hudsonius, Pallas, Glires, pp. 1, 30.
Varying Hare, Pennant, Arct. Zool., vol. i., p. 95.
Lepus Virginianus, Harlan, Fauna, p. 196. 1825.
" Variabilis, var. Godman. Nat. Hist., vol. ii., p. 164.
American Varying Hare, Doughty, Cabinet Nat. Hist., vol. i., p. 217, pl. 19, Autumn pelage.
The Northern Hare, Audubon, Ornithological Biog., vol. ii., p. 469. Birds of America, pl. 181 (in the talons of the Golden Eagle), Winter pelage.
Lepus Americanus, Richardson, Fauna Boreali A., p. 217.
" Virginianus, Bach, Acad. Nat. Sciences, Philadelphia, vol. vii., p. 301.
" Americanus, Bach, Ib., p. 403, and Ib., vol. viii., p. 76.
" Americanus, Dekay, Nat. Hist. State of New-York, p. 95, pl. 26.

DESCRIPTION.

Incisors, pure white, shorter and smaller than in *L. Glacialis*; upper ones moderately grooved; the two posterior upper incisors very small. The margins of the orbits project considerably, having a distinct depression in the frontal bone ; this is more conspicuous in the old than in the younger animals. Head rather short; nose blunt, eyes large and prominent; ears placed far back, and near each other; whiskers, long and numerous; body, elongated, thickly clothed with long loose hair, with a soft downy fur beneath; legs, long; hind-legs, nearly twice the length of the fore-legs; feet, thickly clothed with hair, completely concealing the nails, which are long, thin, very sharp, and slightly arched. So thickly are the soles covered with hair, that an impression by the nails is not generally visible in their tracks made while passing over the snow, unless when running very fast. Tail, very short, covered with fur, but not very bushy. The form of this species is on the whole not very elegant; its long hind legs, although remarkably well adapted for rapid locomotion, and its diminutive tail, would lead the spectator at first sight to pronounce it an awkward animal; which is, nevertheless, far from being the fact. Its fur never lies smooth and compact, either in winter or summer, as does that of many other species, but seems to hang loosely on its back and sides, giving it a somewhat shaggy appearance. The hair on the body is in summer about an inch and a half long, and in winter a little longer.

COLOUR.

In summer, the whole of the upper surface is reddish-brown, formed by hairs that are at their roots and for two-thirds of their length of a blue-ish ash colour, then reddish-yellow, succeeded by a narrow line of dark-brown, the part next the tips or points, reddish-brown, but nearly all the hairs tipped with black—this colour predominating toward the rump. Whiskers, mostly black, a few white, the longest reaching beyond the head; ears, brown, with a narrow black border on the outer margin, and a slight fringe of white hairs on the inner. In some specimens there is a fawn, and in others a light-coloured, edge around the eyes, and a few white hairs on the forehead. The pupil of the eye is dark, the iris light silvery-yellow ; point of nose, chin, and under the throat, white; neck, yellowish-brown. Inner surface of legs, and under surface of body, white: between the hind-legs, to the insertion of the tail, white; upper surface of the tail, brown, under surface white. The summer dress of this species is assumed in April, and remains without much change till about the beginning of November in the latitude of Quebec, and till the middle of the same month in the State of New-York and the western parts of Pennsylvania ; after which season the animal gains its winter pelage. During winter, in high Northern latitudes, it becomes nearly pure white, with the

No. 3. Plate XII

Northern Hare

Winter pelage.

Drawn From Nature by J. J. Audubon F.R.S. F.L.S.

Drawn on Stone by R. Trembly.

Printed by Nagel & Weingärtner N.Y.

Colored by J. Lawrence

exception of the black edge on the outer borders of the ears. In the latitude of Albany, New-York, it has always a tinge of reddish-brown, more conspicuous in some specimens than in others, giving it a wavy appearance, especially when the animal is running, or when the fur is in the least agitated. In the winter season the hair is plumbeous at base, then reddish, and is broadly tipped with white. The parts of the body which are the last to assume the white change, are the forehead and shoulders; we have two winter-killed specimens before us that have the forehead, and a patch on the shoulders, brown. On the under surface, the fur in most specimens is white, even to the roots. A few long black hairs arise above and beneath the eyes, and extend backwards. The soles have a yellowish soiled appearance.

We possess a specimen of the young, about half grown, which in its general aspect resembles the adult; the colour of the back, however, is a shade darker, and the under surface an ashy white. The black edge is very conspicuous on the outer rim of the ear, and some of the whiskers are of unusual length, reaching beyond the head to the middle of the ear. The tail is very short, black above, and grayish-white beneath. The young become white in the autumn of the first year, but assume their winter colouring a little later in the season than the adults. We have met with some specimens in the New-York markets, late in January, in which the change of colour was very partial, the summer pelage still predominating.

DIMENSIONS.

The size and weight of the Northern hare we have found to vary very much. The measurements hitherto given were generally taken from stuffed specimens, which afford no very accurate indications of the size of the animal when living, or when recently killed. Dr. Godman, on the authority of Prince Charles Lucien Bonaparte, gives the measurement of a recent specimen as thirty-one inches, and Dr. Harlan's measurement of the same specimen after it had been stuffed was sixteen inches. We think it probable that the Prince and the Doctor adopted different modes of measuring. All stuffed specimens shrink very much; of a dozen now in our collection, there is not one that measures more than eighteen inches from point of nose to root of tail, and several white adults measure but fifteen inches.

The following measurements are from the largest specimen we have procured, taken when the animal was recently killed.

DIMENSIONS.

From point of nose to root of tail - -	19¼	inches.
Tail (vertebræ) - - - - -	1¼	do.
Do. to end of hair - - - - -	2¼	do.
From heel to end of middle claw - - -	5½	do.
Height of ear - - - - - - - -	3½	do.

Another specimen of moderate size.

From point of nose to root of tail - - -	16	do.
Tail (vertebræ) - - - - - - -	1½	do.
Do. to end of hair - - - - - - -	2½	do.
From heel to end of middle claw - - -	5¼	do.
Height of ear - - - - - - - -	3½	do.

Weight:—This species in the beginning of winter varies from three to six and a half pounds, but we consider 5½ pounds to be the average weight of a full-grown animal in good condition.

HABITS.

Our different species of Hares, and more especially the present one and the little gray rabbit, have been so much mixed up in the accounts of authors, that great confusion exists in regard to their habits, and their specific identity. The assertion of WARDEN, that the American Hare retreats into hollow trees when pursued, applies to the gray rabbit, for which it was no doubt intended, but not to the Northern Hare. We are not aware that the latter ever takes shelter either in a hole in the earth, or in a hollow tree. We have seen it chased by hounds for whole days, and have witnessed the repetition of these hunts for several successive winters, without ever knowing it to seek concealment or security in such places. It depends on its long legs, and on the thickness of the woods, to aid it in evading the pursuit of its enemies. When hunted, it winds and doubles among thick clusters of young pines and scrub-oaks, or leads the dogs through entangled patches of hemlock and spruce fir, until it sometimes wearies out its pursuers; and unless the hunter should appear, and stop its career with the gun, it is almost certain to escape.

In deep snows, the animal is so light, and is so well supported by its broad furry-feet, that it passes over the surface making only a faint impression, whilst the hounds plunge deep into the snow at every bound, and soon give up the hopeless pursuit. It avoids not only open grounds, but even open woods, and confines itself to the densest and most impenetrable forests. Although it wanders by night in many directions in search of its appropriate food, we have scarcely ever seen its tracks in the open fields; it seems cautiously to avoid the cabbage and turnip fields of the farmer, and seldom even in the most retired places makes an encroachment on his cultivated grounds.

The food of this species in summer consists of various kinds of juicy and tender grasses, and the bark, leaves, and buds, of several small shrubs; and these Hares seem to be particularly fond of the young twigs of the wild allspice (*Laurus benzoin*), but in winter, when the earth is covered with snow, they gain a precarious subsistence from the buds and bark of such trees as are suited to their taste. Sometimes they scratch up the snow to feed on the leaves and berries of the various species of *Pyrola*, found in the Northern States. The bark of the willow, birch and poplar, and the buds of young pines, are sought after by them with avidity. We have seen persons in the Northern part of the State of New-York, who were desirous of shooting these animals by moonlight, watching near American black-poplar trees (*Populus Hudsonica*), which they had cut down for the purpose of attracting them to feed on their buds and tender twigs, in which they were often successful. Some of these Hares which we had in a domesticated state, were fed on cabbage leaves, turnips, parsnips, potatoes and sweet apples. During one very cold winter, when these could not be conveniently obtained, they were frequently supplied with clover-hay, to which, when more agreeable food was not given them, they did not evince any aversion; from time to time also, outer branches of willow, poplar or apple trees, were thrown into their enclosure, the bark of which seemed to be greatly relished by them.

The Northern Hare, like most others of the genus, seeks its food only by night or in the early part of the evening. To this habit it is more exclusively confined during autumn and winter, than in spring and summer. In the latter seasons, especially in spring, these animals are frequently observed in the morning, and as the sun is declining, in the afternoon, cautiously proceeding along some solitary by-path of the forest. Two or three may often be seen associated together, appearing full of activity and playfulness. When disturbed on these occasions, they stamp on the ground, making a noise so loud that it can be heard at some distance, then hopping a few yards into the thicket, they sit with ears erect, seemingly listening, to ascertain whether they are pursued or not. This habit of thumping on the earth is common to most hares and rabbits. We have particularly noticed it in the domesticated rabbit (*L. cuniculus*), and in our common gray rabbit. They are more particularly in the habit of doing it on moonlight nights; it is indicative either of fear or anger, and is a frequent action among the males when they meet in combat. During cold weather this Hare retires to its form at early dawn, or shelters itself under the thick foliage of fallen tree tops, particularly those of the pine and hemlock. It occasionally retires to the same cover for a number of nights in succession, but this habit is by no means common; and the sportsman who expects on some succeeding day to find this animal in the place from which it was once started, is likely to be disappointed; although we are not aware, that any other of our species of hare are so attached to particular and beaten paths through the woods, as the one now under consideration. It nightly pursues these paths, not only during the deep snows of winter, but for a period of several years, if not killed or taken, wandering through them even during summer. We have seen a dozen caught at one spot in snares composed of horse-hair or brass wire, in the course of a winter, and when the snow had disappeared and the spring was advanced, others were still captured in the same way, and in the same paths.

The period of gestation in this species is believed to be, (although we cannot speak with positive certainty,) about six weeks. Two females which we domesticated, and kept in a warren, produced young, one on the tenth and the other on the fifteenth of May; one had four, and the other six leverets, which were deposited on a nest of straw the inside of which was lined with a considerable quantity of hair plucked from their bodies. They succeeded in rearing all their young but one, which was killed by the male of a common European rabbit. They were not again gravid during that season. Ill health, and more important studies, required us to be absent for six months, and when we returned, all our pets had escaped to the woods, therefore we could not satisfactorily finish the observations on their habits in confinement, which had interested and amused us in many a leisure hour.

We, however, think it probable that the females in their wild state may produce young twice during the season. Those referred to above were much harassed by other species which were confined in the same warren, and might therefore have been less prolific than if they had enjoyed their liberty undisturbed, amid the recesses of their native woods. We have frequently observed the young of the Northern Hare in May, and again in July. These last must have been either from a second litter, or the produce of a young female of the previous year. The young, at birth, were able to see. They were covered with short hair, and appeared somewhat darker in colour than the adults, at that season. They left their nest in ten or twelve days, and from that time seemed to provide for themselves, and to derive little sustenance or protection from their mothers. The old males at this period seemed to be animated with renewed courage; they had previously suffered themselves to be chased and worried by the common English rabbit, and even retreated from the attacks of the gray rabbit; but they now stood their ground, and engaged in fierce combats with the other prisoners confined with them, and generally came off victorious. They stamped with their feet, used their teeth and claws to a fearful purpose, and in the fight tore off patches of skin and mutilated the ears of their former persecutors, till they were left in undisturbed possession of the premises!

The males did not evince the vicious propensity to destroy their young which is observed in the domesticated English rabbit; on the contrary, they would frequently sit beside their little family, when they were but a day or two old, seeming to enjoy their playfulness and to watch their progress to maturity.

The Northern Hare seems during summer to prefer dry and elevated situations, and to be more fond of grounds covered with pines and firs, than of those that are overgrown with oak or hickory. The swamps and marshes soil their feet, and after having been compelled to pass through them, they are for hours employed in rubbing and drying their paws. In winter, however, when such places are hardened by the frost, they not only have paths through them in every direction, but occasionally seek a fallen tree-top as a hiding or resting place, in the centre of a swamp. We have observed them in great numbers in an almost impenetrable thicket of black larch, or hackmatack, (*Larix pendula,*) considerable portions of which were during summer a perfect morass. In what are called the "bark clearings," places where hemlock trees have been cut down to procure tan bark, this species is sometimes so abundant that twenty or thirty of them may be started in a day's walk.

As an article of food, this is the most indifferent of all our species of Hares; its flesh is hard, dry, almost juiceless, possessing none of the flavour of the English hare, and much inferior to that of our gray rabbit. Epicures, however, who often regard as dainties dishes that are scarce, and who, by the skilful application of the culinary art possess means of rendering things savoury that are of themselves insipid, may dispute this point with us.

The Northern Hare, as is proverbially the case with all the species, has many enemies. It is pursued by men and dogs, by carnivorous beasts of the forest, by eagles, by hawks, and by owls. In the northern parts of Maine, in Canada, and in the countries farther north, their most formidable enemies are the Canada lynx, (*Lynx Canadensis.*) the jer falcon, (*Falco Islandicus,*) and the snowy owl, (*Surnea nyctea.*) In the New England States, however, and in New-York, the red-tailed hawk, (*Butea*

Borealis,) is occasionally seen with one of these species in its talons. But its most formidable enemy is the great horned owl, (*Bubo Virginianus.*) We have also, on one occasion, observed a common house-cat dragging a full grown Northern Hare from the woods, to feed her young. Lads on their way to school, entrap them with snares attached to a bent twig, placed along the paths they nightly resort to. The hunter finds recreation in pursuing them with hounds, whilst he places himself in some wood-path where they were last seen to pass. The Hare runs from fifty to a hundred yards ahead of the dogs, and in its windings and turnings to escape from them frequently returns to the spot where the hunter is stationed, and falls by a shot from his gun.

The Northern Hare, when rapidly pursued, makes such great efforts to escape, that the poor creature (as we have said already) is occasionally successful, and fairly outruns the hounds, whilst the hunter is cunningly avoided by it when doubling. After one of these hard chases, however, we have known the animal die from the fatigue it had undergone, or from having been overheated. We once saw one, which had been closely pressed by the dogs nearly all the afternoon, return to a thicket after the hounds had been called off and the sportsmen had given up the vain pursuit. Next morning we examined the place to which it had retired, and to our surprise, discovered the hare sitting in its form, under a dwarfish, crooked, pine-bush; it was covered with snow and quite dead. In this instance the hare had no doubt been greatly overheated by the race of the preceding day, as well as exhausted and terrified; and the poor thing being in that condition very susceptible of cold was probably chilled by the night air and the falling snow, until its palpitating heart, gradually impelling the vital fluid with fainter and slower pulsations, at length ceased its throbbings forever.

Sometimes we have found these Hares dead in the woods after the melting of the snow in the Spring, and on examination we found they were entangled in portions of wire snares, frequently entwined round their necks, from which they had been unable to extricate themselves.

This species when caught alive cannot be taken into the hand like the gray rabbit, with impunity; the latter, when seized by the ears or hind-legs soon becomes quiet and is harmless; but the Northern Hare struggles to escape, and makes a formidable resistance with its teeth and nails. On one occasion a servant who was expert at catching the gray rabbit in traps, came to us with a rueful countenance holding a hare in his hands, exhibiting at the same time sundry severe scratches he had received, showing us his torn clothes, and a place on his leg which the animal had bitten, and declaring that he had caught "a rabbit as cross as a cat." We ascertained it to be a Northern Hare in its summer dress, and although its captor had not been able to distinguish it from the gray rabbit by its colour, he certainly received a practical lesson in natural history which he did not soon forget.

A living individual of this species, which we have in Charleston in a partially domesticated state, for the purpose of trying to ascertain the effect of a warm climate on its changes of colour, is particularly cross when approached by a stranger. It raises its fur, and springs at the intruder with almost a growl, and is ready with its claws and teeth to gratify its rage, and inflict a wound on the person who has aroused its ire. When thus excited, it reminded us by its attitudes of an angry racoon.

The skin of the Northern Hare is so tender and easily torn, and the fur is so apt to be spoiled and drop off on being handled, that it is difficult to prepare perfect specimens for the naturalist's cabinet. The pelt is not in much request among the furriers, and is regarded by the hatter as of little value. The hind-feet, however, are used by the latter in a part of the process by which the soft, glossy, surface is imparted to his fabric, and answer the purpose of a soft hat-brush.

GEOGRAPHICAL DISTRIBUTION.

This species is found in portions of the British possessions, as far as the sixty-eighth parallel of North latitude. It is, however, confined to the Eastern portion of our Continent; RICHARDSON, who represents it as "a common animal from one extremity of the Continent to the other," seems to have mistaken for it another species which replaces it on the North West coast. Although it does not range as far to the North as the Polar hare, it is decidedly a Northern species; it is found at Hudson's Bay, in Newfoundland, Canada, all the New-England States, and in the Northern portions of New-York, Pennsylvania, and Ohio. Mr. DOUGHTY informed us that he procured a specimen on the Alleghany Mountains in the Northern part of Virginia, Lat. 40° 29′, where it had never before been observed by the inhabitants. On seeking for it afterwards in the locality from which he obtained it, we were unsuccessful, and we are inclined to believe that it is only occasionally that some straggler wanders so far South among these mountains, and that its Southern limit may be set down at about 40°.

GENERAL REMARKS.

The history of this Hare has been attempted from time to time, by early and recent travellers and naturalists, and most of their accounts of it are only sources of perplexity, and additional difficulties in the way of the naturalist of the present day. Strange mistakes were committed by some of those who wrote on the subject, from PENNANT down to HARLAN, GODMAN, and others still later; and one error appears to have led to another, until even the identity of the species meant to be described by different authors, was finally involved in an almost inextricable web of embarrassment.

As far as we have been able to ascertain, the Northern Hare was first noticed by SAGARD THEODAT, (Hist. de Canada,) in 1636. KALM, (who travelled in America from 1748 to 1751, and whose work was published in the Swedish language, and soon after translated into German and English,) speaks of this species as follows:—"Hares are likewise said to be plentiful even in Hudson's Bay, and they are abundant in Canada, where I have often seen, and found them perfectly corresponding with our Swedish hares. In summer they have a brownish-gray, and in winter a snowy-white colour, as with us." (KALM's Travels, &c., vol. ii., p. 45. English translation.)

This judicious and intelligent traveller, undoubtedly here referred to the Northern Hare. He supposed it to be identical with the Alpine or variable Hare, (*Lepus variabilis*,) which is found in Sweden and other Northern countries of Europe. That species is a little larger than the Northern Hare, and the tips of its ears are black; but although it is a distinct species, it so nearly resembles the latter, that several authors, GODMAN not excepted, were induced to regard these two species as identical. KALM, (see vol. i., p. 105, Eng. trans.,) whilst he was in the vicinity of Philadelphia, where the Northern Hare never existed, gave a correct account of another species, the American gray rabbit, which we will notice more in detail when we describe that animal. It is very evident that in these two notices of American hares, KALM had reference to two distinct species, and that he pointed out those distinctive marks by which they are separated. If subsequent authors confounded the two species, and created confusion, their errors evidently cannot be owing to any fault of the eminent Swedish traveller.

The first specimens of the Northern Hare that appeared in Europe, were sent by the servants of the Hudson's Bay Company to England in 1771, (see Phil. Trans., vol. lxii., p. 13.) There were four specimens in the collection, exhibiting the various gradations of colour. In addition to these, a living animal of the same species was received about the same time, probably by the same ship. It was brought to the notice of the Philosophical Society, in a letter from the Hon. DAINES BARRINGTON, read 16th January, 1772. This letter is interesting, since it gives us some idea of the state of natural science in England, at that early day. The animal had for some time remained alive, but had died in the previous November. It had at that time already changed its summer colour, and become nearly white. It was *boiled*, in order to ascertain whether it was a hare or a rabbit, as according to RAY, if the flesh was brown it was a hare, if white a rabbit. It proved to be brown, and was declared to be a hare. The test was strange enough, but the conclusion was correct. In May of the same year, J. R. FORSTER, Esq., F. R. S., described this, among twenty quadrupeds, that had been sent from Hudson's Bay. After giving an account of the manner in which it was captured by snares made of brass wire and pack thread, he designates its size as "bigger than the rabbit, but less than the Alpine hare." In this he was quite correct. He then goes on to show that its hind-feet are longer in proportion to the body than those of the rabbit and common hare, &c. He finally speaks of its habits, and here his first error occurs. KALM's accounts of *two* different species were supposed by him to refer to one species only, and whilst the Northern Hare was *described*—some o' the *habits* of the American gray rabbit were incorrectly referred to it.

As, however, FORSTER gave it no specific name and his description on the whole was but a loose one, it was left to another naturalist to give it a scientific appellation.

In 1777, ERXLEBEN gave the first scientific description of it, and named it *Lepus Americanus*. SCHREBER, (as we are prepared to show in our article on *Lepus sylvaticus*,) published an account of it immediately afterwards, under the name of *Lepus nanus*.

This description, as may easily be seen, was principally taken from FORSTER. SCHŒPFF about the same period, and PALLAS in 1778, under the name of *L. Hudsonicus*, and PENNANT in 1780, under that of *American hare*, followed each other in quick succession.

In GMELIN'S LINNÆUS, (1788,) it is very imperfectly described in one single line. All these authors copied the error of FORSTER in giving to the Northern Hare the habits of the American gray rabbit.

In the work of DESMAREST, (Mammalogie, ou description des espèces de Mammifères, p. 351, Paris, 1820,) a description is given of "Esp. Lièvre d'Amérique, Lepus Americanus." This, however, instead of being a description of the true *L. Americanus* of all previous authors, is in most particulars a pretty good description of our gray rabbit. HARLAN, who published his Fauna in 1825, translated and published this description very literally, even to its faults, (see Fauna Americana, p. 196.) Having thus erroneously disposed of the gray rabbit under the name of *L. Americanus*, the true *Lepus Americanus* was named by him *L. Virginianus!* The following year, Dr. GODMAN gave a description of the Northern Hare, referring it to the *Lepus variabilis* of Europe!

After Dr. RICHARDSON's return from his perilous journey through the Polar regions, he prepared in England his valuable Fauna Boreali Americana, which was published in 1829. Specimens labelled *L. Americanus* of ERXLEBEN, were still in the British Museum, and he published descriptions of his own specimens under that name. The gray rabbit did not come within the range of his investigations, but having received a hunter's skin from the vicinity of the Columbia river, he supposed it to be the *L. Virginianus* of HARLAN, and described it under that name. This skin, however, has since proved to belong to a different species; the Northern Hare not being found in the regions bordering that river.

In 1837, having several new species of Hare to describe, we began to look into this subject, and endeavoured to correct the errors in regard to the species, that had crept into the works of various authors.

We had not seen ERXLEBEN's work, and supposing that the species were correctly designated, we published our views of the habits, &c., of the two species, (whose identity and proper cognomen we have, we hope, just established,) under the old names of *L. Virginianus* and *L. Americanus*, (see Jour. of Acad. of Nat. Sciences of Phila., vol. vii., pl. 2. p. 282.) The article had scarcely been printed, before we obtained a copy of ERXLEBEN, and we immediately perceived and corrected the errors that had been committed, giving the Northern Hare its correct name, *L. Americanus*, and bestowing on the gray rabbit, which, through the mistakes we have already described had been left without any name, that of *Lepus sylvaticus*, (Jour. Acad. Nat. Sciences of Phil., vol. vii., p. 403.) The reasons for this arrangement were given in our remarks on the genus LEPUS, in a subsequent paper, (Jour. Acad. Sc., vol. viii., pl. 1, p. 75,) where we characterized a number of additional new species. In 1842, Dr. DEKAY, (see Nat. Hist. of New-York, p. 95,) acceding to this arrangement of the Northern Hare under the specific name of *L. Americanus*, remarks, "This Hare was first vaguely indicated by ERXLEBEN in 1777." In a spirit of great fairness, however, that author's original description was published at the foot of the article.

In order to set this matter at rest, remove this species from the false position in which it has so long stood, and give its first describer the credit to which he is entitled, we will here insert the description above alluded to.

"Lepus Americanus, L. cauda abbreviata; pedibus posticis corpore dimidio longioribus; auricularum caudoque apicibus griseis.

"Die Hasen—KALM, Hudson's Bay Quadrup., BARRINGTON, Phil. Trans.

vol. lxii., p. 376. Magnitudine medius inter L. cuniculum et timidum Alpinum, (sc. L. timidus, FORSTER, Phil. Trans. vol. lxii., p. 375.) Auriculanum et caudæ apices perpetuo grisei — Pedes postici longiores quam in L. timido et cuniculo, color griseo-fuscus; Hieme in frigidioribus albus.

"Habitat in America boreali ad fretum Hudsoni copiosissimus, nocturnus. Non fœdit, degit sub arborum radicibus, inque cavis arboribus. Parit bis vel semel in anno; pullos quinque ad septem; caro bona, colore L. timidi."

In great deference, we would submit whether the above is not more than a "*vague indication*" of a species. To us it appears a tolerably full description for the era in which the author lived and considering the few species of Hare then known.

There were at that early period but three Hares with which naturalists were familiar:—*L. timidus*, the common European Hare; *L. variabilis*, the variable Hare; and *L. cuniculus*, the European burrowing rabbit. With these ERXLEBEN compares this species in size and colour. With the exception of one of the habits he mentions, this description appears to us creditable to him. There have been many occasions, when, perplexed in guessing at the species intended to be described by old authors, (the Father of natural history, LINNÆUS himself, not excepted,) we would have hailed a description like this, as a light in darkness. The species ERXLEBEN had in view cannot be mistaken; he describes it very correctly as "*magnitudine medius inter L. cuniculum et timidum Alpinum.*" Our American gray rabbit, instead of being intermediate between *L. cuniculus* and the Alpine hare, is smaller than either. "*Pedes postici longiores quam in L. timido et cuniculo.*" The long hind-feet are distinctive marks of the Northern Hare; but those of our gray rabbit are much shorter than those of *L. timidus*, or common hare of Europe. "*Hieme in frigidioribus albus.*" Our gray rabbit, contrary to the assertion of most authors, does not become white in winter in any latitude. "*Habitat in America boreali ad fretum Hudsoni copiosissimus.*" Dr. RICHARDSON, and every Northern traveller with whom we have conversed, have assured us that our gray rabbit does not exist at Hudson's Bay, where the Northern Hare is quite abundant, and where that and the Polar hare, (the last named species existing still further North,) are the only species to be found. We have examined and compared the original specimen described by Dr. RICHARDSON, and also those in the British Museum that have successively replaced the specimens first sent to England, and find that they all belong to this species. In fact our gray rabbit is very little known in England or Scotland; since, after an examination of all the principal Museums in those countries, we met with but two specimens, one of which was not named, and the other was not improperly labelled, "Lepus Americanus Harlan, non Erxleben."

The rigid rule of priority will always preserve for the Northern Hare the name of *L. Americanus*, whilst *L. nanus*, *L. Hudsonicus*, and *L. Virginianus*, must be set down merely as synonymes.

GENUS FIBER.—Illiger.

DENTAL FORMULA.

$$\textit{Incisive}\ \frac{2}{2};\quad \textit{Molar}\ \frac{3-3}{3-3} = 16.$$

Lower incisors, sharp-pointed, and convex in front; molars, with flat crowns, furnished with scaly transverse zig-zag laminæ. Fore-feet with four toes and the rudiment of a thumb; hind-feet, with five toes, the edges furnished with stiff hairs, which assist the animal in swimming, instead of the feet being palmated or webbed; hind-toes, slightly palmated. Tail, long, compressed, granular, nearly naked, having but a few scattered hairs. Glands, near the origin of the tail, which secrete a white, musky, and somewhat offensive fluid. Mammæ six, abdominal.

This genus differs from the Arvicolæ in its dentition; the first inferior molar has one point more than the corresponding tooth in the latter, and all the molars acquire roots immediately after the animal becomes an adult. We have frequently heard complaints made by students of natural history, of the difficulties they had to encounter at the very outset, from the want of accuracy and uniformity in the works of authors, when stating the characters by which they defined the genera they established. The justness of these complaints may be well illustrated by examining the accounts of the present genus as given by several well-known writers.

Illiger says it has four molars on each side, (*Utrinqui quaterni*,) see Prodromus systematis mammaliarum et avum, making in all twenty teeth. Wiegman and Ruthe have given the same dental arrangement, see Handbuch der Zoologie, Berlin, 1832. F. Cuvier, who has been followed by most authors, has given it—Incisive $\frac{2}{2}$; Canine $\frac{3}{3}=\frac{3}{3}$, = sixteen teeth. Griffith, Animal Kingdom, vol. iii., p. 106, describes it as having—Incisive $\frac{2}{2}$; Canine $\frac{4}{4}=\frac{4}{4}$ = twenty teeth; and in his synopsis of the species of mammalia, (sp. 532,) its dental arrangement is thus characterized—Incisive $\frac{2}{2}$, Canine $\frac{3}{3}=\frac{3}{3}$, Cheek-teeth, $\frac{3}{3}=\frac{3}{3}$, giving to it the extravagant number of twenty-eight teeth. This last statement is most probably only a typographical error. A correct examination and description of the teeth of this genus requires a considerable degree of labour, besides great attention and care, as they are placed so close to each other that without a good magnifying glass it is difficult to find the lines of separation, and almost impossible to ascertain their number without extracting them one by one.

The descriptions and figures of their dental arrangement, by Baron CUVIER and F. CUVIER, are correct: see Ondatras, dents des mammifères, pl. 53, p. 157, and Recherches sur les ossemens fossiles, t. 5, p. 1.

ILLIGER's generic name, Fiber, is derived from the Latin word, *Fiber*, a beaver. There is only one species described as belonging to this genus.

FIBER ZIBETHICUS.—LINN.

MUSK-RAT.—MUSQUASH.

PLATE XIII.—OLD, AND YOUNG.

F. supra, rufo-fuscus; subtus cinereus; Leporem sylvaticum magnitudine sub æquans.

CHARACTERS.

General colour, reddish-brown above, cinereous beneath; about the size of the American gray rabbit.

SYNONYMES.

MUSSASCUS, Smith's Virginia, 1626. (Pinkerton's Collection of Voyages and Travels, vol. xiii., p. 31.)
RAT MUSQUÉ, Sagard Theodat, Canada, p. 771.
CASTOR ZIBETHICUS, Linn. Syst. Nat., xii. ed., vol. 1, p. 79.
L'ONDATRA, Buffon, Tom 10, p. 1.
MUSKRAT, Lawson, Carolina, p, 120.
MUSK BEAVER, Pennant, Arc. Zool., vol. i., p. 106.
MUSQUASH, Hearne, Journey, p. 379.
MUS ZIBETHICUS, Linn., Gmel., vol. i., p. 125.
FIBER ZIBETHICUS, Sabine, Franklin's Journey, p. 659.
MUSK RAT, Godman's Nat. Hist., p. 58.
ONDATHRA, Huron Indians.
MUSQUASH, WATSUSS, or WACHUSK; the animal that sits on the ice in a round form. Cree Indians, (Richardson.)

DESCRIPTION.

Body, of a nearly cylindrical shape, resembling that of the Norway rat. Head, short; neck, very short, and indistinct; legs, short; thighs

No 3. Plate XIII

Mush Rat-Musquash

Old & Young

Drawn from Nature by J.J.Audubon F.R.S. F.L.S.

Drawn on Stone by R. Trembly

Printed by Nagel & Weingaertner N.Y.

Colored by J.Lawrence

hid in the body. Tail, two-thirds the length of the body, compressed, convex on the sides, thickest in the middle, tapering to an acute point at the extremity; covered with small scales, which are visible through the thinly scattered hairs. Incisors, large; upper ones, a little rounded anteriorly without grooves, truncated on the cutting edge; lower ones, a little the longest; nose, thick, and obtuse; whiskers, moderate in length, seldom reaching beyond the ear; eyes, small, and lateral, nearly concealed in the fur; ears, short, oblong, covered with hair, and hidden by the fur.

On the fore-legs, the wrist and fingers only are visible beyond the body, they are covered with a short shining coat of hair.

The thumb has a conspicuous palm, and is armed with a nail, as long as the adjoining finger nails. Hind-legs, as short as the fore-legs, so that the body when the animal is walking touches the ground.

The hind-feet are turned obliquely inwards, and at first sight remind us of the foot of a duck. The two middle toes may be called semi-palmated, and there is also a short web between the third and fourth toes. The margins of the soles, and toes, are furnished with an even row of rigid hairs, curving inwards; under-surface of feet, naked; claws, conical, and slightly arched.

The whole body is clothed with a short, downy, fur, intermixed with longer and coarser hairs. In many particulars the skin resembles that of the beaver, although the fur is far less compact, downy and lustrous.

COLOUR.

Fur, on the upper parts a third longer than beneath; from the roots to near the extremities, bluish-gray, or lead-colour, tipped with brown; on the under surface it is a little lighter in colour, and the hairs are tipped with brownish-gray. This species, when viewed from above, appears of a general dark-brown colour, with a reddish tint visible on the neck, sides, and legs; chin, throat, and under-surface, grayish-ash; tail, dark-brown. Incisors, yellow; nails, white. The colour of this animal so much resembles that of the muddy banks on which it is frequently seated, that we have often, when looking at one from a little distance, mistaken it for a lump or clod of earth, until it moved.

DIMENSIONS.

Length of head and body - - - - -	15	inches.
" of tail - - - - - - - -	10	do.
From heel to longest nail - - - - -	3	do.
Height of ear - - - - - - -	½	do.

HABITS.

Reader! if you are a native of, or have sojourned in any portion, almost, of our continent, and have interested yourself in observing the "beasts of the field" in our woods or along our streams, to the slightest degree, you have probably often seen the Musk-Rat; or should you have been confined to the busy marts of commerce, in our large cities, you may even there have seen his *skin*, and thought it a beautiful fur. It is, in fact, when the animal is killed in good season, superior to very many other materials for making *beaver* (?) hats, as well as for other purposes, and thousands of Musk-Rat skins are annually used in the United States, while still greater numbers are shipped to Europe, principally to Great Britain.

This species is nocturnal, and consequently its manners and customs cannot be correctly ascertained from the occasional glimpses of it which we obtain by day-light, as it may chance to pass rapidly through the water seeking to conceal itself under the root of some large tree projecting into the deep pool, or as it dives suddenly to the mouth of its hole under the shelter of the steep or over-hanging bank of the stream, into which it hastily retires when our appearance has alarmed it.

We have often, in the Northern part of the State of New-York, or on the Schuylkill, or near Frankford, in Pennsylvania, gone during the day to look for and observe these animals, to places where we knew they abounded; but although we might patiently wait for hours, with book in hand to beguile the time, we could rarely see one, and should one appear, it was only for an instant. But at such places, so soon as the last rays of the setting sun have ceased to play upon the smooth water, and when the last bright sparkling tints he has thrown as a "farewell till to-morrow," upon rock tree and floweret, are succeeded by the deep quiet gray of twilight; the placid surface of the stream is agitated in every direction, many a living creature emerges from its diurnal retreat and may be observed in full activity above or beneath the water, and first to appear is the Musk-Rat—which may perchance dart out from underneath the very old stump on which we have been so patiently seated! We are perhaps startled by an unexpected noise and plash —and two seconds after, up comes the head of the animal to the surface, at least five yards off—and, if we happen not to be observed, we may look on, and see him swimming merrily with his companions, or seeking his "breakfast," for his day has just begun!

When we were about seventeen years of age, we resided on our farm, "Mill-Grove," situated at the confluence of the Schuylkill river and the Perkioming creek.

On the latter, above a mill-dam which then existed, there was an island divided from the shore on the southerly side by a small channel not more than twenty-five or thirty feet in width, in which we had occasionally observed Musk-Rats swimming. Having a friend at our house for a few weeks, we one evening persuaded him to accompany us to this spot, with the view of procuring a few of these animals. Accordingly, after due preparation we made our way toward the creek. We approached the bank quietly, and seated ourselves on some moss-covered stones without disturbing the silence of the night, the only interruption to which was the gentle ripple of the pure stream, which, united with the broader Schuylkill, still flows onward, and conveys to the now great city of Philadelphia, that inestimable treasure *pure water*. Here then we waited, long and patiently—so long, that our companion became restless, said that he would like to smoke a cigar, and accordingly lighted a "fragrant Havana." We remained watching, but saw no Musk-Rats that evening, as these cunning animals no doubt observed the light at the end of my friend's cigar. We have since that time known many a sportsman lose a shot at a fine buck, by indulging in this relaxation, while at a "*stand*," as it is generally termed. To return to our Musk-Rats, we went home disappointed, but on the next evening proceeded to the same spot, and in less than an hour shot three, which we secured. Next day we made a drawing of one of them, which was afterwards lost. We have now in our possession only two drawings of quadrupeds made by us at this early period; one of which represents the American otter, and the other a mink. They were drawn with coloured chalks and crayons, and both are now quite rubbed and soiled, like ourselves having suffered somewhat from the hand of time, and the jostling we have encountered.

We have sometimes, when exami ing or describing one of our well-known animals, allowed ourselves to fall into a train of thought as we turned over the pages of some early writer, which carried us back to the period of the discovery of our country, or still earlier explorations of wild and unknown regions. We have endeavoured to picture to ourselves the curiosity eagerly indulged, the gratified hopes, and the various other feelings that must have filled the minds of the adventurous voyagers that first landed on America's forest-margined coast. What were their impressions on seeing the strange objects that met their eyes in all directions? what thought they of the inhabitants they met with? and what were their ideas on seeing birds and quadrupeds hitherto unheard of and unknown? The most indifferent or phlegmatic temperament must have been aroused, and the traveller, whatever his profession—whether soldier, sailor, trader, or adventurer—at such times, doubtless, would pause for awhile, conceal himself, and noiselessly observe the strange movements of the *wonderful* creature he has just for the first time seen—for all the Creator's works *are* wonderful—and it is only because we behold many of them continually, that we finally cease to marvel at the conformation of the most common domesticated species.

Something in this way were our reflections directed while turning over the pages of Captain John Smith, whose life was preserved by the fair and heroic Pocahontas. This gallant soldier was, as well as we can learn, the first person who gave any account of the Musk-Rat. His "General History of Virginia, New England, and the Summer Isles," was published in London, in 1624, folio; he styles himself, "sometime Governor in those Countries, and Admiral of New England."

Smith, in this account of Virginia, &c., says of this animal—"A Mussascus is a beast of the form and nature of our water-rat, but many of them smell exceedingly strong of musk."

La Hontan, in a letter dated Boucherville, May, 1687, (see Trav. in Canada,) says—"In the same place we killed some Musk-Rats, or a sort of animals which resemble a rat in their shape, and are as big as a rabbit. The skins of these rats are very much valued, as differing but little from those of beavers." He goes on to describe the manner in which the "strong and sweet smell" of musk is produced; in which he so much betrays his ignorance of natural history, that we will not expose the vulgar error by repeating it here. But if one Frenchman of the 17th century committed some errors in relating the habits of this species, another, early in the 18th, (1725,) made ample amends, by giving us a scientific description of its form, internal structure, and habits, that would do credit to the most careful investigator of the present day. This accomplished naturalist was Mons. Sarrasin, King's Physician at Quebec, and correspondent of the French Academy; in honour of whom Linnæus named the genus *Sarrasenia*. He dissected a number of Musk-Rats, described the animal, gave an account of the "follicles which contain the perfume," and noted its habits.

To this intelligent physician, Buffon was principally indebted for the information which enabled him to draw up his article on the Canadian Musk-Rat.

In 1789, Kalm, (Beschreibung der Reise nach dem Noerdlichen America,) gives a very correct account of the characteristics and habits of this species.

Musk-Rats are lively playful animals when in their proper element the water, and many of them may be occasionally seen disporting on a calm night in some mill-pond or deep sequestered pool, crossing and re-crossing in every direction leaving long ripples in the water behind them, whilst others stand for a few moments on little knolls or tufts of grass, or on stones or logs, on which they can get footing above the water, or on the banks of the pond, and then plunge one after another into the water; at times, one is seen lying perfectly still on the surface of the pond or stream, with its body widely spread out, and as flat as it can be. Suddenly it gives the water a smart flap with its tail, somewhat in the manner of the beaver, and disappears beneath the surface instantaneously—going down head foremost—and reminding one of the quickness and ease with which some species of ducks and grebes dive when shot at. At the distance of ten or twenty yards, the Musk-Rat comes to the surface again, and perhaps joins its companions in their sports; at the same time others are feeding on the grassy banks, dragging off the roots of various kinds of plants, or digging underneath the edge of the bank. These animals thus seem to form a little community of social playful creatures, who only require to be unmolested in order to be happy. Should you fire off a fowling-piece whilst the Musk-Rats are thus occupied, a terrible fright and dispersion ensues—dozens dive at the flash of the gun, or disappear in their holes; and although in the day-time, when they see imperfectly, they may be shot whilst swimming, it is exceedingly difficult to kill one at night. In order to insure success the gunner must be concealed, so that the animal cannot see the flash, when he fires even with a percussion lock.

The burrows and houses of this species are not constructed on such admirable architectural principles as those of the beaver, but are, nevertheless, curious, and well-adapted for the residence of the animal. Having enjoyed opportunities of examining them in several portions of the Northern States, and having been present when hundreds of Musk-Rats were taken, either by digging them out or catching them in traps, we will endeavour to describe their nests, and the manner in which the hunters generally proceed in order to procure the animals that are in them.

In different localities the Musk-Rat has very opposite modes of constructing its winter domicil. Where there are overhanging clayey or loamy banks along the stream or pond, they form a winter retreat in the side of the bank, with openings under the water, and their galleries run sometimes to the distance of fifteen or twenty yards from the shore, inclining upward, so as to be above the influence of the high waters on the breaking up of the ice in spring, or during freshets. There are usually three or four entrances from under the water, which all, however, unite at a point some distance from the water, and sufficiently high to be secure from inundation, where there is a pretty large excavation. In this "central hall" we have seen nests that would fill a bushel basket. They were composed of decayed plants and grasses, principally sedge, (*Carex*,) the leaves of the arrow-head, (*Sagittaria*,) and the pond-lily, (*Nymphæa.*) They always contained several dried sticks, some of them more than a foot in length; these were sometimes arranged along the sides, but more frequently on the top of the nests. From these nests there are several galleries extending still farther from the shore; into the latter the animals retreat, when, after having been prevented from returning to the water, by stopping the entrances, they are disturbed in their chamber. Sometimes we have found their subterranean strongholds leading into others by transverse galleries. These were never so far beneath the surface as those of the fox, marmot, or skunk. On passing near the burrows of the Musk-Rat, there is always sufficient evidence of their existence in the vicinity: the excrement of the animal, resembling that of the Norway rat, being deposited around, and paths that they have made through the rushes and aquatic plants, that grow in thick profusion in the immediate neighbourhood, being easily traced; but it is not so easy to discover the entrances. The latter are always under the water, and usually where it is deepest near the shore. When the Musk-Rat is about to retire to its hole, it swims to within a few feet of the shore, and then dives suddenly and enters it. If you are standing on the bank directly above the mouth of the hole, the rumbling noise under your feet, if you listen attentively, will inform you that it has entered its burrow. It seldom, however, immediately retreats far into its hole, but has small excavations and resting-places on the dry ground a little beyond the reach of the water.

There are occasionally very differently constructed nests of the Musk-Rat; we have seen some of them, in the town of Clinton Dutchess county, and along the margins of swamps in the vicinity of Lake Champlain, in the State of New-York; and others in several localities in Canada. A pond supplied chiefly if not entirely by springs and surrounded by low and marshy ground, is preferred by the Musk-Rats; they seem to be aware that the spring-water it contains, probably will not be solidly frozen, and there they prepare to pass the winter. Such a place, as you may well imagine, cannot without great difficulty be approached until its boggy and treacherous foundation has been congealed by the hard frosts and the water is frozen over; before this time the Musk-Rats collect coarse grasses and mud, with which, together with sticks, twigs, leaves, and any thing in the vicinity that will serve their purpose, they raise their little houses from two to four feet above the water; the entrance being always from below. We have frequently opened these nests and found in the centre a dry comfortable bed of grass, sufficiently large to accommodate several of them. When the ponds are frozen over, and a slight fall of snow covers the ground, these edifices resemble small hay-cocks. There is another peculiarity that, it appears to us, indicates a greater degree of intelligence in the Musk-Rat than we are usually disposed to award to it. The animal seems to know that the ice will cover the pond in winter, and that if it has no places to which it can resort to breathe, it will be suffocated. Hence you here and there see what are called breathing places. These are covered over with mud on the sides, with some loose grass in the centre to preserve them from being too easily frozen over. We have occasionally seen these winter-huts of the Musk-Rat, in the vicinity of their snug summer retreats in some neighbouring river's bank, and have sometimes been half inclined to suppose, that for some cause or other they gave a preference to this kind of residence. We are not, however, aware that these nests are made use of by the Musk-Rat in spring for the purpose of rearing its young. We believe these animals always for that purpose resort to holes in the sides of ponds, sluggish streams, or dykes.

In such situations we have frequently observed the young, which when they first make their appearance are seen emerging from a side gallery leading to the surface, so that they are not of necessity obliged to "take a dive" until they have had a little acquaintance with the liquid element. They are at this time very gentle, and we have on several occasions taken them up with the hand without their making any violent struggles to escape, or attempts to punish us with their teeth.

The fur of this species was formerly a valuable article of commerce, and is still in some demand. But since so many new inventions are supplying the public with cheap hats, and the Nutria skin has been extensively introduced from South America, the Musk-Rat is less sought after, and in some of our most thickly populated districts has greatly increased in numbers. The country-people, however, continue to destroy it to prevent its becoming so numerous as to cause loss, by making holes in the mill-dams, embankments, or ditches, that happen to be inhabited by it, and allowing the water to flow through, when frequently much mischief results. The Musk-Rat has little of the cunning of the fox, the beaver, or even the common Norway rat, and may be easily taken in almost any kind of trap; and although it is very prolific, it might by proper attention be so thinned off in a single year as to cease to be a nuisance. A dozen common rat-traps carefully and judiciously attended to would go far toward reducing if not exterminating these pests in a small neighbourhood, in the course of one or two seasons. The traps should be set in shallow water near the edge of the stream or pool, or on a log sunk about an inch under the water, with a cord ten or twelve feet long so as to prevent the animals from running away with the traps when they have been caught; one or two slices of parsnips or sweet apples may be stuck upon small twigs so that they will hang about six inches above the traps. The animal, having evidently a good nose, whilst swimming at some little distance from the traps when thus set, suddenly turns as it scents the bait, swims along the shore toward it, and reaching up to seize it, is caught by the foot, and being of course greatly alarmed, jerks the trap off the log or pulls it into the water, where the weight of the trap soon drowns it. The Musk-Rat also readily enters and is easily taken in a box-trap, but it ought to be lined with tin or sheet iron, for its formidable incisors otherwise enable the animal to make its escape by gnawing a hole in the box. We have sometimes seen it taken between two boards in what is called a figure of 4 trap, with a heavy weight on the upper board.

The following mode of hunting the Musk-Rat frequently affords a considerable degree of amusement. A party is made up to go: a spade, an axe, and a hoe, are carried along, and a spear, or in lieu of it, a pitchfork; in addition to these, a hoop-net is sometimes wanted, but what is most important, and regarded as a *sine quâ non*, is a dog accustomed to hunting these aquatic animals. The season which promises most success in this way of hunting them is the autumn, before the heavy rains have swelled the waters. The party go to some sluggish stream that winds through a meadow or across a flat country where the banks are not so high as to render the "digging" that has to be done too laborious. The little islands which in such places rise but a few feet above the water, are sometimes perforated by the Musk-Rats, and their holes and excavations undermine them in a great degree, so that it is difficult to find and stop all the mouths of these galleries, and thereby render success tolerably certain. But as these are the very places in which the greatest number of these animals are to be found, it is quite important to "invest" them. It is necessary to be very cautious in digging down along the banks of these islets, in order to reach and stop up the holes, and it usually happens that notwithstanding every precaution is taken, the animals find some way to escape. No sooner is their ancient domicil disturbed, than they issue forth from their holes under the water, to seek some safer retreat along the banks of the main-land; one after another is seen, alternately rising and diving, and making for the shore. If it is ascertained that it is not possible to prevent their escape, the hunters resolve to drive them all from the little islet. A hole is dug in the centre of the place, and the dog encouraged to go in; the few remaining Musk-Rats, at this last and worst alarm, scamper out of the burrows with all haste, and the island is left in possession of the allied forces. All this time the hunters have been sharply looking out to observe to what spot the greatest number of Musk-Rats have retired. They have marked the places in front of which they were seen to dive, well knowing that they are closely concealed in some of the holes along the bank. The animals have now retreated far up into their burrows, and are not very apt to make for the water. The ground is struck with a stick in different places, and where a hollow sound is heard, the hunters know there is an excavation, and at once dig down to it. In this way several holes are found, and are successively stopped to prevent the return of the Musk-Rats to the water. The digging is then continued till the hunters reach the nest, which being laid open, is entered by the dog, in order that the sagacious animal may ascertain the gallery into which the Musk-Rats have retired, as a last resort. The digging is seldom fatiguing, as the holes run very near the surface. A net to catch them is now placed at the hole, or in lieu of it, a man stands with a spear touching the mouth of it, placing his foot immediately behind the spear. As the Rat attempts to rush out, the weapon is driven into its neck. Thus these animals are killed one after another until the whole colony is destroyed; sometimes they are knocked on the head with a club, instead of being speared. In some places we have seen more than a dozen killed in one hole, and we have known upwards of fifty to be taken in this manner in a single day.

When the Musk-Rats have gone to their winter huts among the marshes, there is another way of procuring them. The party go to the marshes, when the ice is sufficiently strong to support a man. They proceed cautiously to their nests (the manner of building which we have already described,) where the Rats are snugly ensconsed in their warm beds, within seven or eight inches of the top. A spear with four prongs, about as long as those of a pitch-fork, is used upon the occasion. One of the men strikes the spear into the nest with all the force he is capable of exerting, and if he understands his business and knows where to strike, he is almost sure to pin one, if not two or three, of the animals to the earth with one blow. Another hunter stands by with an axe to demolish the little mud habitation, and aid in securing the Musk-Rats which have been speared by his companion. It often occurs that the water under the ice is shallow and the ice transparent, in which case the animals may be seen making their way through the water, almost touching the ice, and we have frequently seen them stunned by a blow with the axe on the ice above them, (in the manner in which pike and other fish are sometimes killed in our rivers when they are frozen over;) a hole is then cut in the ice, and they are secured without difficulty. The houses of the Musk-Rats which have been broken up by the hunters are soon restored, the repairs commence the following night, and are usually completed by morning!

In regard to the food of the Musk-Rat, our experience induces us to believe, that like its congener, the house-rat, it is omnivorous. In 1813, we obtained two of this species when very young for the purpose of domesticating them, in order that we might study their habits. They became so perfectly gentle that they came at our call, and were frequently carried to an artificial fish-pond near the house, and after swimming about for an hour or two, they would go into their cage, which was left for them at the water's edge. A few years ago, we received from LEE ALLISON, Esq., residing at Aikin, South Carolina, one of this species in a box lined with tin. We have thus had opportunities of ascertaining the kind of food to which they gave the preference. We would, however, remark, that the food taken by an animal in confinement is no positive evidence of what it would prefer when left to its free choice in the meadows, the brooks, and the fields it inhabits in a state of nature. Their food in summer consists chiefly of grasses, roots, and vegetables. We have often watched them early in the morning, eating the young grass of the meadows; they seemed very fond, especially of the timothy, (*Phleum pratense,*) and red-top, (*Agrostis;*) indeed, the few bunches of clover and other kinds of grass remaining in their vicinity gave evidence that the Musk-Rats had been at work upon them. The injury sustained by the farmer from these animals, however, is by the destruction of his embankments and the excavations through his meadows, made in constructing their galleries, rather than from the loss of any quantity of grass or vegetables they may destroy; although their depredations are sometimes carried on to the great injury of vegetable gardens.

An acquaintance who had a garden in the neighbourhood of a meadow which contained a large number of Musk-Rats, sent one day to inquire whether we could aid in discovering the robbers who carried off almost every night a quantity of turnips. We were surprised to find on examining the premises, that the garden had been plundered and nearly ruined by these Rats. There were paths extending from the muddy banks of the stream, winding among the rank weeds and grasses, passing through the old worm fence, and leading to the various beds of vegetables. Many of the turnips had disappeared on the previous night—the duck-like tracks of the Musk-rats were seen on the beds in every direction. The paths were strewn with turnip leaves, which either had dropped, or were bitten off to render the transportation more convenient. Their paths after entering the meadow diverged to several burrows, all of which gave evidence that their tenants had been on a foraging expedition on the previous night. The most convenient burrow was opened, and we discovered in the nest so many different articles of food, that we were for some time under an impression that like the chipping squirrel, chickaree, &c., this species laid up in autumn a store of food for winter use. There were carrots and parsnips which appeared to have been cut into halves, the lower part of the root having been left in the ground; but what struck us as most singular was that ears of corn (maize) not yet quite ripe, had been dragged into the burrow with a considerable portion of the stalk attached.

The corn-stalks then standing in the garden were so tall that the ears could not be reached by the Musk-Rats, and on examining the beds from which they had probably some days previously taken the corn we found in the burrow, we ascertained that the stalks had been gnawed off at the roots.

Professor Lee, who resides at Buncomb, North Carolina, lately informed us that for several summers past his fields of Indian corn, which are situated near a stream frequented by Musk-Rats, have been greatly injured by their carrying off whole stalks at a time, every night for some weeks together. The above, however, are the only instances that have come to our knowledge of their doing any injury to the vegetable garden or to the corn-field, although this may probably be frequently the case where the fields or gardens skirt the banks of water-courses.

These animals walk so clumsily that they seem unwilling to trust themselves any distance from the margin of the stream or dam on which they have taken up their residence. We have supposed, that a considerable portion of their food in the Northern States in some localities, was the root of the common arrow-head, (*Sagittaria, sagittifolia*,) as we have often observed it had been gnawed off, and have found bits of it at the mouths of their holes. We have also seen stems of the common Indian turnip, (*Arum triphyllum*,) which were cut off, portions of which, near the root, appeared to have been eaten. They also feed on the spice wood, (*Laurus benzoin*.) Richardson says, "they feed in the Northern districts on the roots and tender shoots of the bulrush and reed-mace, and on the leaves of various carices and aquatic grasses." Pennant says, "they are very fond of the *Acorus verus*, or *Calamus aromaticus*;" and Kalm speaks of apples being placed in traps as a bait for them. Nearly all our writers on natural history are correct in saying that fresh water mussels compose a portion of their food. Sometimes several bushels of shells may be found in a small space near their nests. Our young friend, SPENCER F. BAIRD, Esq., assures us that in the neighbourhood of Carlisle Pennsylvania, on the Conodoguinet creek, he has often observed large quantities of shells, most of which were so adroitly opened by these animals as not to be at all broken, and would have made very good specimens for the conchologist. He has seen the Musk-Rat eating a mussel occasionally on a log in the water, holding the shell between its fore-paws, as a squirrel holds a nut.

We once placed a quantity of mussels in a cage, to feed some Musk-Rats we had domesticated in the North; they carried them one by one into an inner compartment, where they were hidden from view. Here we heard them gnawing at the shells; we then removed a slide in the cage, which enabled us to see them at work; they were seated, sometimes upright like a squirrel, at other times like a rat, with the shell-fish lying on the floor, holding on to it by their fore-paws, and breaking it open with their lower incisors. In Carolina, we obtained for the same purpose, although for a different family of Musk-Rats, a quantity of mussels of the species *Unio angustatus* and *Anodon cataracta;* some of these were too hard to be immediately opened by the animals with their teeth. They were carried by the Musk-Rats, as usual, into a separate and darkened portion of the cage. We heard an occasional gnawing, but three days afterwards many of the harder species of shell still remained unopened. We did not again examine the cage till after the expiration of ten days, when the shells were all empty. They had probably opened in consequence of the death of the animal within, when their contents were eaten by the Rats. Oysters were placed in the cage, which on account of their saltness we believed would not be relished; but a week afterwards the shells only were left. We procured a pint of a small species of imported snail, (*Bulimus decollatus*, GMEL., *mutilatus*, SAY,) that has become very destructive in many of the gardens of Charleston, and the Musk-Rats immediately began to crush them with their teeth, and in a few days nothing but the broken shells remained. We have therefore come to the conclusion that whilst vegetables are the general food of this species, various kinds of shell-fish form no inconsiderable portion of it. Our Musk-Rats refused fish, but were, like most animals in confinement, fond of bread. They were generally fed on sweet potatoes, parsnips, cabbage, and celery; the sweet flag, (*Acorus calamus,*) they rejected altogether.

Although the Musk-Rat walks awkwardly, and proceeds so slowly that it can scarcely be said to run, it swims and dives well. We regard it as a better swimmer than the mink, and from its promptness in diving at the flash of the gun, it frequently escapes from its pursuers. It may, however, be easily drowned. We once observed several of them which had been driven from their holes, after struggling under the ice for about fifteen minutes rising to the surface; and on taking them out, by cutting holes in the ice, they were found to be quite dead. RICHARDSON speaks of "their being subject at uncertain intervals to a great mortality from some unknown cause." We have no doubt that in very cold winters when the ice reaches to the bottom of the ponds, and they are confined to their holes, they devour each other, since we have seen many burrows opened in autumn, and except in the instances we have already mentioned, we found no provision laid up for winter use. When a Musk-Rat has been caught by one foot in a trap set on the land, it is frequently found torn to pieces and partially devoured; and from the tracks around one might be induced to believe, that, as is the case with porpoises and many other animals, when one is wounded and cannot escape its companions turn upon and devour it. When one is shot and dies in the water it is very soon carried off by the living ones, if there are any in the vicinity at the time, and is dragged into one of their holes or nests. We have frequently found carcasses of these animals thus concealed, but in these cases the flesh had not been devoured. This singular habit reminds us of the Indians, who always carry their dead off the field of battle when they can, and endeavour to prevent their bodies falling into the hands of their enemies.

After a severe winter on a sudden rise in the water before the breaking up of the ice, hundreds of Musk-Rats are drowned in their holes, especially where there are no high shelving banks to enable them to extend their galleries beyond the reach of the rising waters. During these occasional freshets in early spring, the Musk-Rats that escape drowning are driven from their holes, and swim about from shore to shore without shelter and without food, and may be easily destroyed. We remember that two hunters with their guns, coursing up and down opposite sides of a pond on one of these occasions, made such fearful havoc among these animals that for several years afterwards we scarcely observed any traces of them in that locality. Many rapacious birds as well as quadrupeds seize and devour the Musk-Rat. When it makes its appearance on land, the fox and the lynx capture it with great ease. One of our young friends at Dennisville in the State of Maine, informed us that his greatest difficulty in procuring this species in traps arose from their being eaten after they were caught, by the snowy owl and other birds of prey, which would frequently sit and watch the traps, as it were keeping guard over them, until the poor Musquash was in the toils, on seeing which they descended and made a hearty meal at the trapper's expense, taking good care meanwhile not to expose themselves to his vengeance, by keeping a sharp look out for him in every direction. Our friend, however, got the better of these wary thieves by occasionally baiting his traps with meat instead of apples or vegetables, by which means he often caught an owl or a hawk, instead of a Musk-Rat. Although this species has such a long list of enemies, it is so prolific that, like the common rat, (*Mus decumanus,*) it continues to increase and multiply in many parts of the country, notwithstanding their activity and voracity.

The Musk-Rat has occasionally been known to leave its haunts along the streams and ponds, and is sometimes found travelling on elevated grounds. We were informed by our friend Mr. BAIRD, that one was caught in a house near Reading, in Pennsylvania, three-quarters of a mile from the water; and the late Dr. WRIGHT of Troy once discovered one making its way through the snow, on the top of a hill near that city.

The number of young produced at a litter varies from three to six. RICHARDSON states that they sometimes have seven, which is by no means improbable. They usually have three litters in a season.

Although the Musk-Rat does not seem to possess any extraordinary instincts by which to avoid or baffle its pursuers, we were witnesses of its sensibility of approaching danger arising from a natural cause, manifested in a way we think deserving of being recorded. It is a well-known fact, that many species of quadrupeds and birds are endowed by Nature with the faculty of foreseeing or foreknowing the changes of the seasons, and have premonitions of the coming storm. The swallow commences its long aerial voyage even in summer, in anticipation of the cold. The sea-birds become excessively restless: some seek the protection of the land, and others, like the loon, (*Colymbus glacialis,*) make the shores re-echo with their hoarse and clamorous screams, previous to excessively cold weather; the swine also are seen carrying straw in their mouths and enlarging their beds. After an unusual drought, succeeded by a warm Indian-summer, as we were one day passing near a mill-pond inhabited by some families of Musk-Rats, we observed numbers of them swimming about in every direction, carrying mouthfuls of withered grasses, and building their huts higher on the land than any we had seen before. We had scarcely ever observed them in this locality in the middle of the day, and then only for a moment as they swam from one side of the pond to the other; but now they seemed bent on preparing for some approaching event, and the successive reports of several guns fired by some hunters only produced a pause in their operations for five or ten minutes. Although the day was bright and fair, on that very night there fell torrents of rain succeeded by an unusual freshet and intensely cold weather.

This species has a strong musky smell; to us this has never appeared particularly offensive. It is infinitely less unpleasant than that of the skunk, and we are less annoyed by it than by the smell of the mink, or even the red fox. We have, however, observed in passing some of the haunts of this Rat at particular periods during summer, that the whole locality was strongly pervaded by this odour.

It is said, notwithstanding this peculiarity, that the Musk-Rat is not an unpalatable article of food, the musky smell not being perceptible when the animal has been properly prepared and cooked; we have, indeed, heard it stated that Musk-Rat suppers are not unfrequent among a certain class of inhabitants on the Eastern shore of Maryland, and that some persons prefer them, when well dressed, to a wild duck. Like the flesh of the bear and some other quadrupeds, their meat somewhat resembles fresh pork, and is too rich to be eaten with much relish for any length of time.

By what we may almost look at as a merciful interposition of Providence, the Musk-Rat is not found on the rice plantations of Carolina; it approaches within a few miles of them, and then ceases to be found. If it existed in the banks and dykes of the rice fields, it would be a terrible annoyance to the planter, and possibly destroy the reservoirs on which his crops depend. Although it reaches much farther South, and even extends to Louisiana, it is never found on the alluvial lands within seventy miles of the sea, either in Carolina or Georgia.

The skins of the Musk-Rat are no longer in such high repute as they enjoyed thirty-five years ago, and they are now only worth from six and a quarter to twenty-five cents each.

Dr. Richardson states, (in 1824,) that between four and five thousand skins were annually imported into Great Britain from North America.

GEOGRAPHICAL DISTRIBUTION.

The Musk-Rat is found as far North as the mouth of the Mackenzie river, in latitude 69°, on the Rocky mountains, on the Columbia river, and on the Missouri. With the exception of the alluvial lands in Carolina, Georgia, Alabama, and Florida, it abounds in all parts of the United States north of latitude 30°. It exists, although not abundantly, in the mountains of Georgia, and the higher portions of Alabama. In South Carolina, we have obtained it from Aikin, and St. Matthew's parish, on the Congaree river, but have never found traces of it nearer the sea than seventy miles from Charleston.

GENERAL REMARKS.

The Musk-Rat, although the only species in the genus, was moved about among several genera before it found a resting place under its present name. SCHREBER placed it under MUS. GMELIN, and F. CUVIER described it as a LEMMUS. LINNÆUS and ERXLEBEN arranged it with the beaver, and referred it to the genus CASTOR. LESSON, LACEPEDE and CUVIER, under ONDATRA. In 1811, ILLIGER proposed changing its specific into a generic name. As LINNÆUS had called it *Castor Fiber*, he then established for it the genus FIBER.

N.º 3 Plate XIV.

Hudson's Bay Squirrel - Chickaree - Red Squirrel.

SCIURUS HUDSONIUS.—Pennant.

Hudson's Bay Squirrel.—Chickaree.—Red-Squirrel.

PLATE XIV.—Male and Female.

S. cauda corpore breviore, auriculis apice sub-barbatis; corpore supra subrufo, subtus albo; S. migratorii tertia parte minore.

CHARACTERS.

A third smaller than the Northern Gray-Squirrel, (Sc. migratorius;) tail shorter than the body; ears, slightly tufted; colour, reddish above, white beneath.

SYNONYMES.

Ecureuil Commun, ou Aroupen, Sagard Theodat, Canada, p. 746.
Common Squirrel, Forster, Phil. Trans., vol. lxii., p. 378, 1772.
Sciurus Vulgaris, var. E. Erxleben, Syst., An. 1777.
Sciurus Hudsonicus, Pallas, Glir., p. 377.
Sciurus Hudsonicus, Gmel., Linn., ——— 1788.
Hudson's Bay Squirrel, Penn. Arctic Zool., vol. i., p. 116.
" " " " Hist. Quadrupeds, vol. ii., p. 147.
Common Squirrel, Hearnes' Journey, p. 385.
Red Squirrel, Warden's Hist. U. S., vol. i., p. 330.
Red Barking Squirrel, Schoolcraft's Journal, p. 273.
Sciurus Hudsonicus, Sabine, Franklin's Journey, p. 663.
" " " Godman, vol. ii., p. 138.
" " " Fischer, Mam., p. 349.
Ecureuil de la Baie d'Hudson, F. Cuvier, Hist. Nat. des Mammifères.
Sciurus Hudsonicus, Bach. Trans. Zool. Soc., London, 1839.
" " " Dekay, Nat. Hist. New-York, 1842.

DESCRIPTION.

On examining the teeth of this species, we do not find the small and usually deciduous molar that exists in all the other species of Sciurus with which we are acquainted; it is possible, however, that it may be found in very young animals. It will be perceived, on referring to the dental formula of the genus, (which we have given at p. 38,) that the molars are set down as $\frac{4}{4}$—$\frac{4}{4}$ or $\frac{5}{4}$—$\frac{5}{4}$; and we will for the present assign the former arrangement to this species. Forehead, very slightly arched; nose, somewhat obtuse; eyes, of moderate size; ears, broad, rounded, clothed on both sides with short hairs, not distinctly tufted like those of the European Squirrel, (*Sc. vulgaris*,) although the hairs, when the animal has its winter pelage, project beyond the margins, and resemble tufts; whiskers, a little longer than the head; the body presents the appearance of lightness and agility; the tail is somewhat depressed, and linear, not as bushy as in most other squirrels, but capable of a distichous arrangement; limbs, robust; claws, compressed, sharp, slightly hooked; third toe a little the longest; palms, and under surface of the toes, naked; soles of hind-feet, clothed with hair, except on the tubercles at the roots of the toes.

COLOUR.

This species exhibits some shades of difference in colour, and we have sometimes, although very rarely, found a specimen that might be regarded as a variety. General colour, deep reddish-brown on the whole of the upper surface; short fur beneath, plumbeous, mixed with so large a quantity of longer hairs, that the colour of the fur does not show on the surface. These long hairs are dark at the roots, then brown, and are slightly tipped with black. In most specimens there is an orange hue on the outer surface of the fore-legs, running up to the shoulder; this colour is also frequently visible on the upper surface of the hind-feet, and behind the ears. Whiskers, black; tail, on the upper surface, deep reddish-brown; the hair on the sides may be so arranged as to present a line of black near the outer borders; on the under side it has two or three annulations of light-brown and black; lips, chin, throat, inside of legs, and belly, white; in some specimens the hairs on these parts of the body are plumbeous at the roots, and white to the tips, giving it a light, grayish-white appearance. There is in a great many specimens a black line, running from near the shoulders along the sides to within an inch of the thighs.

DIMENSIONS.

Recent specimen.

	Inches.	Lines.
Length from nose to root of tail - - - -	8	0
Tail (vertebræ) - - - - - - - -	3	7
Tail to end of hair - - - - - - -	6	5

HABITS.

The genus Sciurus is illustrated in North America by a greater variety of species than any other among the various genera we shall have the pleasure of introducing to our readers:—Permit us to dwell for a moment on the subject, and to relate the following anecdote:

When we began the publication in Great Britain of the "Birds of America," we were encouraged by the approbation of many excellent friends, and by the more essential, although less heartfelt favours, bestowed by those noblemen and gentlemen who kindly subscribed to the work, and without whose aid, it is frankly acknowledged, it could never have been completed. Among those whom we then had the honour of calling patrons, we found as many varieties of character as among the beautiful feathered inhabitants of our woods, lakes, and sea-shores, themselves; and had we time just now to spare, we might undertake to describe some of them. We published as the first plate of the first number of "The Birds of America," the Wild Turkey Cock, and gave the Turkey Hen and Young, as the first plate of the second number. We need not stop to enumerate the other species of birds that completed those two numbers; but judge of our surprise, on being told gravely, by a certain noble subscriber, that, "as the work was to consist of *Turkeys* only, he begged to be allowed to discontinue his subscription!"

Now, kind reader, we are obliged to follow Nature in the works of infinite wisdom, which we humbly attempt to portray; and although you should find that more Squirrels inhabit our forests than you expected or desired to be figured in this work, we assure you it would give us pleasure to discover a new species at any time! We are not, however, wanting in a due knowledge of the sympathy and kindness that exist among our patrons toward us, and we hope you will find this really beautiful genus as interesting as any other among the quadrupeds we desire to place before you.

The Chickaree, or Hudson's Bay Squirrel, is the most common species of this numerous genus around New-York and throughout the Eastern States. It is a graceful, lively animal, and were you to walk with us through the woods in the neighbourhood of our great commercial metropolis, where boys and sportsmen (?) for years past have been hunting in every direction, and killing all the game left in the vicinity; where woodcocks are shot before the first of July, and quails (Virginian partridges) when they are half-grown, in defiance of the laws for their preservation, you would be glad to find the comparative silence which now reigns amid the trees, interrupted by the sprightly querulous cry of the Chickaree, and would pause with us to look at him as he runs along the rocky surface of the ground, or nimbly ascends some tree; for in these woods, once no doubt abounding in both beasts and birds, it is now a hard task to start anything larger than a robin, or a *High-hole*, (*Picus auratus*.) The Hudon's Bay Squirrel is fearless and heedless, to a great degree, of the presence of man; we have had one occasionally pass through our yard, sometimes ascending an oak or a chesnut, and proceeding leisurely through our small woody lawn. These little animals are generally found singly, although it is not uncommon for many to occupy the same piece of wood-land, if of any extent. In their quick graceful motions from branch to branch, they almost remind one of a bird, and they are always neat and cleanly in their coats, industrious, and well provided for the cold of winter.

In parts of the country, the Chickaree is fond of approaching the farmer's store-houses of grain, or other products of the fields, and occasionally it ventures even so far as to make a nest for itself in some of his outbuildings, and is not dislodged from such snug quarters without undergoing a good deal of persecution.

One of these Squirrels made its nest between the beams and the rafters of a house of the kind we have just spoken of, and finding the skin of a peacock in the loft, appropriated the feathers to compose its nest, and although it was destroyed several times, to test the perseverance of the animal, it persisted in re-constructing it. The Chickaree obtained this name from its noisy chattering note, and like most other Squirrels, is fond of repeating its cries at frequent intervals. Many of the inhabitants of our Eastern States refuse to eat Squirrels of any kind, from some prejudice or other; but we can assure our readers that the flesh of this species, and many others, is both tender and well-flavoured, and when nicely broiled, does not require a hunter's appetite to recommend it.

The habits of this little Squirrel are, in several particulars, peculiar; whilst the larger Gray Squirrels derive their sustenance from buds and nuts, chiefly inhabit warm or temperate climates, and are constitutionally fitted to subsist during winter on a small quantity of food, the Chickaree exhibits the greatest sprightliness and activity amidst the snows and frosts of our Northern regions, and consequently is obliged, during the winter season, to consume as great a quantity of food as at any other. Nature has, therefore, instructed it to make provision in the season of abundance for the long winter that is approaching; and the quantity of nuts and seeds it often lays up in its store-house, is almost incredible. On one occasion we were present when a bushel and a half of shell-barks (*Carya alba*), and chesnuts, were taken from a hollow tree occupied by a single pair of these industrious creatures; although generally the quantity of provision laid up by them is considerably less. The Chickaree has too much foresight to trust to a single hoard, and it often has several, in different localities among the neighbouring trees, or in burrows dug deep in the earth. Occasionally these stores are found under leaves, beneath logs, or in brush-heaps; at other times they are deposited in holes in the ground; and they are sometimes only temporarily laid by in some convenient situation to be removed at leisure. When, for instance, nuts are abundant in the autumn, large quantities in the green state, covered by their thick envelope, are collected in a heap near the tree whence they have fallen; they are then covered up with leaves until the pericarp, or thick outer covering, either falls off or opens, when the Squirrel is able to carry off the nuts more conveniently. In obtaining shell-barks, butter-nuts, (*Juglans cinerea*,) chesnuts, hazel-nuts, &c., this Squirrel adopts the mode of most of the other species. It advances as near to the extremity of the branch as it can with safety, and gnaws off that portion on which the nuts are dependent. This is usually done early in the morning, and the noise occasioned by the falling of large bunches of chesnut burrs, or clusters of butter-nuts, hickory, or beech-nuts, thus detached from the parent stem, may be heard more than a hundred yards off. Some of the stems attached to the nuts are ten inches or a foot in length. After having thrown down a considerable quantity, the Squirrel descends and drags them into a heap, as stated above.

Sometimes the hogs find out these stores, and make sad havoc in the temporary depot. But Providence has placed much food of a different kind within reach of the Red-Squirrel during winter. The cones of many of our pines and firs in high northern latitudes are persistent during winter; and the Chickaree can be supported by the seeds they contain, even should his hoards of nuts fail. This little Squirrel seems also to accommodate itself to its situation in another respect. In Pennsylvania and the southern part of New-York, where the winters are comparatively mild, it is very commonly satisfied with a hollow tree as a winter residence; but in the latitude of Saratoga, N. Y., in the northern part of Massachusetts, in New Hampshire, Maine, Canada, and farther north, it usually seeks for additional protection from the cold by forming deep burrows in the earth. Nothing is more common than to meet with five or six Squirrel-holes in the ground, near the roots of some white pine or hemlock; and these retreats can be easily found by the vast heaps of scales from the cones of pines and firs, which are in process of time accumulated around them. This species can both swim and dive. We once observed some lads shaking a Red-Squirrel from a sapling that grew on the edge of a mill-pond. It fell into the water and swam to the opposite shore, performing the operation of swimming moderately well, and reminding us by its movements of the meadow-mouse, when similarly occupied. It was "headed" by its untiring persecutors on the opposite shore, where on being pelted with sticks, we noticed it diving two or three times, not in the graceful curving manner of the mink, or musk-rat, but with short and ineffectual plunges of a foot or two at a time.

We have kept the Chickaree in cages, but found it less gentle and more difficult to be tamed than many other species of the genus.

Richardson informs us that in the fur countries, "the Indian boys kill many with the bow and arrow, and also take them occasionally with snares set round the trunks of the trees which they frequent." We have observed that during winter a steel-trap baited with an ear of corn, (maize,) placed near their burrows at the foot of large pine or spruce trees, will secure them with the greatest ease.

GEOGRAPHICAL DISTRIBUTION.

The limits of the northern range of this species are not precisely determined, but all travellers who have braved the snows of our Polar regions, speak of its existence as far north as their journeys extended. It has been observed in the 68th or 69th parallel of latitude; it also exists in Labrador, Newfoundland and Canada. It is the most common species in New-England and New-York, and is by no means rare in Pennsylvania and New Jersey, especially in the hilly or mountainous portions of the latter State. It is seen, in diminished numbers, in the mountains of Virginia, although in the alluvial parts of that State it is scarcely known; as we proceed southwardly it becomes more rare, but still continues to be met with on the highest mountains. The most southern locality to which we have traced it, is a high peak called the Black mountain, in Buncombe county, N. Carolina. The woods growing in that elevated situation are in some places wholly composed of balsam-fir trees, (*Abies balsamea,*) on the cones of which these Squirrels feed. There this little animal is quite common, and has received a new English name, viz., that of, "Mountain boomer." Toward the west we have traced it to the mountains of Tennessee; beyond the Rocky mountains, it does not exist. In the Russian settlements on the Western coast, it is replaced by the Downy Squirrel, (*Sc. lanuginosus.*) In the vicinity of the Columbia, and for several hundred miles along the mountains South of that river, by Richardson's Columbian Squirrel; and in the mountainous regions bordering on California, by another small species much resembling it, which we hope, hereafter, to present to our readers.

GENERAL REMARKS.

Although this species, from its numbers and familiarity as well as from its general diffusion, has been longer known than any other of our Squirrels, and has been very frequently described, it has, with few exceptions, retained its name of *Hudsonius*. ERXLEBEN supposed it to be only a variety of the common Squirrel, *S. vulgaris* of Europe, and so described it. The *Sciurus Hudsonius* of GMELIN is a flying Squirrel, (*Pteromys sabrinus*,) and the Carolina Gray Squirrel, which in Shaw's General Zoology, vol. ii., p. 141, is given as a variety of *Sciurus Hudsonius*, is our own species, (*Sc. Carolinensis*.) This species was unknown to LINNÆUS. PALLAS appears to have been the first author who gave the specific name of *Hudsonius*, (see Pall. Glir., p. 377, A. D. 1786,) and GMELIN, in 1788, adopted his name.

In examining the form and inquiring into the habits of this species, we cannot but observe a slight approach to TAMIAS, and a more distant one to SPERMOPHILUS. Its ears are placed farther back than in the Squirrels generally, its tail is only sub-distichous, and withal it often digs its own burrow, and lives indiscriminately in the ground and on trees. In all these particulars it appears, in connexion with the Downy Squirrel, (*Sc. lanuginosus*,) to form a connecting link between SCIURUS and TAMIAS. It has, however, no cheek pouches, and does not carry its food in its cheeks in the manner of the TAMIÆ and SPERMOPHILI, but between its front teeth, like the rest of the squirrels.

GENUS PTEROMYS.—Illiger.

DENTAL FORMULA.

$$\textit{Incisive}\ \frac{2}{2};\quad \textit{Molar}\ \frac{5-5}{4-4} = 22.$$

Dentition similar to that of the genus Sciurus. Head, round; ears, round; upper lip, divided; eyes, large; fore-feet, with four elongated toes, furnished with compressed, sharp, talons, with the rudiment of a thumb having an obtuse nail: hind-feet, with five long toes, much divided, and fitted for seizing or climbing; tail, long, villose; skin of the sides, extending from the anterior to the posterior extremities, forming a thin membrane, by the aid of which, when extended, the animal sails through the air in a descending curve from a tree or any elevated point, occasionally for some distance.

The generic name, *pteromys*, is derived from two Greek words, πτερον, (*pteron*,) a wing, and μυς, (*mus*,) a mouse.

There are thirteen well-determined species belonging to this genus. One is found in the north of Europe, four in North America, and the remainder in Asia and other parts of the old world.

PTEROMYS OREGONENSIS.—Bachman.

Oregon Flying Squirrel.

PLATE XV.—Male and Female.

P. magnitudine inter P. volucellam et P. sabrinum medius, supra fuscus, subtus luteo-albus; auribus P. sabrini auriculis longioribus; vellere densiore, membrana volatica largiore, pedibus grandioribus.

CHARACTERS.

Intermediate in size between P. volucella, and the Northern species, P. sabrinus; ears, longer than in the latter, and far more compact; lobe of the flying membrane joining the fore-feet, much longer in proportion, making that membrane broader. Foot larger; general colour above, brown; beneath, yellowish-white.

Oregon Flying Squirrel.

Drawn from Nature by J.J. Audubon F.R.S. F.L.S.

Drawn on Stone by R. Trembly

SYNONYME.

PTEROMYS OREGONENSIS; Oregon Flying Squirrel, Bach., Jour. Acad. of Nat. Sciences, Phil., vol. viii., pt. i., p. 101.

DESCRIPTION.

This species differs from *P. sabrinus,* in several very striking particulars; the arm which supports the flying membrane is 11¾ lines in length, whilst that of the latter is only 9. Thus the smaller of the two has the largest flying membrane.

The fur of *P. sabrinus* is much the longest, and is white, whilst that of *P. Oregonensis* has a yellowish tinge. The hairs on the tail of the former are only slightly tinged with lead-colour at the roots, whilst in the latter that colour extends outwardly, (towards the tips,) for half their length. The different shape of the ear, it being longer and narrower in our present species than in *P. sabrinus,* is a sufficient distinctive character. *P. Oregonensis* differs from the common flying squirrel (*P. volucella*) so entirely, that it is hardly necessary to give a particular comparison. Besides being much larger than the latter, and not possessing the beautiful downy-white on the belly, it may be distinguished from *P. volucella* by the hairs on that species being white to the roots, which is not the case with the *Oregonensis.* Whiskers, numerous, and very long.

COLOUR.

Fur, deep gray at the base, on the back tipped with yellowish-brown; tail, pale-brown above, dusky toward the extremity; beneath, brownish-white; whiskers, chiefly black, grayish at the tips. Hairs covering the flying membrane, mostly black, slightly tipped with pale-brown; feet, dusky; around the eyes, blackish; ears, with minute adpressed brown hairs externally, and brownish-white internally.

DIMENSIONS.

	Inches.	Lines.
Length from point of nose to root of tail - -	6	8
Tail, to point of fur - - - - - - -	6	0
Height of ear posteriorly - - - - - -	0	7
Breadth between the outer edges of the flying membrane - - - - - - - -	8	0
Longest hind-toe, including nail - - - - -	0	5¼
" fore-toe, " " - - - - -	0	5½
From heel to point of nail - - - - - -	1	6½
From nose to ear - - - - - - -	1	6

HABITS.

The habits of this handsome Flying Squirrel, we regret to say, are almost unknown to us; but from its general appearance, it is undoubtedly as active and volatile as our common little species; and much do we regret that we have never seen it launch itself into the air, and sail from the highest branch of one of the enormous pines of the valley of the Columbia river, to some other tall and magnificent tree. Indeed, much should we like to know the many works of the Creator that yet remain to be discovered, examined, figured, and described, in the vast mountain-valleys and forests beyond the highest peaks of the great Rocky Chain.

We hope, however, to obtain a good deal of information through various sources ere the conclusion of this work, from the remote portions of our Continent that have not yet been well explored by naturalists, and we shall then perhaps be able to say something more in regard to the subject of this article, of which we can now only add, that Mr. Townsend remarks, that it inhabits the pine woods of the Columbia, near the sea, and has the habits of *P. volucella.*

GEOGRAPHICAL DISTRIBUTION.

Dr. Richardson (Fauna Boreali Americana, p. 195) speaks of a Flying Squirrel which was "discovered by Mr. Drummond on the Rocky mountains, living in dense pine-forests, and seldom venturing from its retreats except in the night." This animal he considers a variety of *P. sabrinus,* (var. *B. Alpinus.*) The locality in which it was found, and parts of his description, however, on the whole incline us to suppose that the specimen procured by Mr. Drummond was one of our present species, although of a very large size. Dr. Richardson says, "I have received specimens of it from the head of Elk river, and also from the south branch of the Mackenzie." So that if this supposition be correct, we may conclude that it inhabits a very extensive tract of country, and is, perhaps, most common on, and to the west, of the Rocky Mountains; in which last locality Mr. Townsend met with it in the woods on the shores of the Columbia river.

GENERAL REMARKS.

There are no accounts of this species of Flying Squirrel, or of the larger one, *P. sabrinus,* in Lewis and Clarke's Journal. Those travellers not having, as we suppose, heard of either, although they traversed a considerable portion of the country in which both species have since been found.

We hope, when presenting an account of the habits of *P. sabrinus*, to be able to identify the *variety* above-mentioned, (*P. sabrinus*, var. *B. Alpinus* of RICHARDSON,) and if necessary, correct any error in our account of the geographical distribution of the present species (*P. Oregonensis*.)

LYNX CANADENSIS.—GEOFFROY.

CANADA LYNX.

PLATE XVI.—MALE.

L. magnitudine L. rufum superans; auribus triangularibus, apice pilis crassis nigris erectis barbatis; cauda capite breviore, plantis villosis; supra cinereus, maculis obscuris nebulosus, subtus dilutior.

CHARACTERS.

Larger than F. rufus; ears, triangular, tipt with an upright slender tuft of coarse black hairs; tail, shorter than the head; soles, hairy; general colour, gray above, a little clouded with irregular darker spots, lighter beneath.

SYNONYMES.

LOUP-CERVIER, (anaris qua,) Sagard Theodat, Canada, 744, An. 1636.
" " or LYNX, Dobb's Hudson's Bay, p. 41, An. 1744.
LYNX, Pennant, Arc. Zool., vol. i., p. 50.
" or WILD-CAT, Hearne's Journey, p. 366.
CANADIAN LYNX, Buff., vol. v., suppl. p. 216, pl. 125.
" " Mackenzie's Journey, p. 106.
FELIS CANADENSIS, Geoffroy, An. du Mus.
" CANADENSIS, Sabine, Franklin's Journey, p. 659.
" CANADENSIS, Desm. Mam., p. 225.
NORTHERN LYNX, Godman, Nat. Hist., vol. i., p. 302.
FELIS BOREALIS, Temminck, Monographie, t. i., p. 109.
" CANADENSIS, Rich., F. B. A., p. 101.
" " Reichenbach, Regnum Animale, sp. 551, p. 46, pl. 551, Lipsiæ, 1836.
LYNCUS BOREALIS, Dekay, Nat. Hist., N. Y., p. 50, pl. 10, fig. 2.

DESCRIPTION.

This species has a rounder, broader, and proportionably shorter head than (*L. rufus*) the Bay Lynx; nose, obtuse; eyes, large; teeth, very strong; whiskers, stiff, horizontal, arranged in three oblique series; ears, acute, thickly clothed with hair on both surfaces, tipped by a long and slender tuft of coarse hairs; beneath the ears commences a broad ruff

No. 4 Plate XVI

Drawn on Stone by R. Trembly

Canada Lynx

Male

Drawn from Nature by J.J. Audubon F.R.S.F.L.S.

Printed by Nagel & Weingærtner N.Y.

formed of longer hairs than those on the surrounding parts; this ruff surrounds the throat and reaches the chin, but does not extend around the neck above. The female has the ruff much shorter than the male. Body, robust, thick, and heavy; and from the form, we are inclined to believe that this species is far less fleet than its congener the Bay lynx. The hair has a woolly appearance; under-fur, very dense and soft, mixed with hairs somewhat rigid and two inches in length. On the under surface, the hairs are thinner and a little longer than those above. Thighs, strong; legs, thick and clumsy, presenting a slight resemblance to those of the bear. Toes, thick, so completely concealed by the fur that the tracks made in the snow by this animal do not show distinct impressions of them, like those made by the fox, or the Bay lynx. Their tracks are round, leaving no marks of the nails unless the animal is running, when its toes are widely spread and its nails leave the appearance of slight scratches in the snow. Tail, thickly covered with hair, short, slightly turned upward. Nails, very strong, much larger than those of the Bay lynx, curved, and acuminate.

COLOUR.

Nose, flesh-coloured; pupil of the eye, black; iris amber colour; margin of the lips, and inner surface of the ears, yellowish-brown; face, and around the eyes, light-gray; whiskers, nearly all white, a few black; outer margin of the ear, edged with black, widening as it approaches the extremity, where it is half an inch broad; tuft of ear, black; the ruff under the throat is light-gray, mixed in the centre of the circle with long tufts of black hair. When the hairs on the back are blown aside, they exhibit a dark yellowish-brown colour. The long hairs on the back, black to near the extremity, where there is an annulation of yellowish-brown, finally tipped with black; general colour of the back, gray, with a shade of rufous, and slightly varied with shades of a darker colour; under surface, dull white, with irregular broad spots of dark-brown situated on the inner surface of the fore-legs and extending along the belly; these spots are partially covered by long whitish hairs in the vicinity. In one of our specimens these dark-coloured spots are altogether wanting. The legs are of the colour of the sides; upper surface of the tail, to within an inch of the tip, and exterior portion of the thighs, rufous; beneath yellowish-white; extremity of the tail black.

DIMENSIONS.

The Male represented on the Plate:—Recent.

From nose to root of tail - - - - - - -	33 inches.
Tail (vertebræ) - - - - - - - - -	5 "
Tail, to end of hair - - - - - - -	6 "
Entire length - - - - - - - -	39 "
From nose to end of skull - - - - - -	6 "
" " " root of ears - - - - - -	$4\frac{3}{4}$ "
" " " end of ears laid down - - -	$7\frac{1}{2}$ "
Breadth of ears in front - - - - - -	$3\frac{1}{2}$ "
Height of ears - - - - - - - - -	$2\frac{1}{2}$ "
Length of tufts of hair on the ear - - -	2 "
From nose to hind-foot stretched beyond tail -	45 "
From do. to end of fore-foot stretched beyond nose	$5\frac{1}{2}$ "
Distance between roots of ears anteriorly - -	$3\frac{1}{2}$ "
" " tips of do. - - - - -	$7\frac{1}{2}$ "
Spread of fore-feet, between the claws - -	5 "
Breadth of arm - - - - - - - - -	$2\frac{5}{8}$ "
Height to shoulder from middle of fore-claw -	$13\frac{1}{2}$ "

Weight 16 pounds; extremely lean.

A specimen in the flesh from the Petersburg Mountains, east of Troy:—
Male.

	Inches.	Lines.
From point of nose to root of tail - - - - -	37	0
Tail (vertebræ) - - - - - - - - -	4	4
Tail, to end of fur - - - - - - -	5	4
Height of ear - - - - - - - - -	2	2
Length of tufts on the ears - - - - - -	1	9
From shoulder to extremity of toes on fore-feet -	17	0
From heel to end of hind-claw - - - - -	7	5

Weight 22 pounds.

HABITS.

In some parts of the State Maine, and in New Brunswick, there are tracts of land, formerly covered with large trees but over-run by fires not many years since, now presenting a desolate appearance as you look in every direction and see nothing but tall blackened and charred trunks standing, with only their larger branches occasionally stretching out to the righ' (r left, while many of them are like bare poles, half burnt off near the roots perhaps, and looking as if they might fall to the earth with the slightest breath of air. Into one of these "burnt districts," let us go together. Nature has already begun to replace the stately trees, which the destroying element had consumed or stripped of all beauty and vitality, and we find the new growth already advanced; instead of the light, brittle, and inflammable pine, the solid and hard, maple, oak, or beech, are thickly and rapidly raising their leafy branches to hide from our view the unsightly trunks that, half-destroyed, charred, and prostrate on the ground, are strewn around in almost every direction. We must pursue our way slowly and laboriously, sometimes jumping over, and sometimes creeping under, or walking along a fallen tree, our progress impeded by the new growth, by brambles, holes in the ground, and the necessity of cautiously observing the general direction of our crooked and fatiguing march; here and there we come to a small open space, where the wild raspberry tempts us to pause and allay our thirst, and perhaps whilst picking its ripe fruit, a pack of grouse rise with a whirr-whirr, and attract our attention—they are gone ere we can reach our gun: but we are not alone;—see, under cover of yon thicket, crouched behind that fallen pine tree, is the Canada Lynx—stealthily and slowly moving along—it is he that startled the game that has just escaped. Now he ascends to the lower branch of a thick leaved tree, and closely squatted, awaits the approach of some other prey, to dart upon and secure it, ere the unsuspecting object of his appetite can even see whence the devourer comes. We move carefully toward the concealed prowler—but his eyes and ears are full as good as our own—with a bound he is upon the earth, and in an instant is out of sight amid the logs and brush-wood—for savage and voracious as he may be when pursuing the smaller animals, he is equally cowardly when opposed to his great enemy—man; and as his skin is valuable, let us excuse him for desiring to keep it whole.

The Canada Lynx is more retired in its habits than our common wild cat, keeping chiefly far from the habitations of even the settlers who first penetrate into the depths of the wilderness. Its fine long fur enables it to withstand the cold of our northern latitudes, and it is found both in the wooded countries north of the great lakes, and as far south as the Middle States, dispersed over a great many degrees of longitude; even occasionally approaching the sea-coast. The specimen from which we drew the figure of this animal was sent to us from Halifax, Nova Scotia. It had been taken in a wolf-trap, after having (as we supposed) destroyed several sheep. We kept it alive for a few weeks, feeding it on fresh raw meat; it ate but a small quantity at a time and like all predacious animals, appeared able to support a long fast without inconvenience. The precarious life led by beasts of prey, in fact makes this a wise provision of Nature, but for which many would no doubt soon perish, as occasionally several days may pass without their being able to secure a hearty meal.

The Lynx we have just mentioned, when a dog approached the cage in which it was confined, drew back to the farthest part of it, and with open jaws spit forth like a cat at the intruder. We often admired the brilliancy of its large eyes, when it glared at us from a corner of its prison. When killed, it was extremely poor, and we found that one of its legs had been broken, probably by a rifle-ball, some considerable time previous to its having been captured, as the bone was united again pretty firmly; it was in other respects a fine specimen.

When alarmed, or when pursued, the Canada Lynx leaps or bounds rapidly in a straight direction from the danger; and takes to a tree if hard pressed by the dogs. It is very strong, and possessing remarkably large and powerful fore-legs and claws, is able to climb trees of any size, and can leap from a considerable height to the ground without feeling the jar, alighting on all four feet at the same instant, ready for flight or battle. If dislodged from a tree by the hunter, it is instantly surrounded by the dogs, in which case it strikes with its sharp claws and bites severely.

In crossing the Petersburg mountains east of Albany, more than thirty years ago, we procured from a farmer a male Lynx, the measurement of which was taken at the time, and has just been given by us, (see p. 138.) It had been killed only half an hour before, and was in very fine order. The farmer stated that in hunting for the ruffed grouse, his dog had started this Lynx from a thicket of laurel bushes; it made no doublings, but ran about a quarter of a mile up the side of a hill, pursued by the dog, when it ascended a tree, on which he shot it; it fell to the ground quite dead, after having hung for some time suspended from a branch to which it clung with great tenacity until life was extinct.

It has been stated that the Canada Lynx "is easily destroyed by a blow on the back with a slender stick;" this we are inclined to think a mistake, never having witnessed it, and judging merely by the activity and strength manifested by the animal, although we agree with the farther remarks of the same writer, "that it never attacks man." This indeed is a remark applicable to nearly all the beasts of prey in our country, except in extreme cases of hunger or desperation. It is said by Dr. Richardson, that the Canada Lynx "swims well, and will cross the arm of a lake two miles wide"—this is a habit which is also shared by the more southern species, (*Lynx rufus.*)

The Canada Lynx, like all other animals of its general habits, breeds but once a year, generally having two young; we have heard of an instance, however, of three whelps being littered at a time.

The skin of this animal is generally used for muffs, collars, &c., and is ranked among the most beautiful materials for these purposes. It varies somewhat in colour, and the best are much lighter, when killed in good season, than the specimen from which our drawing was made.

We have been informed by the northern trappers that the Canada Lynx is usually taken in steel-traps, such as are used for the beaver and otter, into which he enters very readily.

The Indians, we are told, regard its flesh as good eating, which may perhaps be ascribed to the excellence of their appetites. HEARNE, (see Journey, p. 366,) who ate of it in the neighbourhood of York Fort, says, "the flesh is white, and nearly as good as that of the rabbit." We think we would give the preference, however, to a buffalo-hump well roasted, for either dinner or supper.

The stories told of the great cunning of this species, in throwing mosses from the trees in order to entice the deer to feed on them, and then dropping on their backs and tearing their throats, may as well be omitted here, as they fortunately require no refutation at the present day.

The food of the Canada Lynx consists of several species of grouse and other birds, the northern hare, gray rabbit, chipping squirrel, and other quadrupeds. It has been mentioned to us, that in the territories to the north of the Gulf of St. Lawrence they destroy the Arctic fox, and make great havoc among the lemmings, (GEORYCHUS.) HEARNE informs us, that in Hudson's Bay they "seldom leave a place which is frequented by rabbits till they have killed nearly all of them." They are said to pounce on the wild goose at its breeding places, and to destroy many marmots and spermophiles, by lying in wait for them at their burrows. At a public house in Canada we were shown the skin of one of these Lynxes, the animal having been found quite helpless and nearly dead in the woods. It appears, that leaping on to a porcupine, it had caught a Tartar, as its head was greatly inflamed, and it was nearly blind. Its mouth was full of the sharp quills of that well-defended animal, which would in a day or two have occasioned its death. We have heard one or two accounts of the Canada Lynx having killed a deer; we are somewhat sceptical in regard to this being a general habit of the species, although when pressed by hunger, which renders all creatures desperate at times, it may occasionally venture to attack a large animal.

HEARNE states that he "once saw a Lynx that had seized on the carcass of a deer just killed by an Indian, who was forced to shoot it before it would relinquish the prize." (See HEARNE'S Journey, p. 372.) Young fawns, as we have ourselves ascertained, are killed by these animals, and farmers in some of the wilder portions of our Northern States, and of Canada, complain of their carrying off their lambs and pigs. The Canada Lynx is, however, by no means so great a depredator in the vicinity of the farm-yard as the wild-cat or Bay lynx, as his more retired habits incline him to keep in the deepest recesses of the forests—and besides, for aught we know, he may prefer "game" to "pigs and poultry."

The slow multiplication of this species proves that it is not intended to be abundant, but to exist only in such moderate numbers as are necessary to enable it to play its part with other carnivora in preventing too fast an increase of many of the smaller animals and birds; if the hare, the squirrel, the rat, and all the graminivorous quadrupeds and birds were allowed to increase their species without being preyed upon by the owl, the hawk, the fox, the lynx, and other enemies, the grass would be cut off, and the seeds of plants destroyed, so that the larger animals would find no subsistence, and in time, from the destruction of the seeds by the teeth of the rodentia, the forest itself would become a wide desert.

There is then a meaning in this arrangement of Providence; and the more we investigate the works of Him who hath created nothing in vain, the more we are led to admire the wisdom of His designs.

GEOGRAPHICAL DISTRIBUTION.

The Canada Lynx is a northern species—it is known to exist north of the great Lakes eastward of the Rocky Mountains; it is found on the Mackenzie river as far north as latitude 66°. It exists in Labrador, and in Canada. It still occurs, although very sparingly, in some of the New England States. It is occasionally met with in the northern part of New-York. We heard of one having been taken some fifteen years ago in the mountains of Pennsylvania. Farther south, we have not traced it. It is not found in Kentucky, or in the valley of the Mississippi. Westward of that river it does not appear to exist. There are Lynxes between the Rocky Mountains and the Pacific Ocean; these seem, however, to be the Bay lynx, or a species so nearly resembling the latter, that they appear to be no more than one of its numerous varieties. There is a specimen in the Museum of the Zoological Society of London, marked *F. borealis*, which is stated to have been brought from California by DOUGLASS, which we did not see, having somehow overlooked it. Its characters and history deserve investigation.

GENERAL REMARKS

The question whether the Canada Lynx is, or is not, identical with any species of the north of Europe, is by no means settled. PENNANT considered it the same as the lynx (*Felis lynx*) of the old world. BUFFON, after pointing out the distinctive marks of each, came to the conclusion that they were mere varieties. These naturalists, however, lived at a period when it was customary to consider the animals of America as mere varieties of those of the Eastern continent. GEOFFROY ST. HILAIRE named our present species, considering it distinct from the Lynxes of Europe; and TEMMINCK described it under the name of *F. borealis*, as existing in the northern parts of both continents, thinking it a species distinct from *Felis lynx* of the north of Europe.

We spent some time with Professor REICHENBACH, in comparing specimens of European and American lynxes which exist in the museum of Dresden. From the general appearance of these specimens, a great similarity between *L. Canadensis* and the Lynx (*Felis lynx*) of the north of Europe may undoubtedly be remarked, and they might be regarded as mere varieties of one species. The forms of animals, however, approach each other in both continents where there is a similarity of climate. Many of the genera of New-York and Pennsylvania plants are largely represented in Germany, and although nearly all the indigenous species are different, they are closely allied. In South Carolina, there are several birds, quadrupeds, and reptiles, which bear a striking resemblance to those found in Egypt, in nearly the same parallel of latitude. The black-winged hawk (*F. dispar*) resembles the *F. melanopterus* so nearly, that BONAPARTE published them as identical. Our alligator is a near relative of the crocodile, our soft-shelled turtle (*Trionyx ferox*) is much like the *T. Ægypticus*, and our fox squirrel (*Sc. capistratus*) has a pretty good representative in *Sc. Madagascariensis*. In a more northern latitude, we may point to the American and European badgers, to *Lepus Americanus*, and *L. variabilis*, and to *Tamias striatus* of Siberia and *T. Lysterii*, as examples of the near approach of distinct species to each other; to which we may add, that the wild sheep of the Rocky Mountains (*Ovis montana*) bears so striking a resemblance to the *Ovis Ammon*, another species existing on the mountains of Asia, that the two have been confounded; and our *Spermophilus Townsendii* is in size and colour so like the Souslik (*Sp. guttatus*) of the mountains of Hungary, that Dr. RICHARDSON published it as a mere variety. Taking these facts into consideration, after a careful examination of *Lynx Canadensis*, and after having compared it with *Felis lynx* of Europe, we pronounce them distinct species without hesitation.

Although the *European* lynx varies considerably in colour, especially specimens killed at different seasons of the year, it is in all the varieties we have seen, of a deeper rufous tint than the Canada Lynx; the spots on the body are more distinct, and the hair, in some specimens from Russia and Siberia, is much shorter than in our animal, while the tail is longer and more tufted. TEMMINCK, a very close observer, and distinguished naturalist, thinks the Canada Lynx is found on both continents—in this he may possibly be correct; we, however, saw no specimens in the museums of Europe that corresponded with the description of *L. Canadensis*, that did not come from America. The name, *F. borealis*, which TEMMINCK bestowed on it, can, however, only be considered a synonyme, as GEOFFROY described the animal previously, giving it the name of *Felis Canadensis*. We have not been able to find in America the European species described by TEMMINCK under the name of *Felis cervaria*, which, as he supposes, exists also in the northern part of our continent.

No. 4 Plate XVII

Cat Squirrel.

Drawn from Nature by J. J. Audubon F.R.S. F.L.S.

Drawn on Stone by R. Trembly

Printed by Nagel & Weingærtner N.Y.

Colored by J. Lawrence

SCIURUS CINEREUS.—Linn., Gmel.

Cat-Squirrel.

PLATE XVII.

S. corpore robusto, S. capistratus minore, S. migratorio majore; cruribus paullum curtis; naso et auribus nunquam albis; cauda corpore paullo longiore.

CHARACTERS.

A little smaller than the fox squirrel, (S. capistratus;) larger than the northern gray squirrel, (S. migratorius;) body, stout; legs, rather short; nose and ears, never white; tail, a little shorter than the body.

SYNONYMES.

Sciurus Cinereus, Ray, Quad., p. 215, A. D. 1693.
Cat-Squirrel, Catesby, Carolina, vol. ii., p. 74, pl. 74, A. D. 1771.
" " Kalm's Travels, vol. ii., p. 409, English trans.
" " Pennant's Arctic Zoology, vol. i., p. 119, 1784.
Sciurus Cinereus, Linn., Gmel., ——— 1788.
Fox-Squirrel, (*S. vulpinus,*) Godman, Nat. Hist., vol. ii., p. 128.
Cat-Squirrel, " " " " " vol, ii., p. 129.
Sciurus Cinereus, Appendix to American Edition of McMurtrie's Translation of Cuvier's Animal Kingdom, vol. i., p. 433.
" " Bach, Monog. Zoological Society, 1838.
Vulgo, Fox-Squirrel, of New-York, Pennsylvania, and New Jersey, distinct from the Fox-Squirrel (*S. capistratus*) of the southern States.

DESCRIPTION.

Head, less elongated than that of *S. capistratus*, (the fox-squirrel,) and incisors rather narrower, shorter, and less prominent than in that species. Ears broad at base and nearly round, thickly clothed on both surfaces with hair; behind the ears the hairs are longer in winter than during summer, and in the former season, extend beyond the margin of the ear. Whiskers, numerous, longer than the head; neck, short; body, stouter than that of *S. capistratus*, or any known species of Squirrel peculiar to our continent. Fur, more woolly, and less rigid than in *S. ca-*

pistratus; not as smooth as in *S. migratorius.* Hinder parts heavy, giving it a clumsy appearance. Tail, long, broad, and flat, rather less distichous than in *S. capistratus,* or *S. migratorius;* feet, shorter than in the former. Nails, strong, compressed, moderately arched, and acute.

COLOUR.

Perhaps none of our squirrels are subject to greater varieties of colour than the present; we have seen specimens in (formerly) PEALE's museum, of every tint, from light-gray almost to black. Two others that came under our observation were nearly white, and had not red or pink eyes, which last are a characteristic mark of that variety in any animal which is commonly called an albino.

Between the varieties of our present species and the almost equally numerous varieties of the fox-squirrel, (*S. capistratus,*) there may be remarked an important difference. In the latter species the varieties are generally permanent, scarcely any specimens being found of intermediate colour between the well-known shades which exist in different localities or families, whilst in the former, every variety of tint can be observed, and scarcely two can be found exactly alike. The prevailing variety, or colour, however, is gray, and one of this colour we will now describe from a specimen before us.

Teeth, orange; nails, dark-brown near the base, lighter to the extremities. On the cheeks, a slight tinge of yellowish-brown, extending to the junction of the head with the neck; inner surface of the ears, yellowish-brown; outer surface of the ear, fur soft and woolly in appearance, extending a little beyond the margin, light cinereous edged with rutsy-brown. Whiskers both black and white, the black ones most numerous; under the throat, inner surface of the legs and thighs, and the whole under-fur, white, producing an iron-gray colour at the surface; tail, less flat and distichous (being rather more rounded, and narrower) than in many other species of this genus, composed of hairs which separately examined are of a dull white near the roots, succeeded by a narrow marking of black, then white, followed by a broad line of black, and broadly tipped with white.

Another specimen is dark-gray on the back and head, with a mixture of black and cinereous on the feet, thighs, and under-surface. Whiskers, nearly all white. The markings on the tail are similar to those of the other specimen. A third specimen, obtained from Pennsylvania, is dark yellowish-brown on the upper-surface; legs and belly, of a bright, orange-colour. A fourth specimen, obtained in the New-York market, is grayish-brown above, and black beneath. The bones of this species are invariably of a reddish-colour—this is strikingly perceptible after the flesh is cooked.

We have represented in the plate three of these Squirrels, all of different colours, but the varieties of tint to be observed in different specimens of the Cat-Squirrel, are so great, that among fifty or more perhaps, we never could find two exactly alike; for which reason we selected for our drawing an orange-coloured one, a gray one, and one nearly black.

DIMENSIONS.

An old male.—Recent.	Inches.
From nose to root of tail	$12\frac{1}{8}$
Length of tail, (vertebræ)	$7\frac{1}{2}$
do. of tail, to end of hair	$11\frac{1}{2}$
do. from fore-claws to hind-claws, stretched out	$18\frac{3}{4}$

Weight, 1 lb. 13 oz.

Female specimen sent to us, by Mr. Baird, of Pennsylvania.

Length of body	13
do. of tail, from root to end of vertebræ	11
do. of tail, " to end of hair	14
do. to end of hind-legs	19
Extent of fore-legs	$13\frac{3}{4}$
Hind-foot	3
Fore-foot	2
Height of ear, anteriorly	$\frac{10}{12}$
do. of " posteriorly	1
do. of " laterally, (inside,)	$\frac{19}{24}$
Nose to occiput	3
Breadth of ear	$\frac{17}{24}$
do. of tail	$5\frac{3}{12}$

Weight, 2 lb. 5 oz.

HABITS.

This Squirrel has many habits in common with other species, residing in the hollows of trees, building in summer its nest of leaves in some convenient fork of a tree, and subsisting on the same kinds of food. It is, however, the most inactive of all our known species; it climbs a tree, not with the lightness and agility of the northern gray squirrel, but with the slowness and apparent reluctance of the little striped squirrel, (*Tamias Lysteri.*) After ascending, it does not immediately mount to the top as is the case with other species, but clings to the body of the tree on the side opposite to you, or tries to conceal itself behind the first convenient branch. We have seldom observed it leaping from bough to bough.

When it is induced, in search of food to proceed to the extremity of a branch, it moves cautiously and heavily, and generally returns the same way. On the ground it runs clumsily and makes slower progress than the gray squirrel. It is usually fat, especially in autumn, and the flesh is said to be preferable to that of any of our other species of squirrel. The Cat-Squirrel does not appear to be migratory in its habits. The same pair, if undisturbed, may be found in a particular vicinity for a number of years in succession, and the sexes seem paired for life.

William Baird, Esq., of Carlisle, Pennsylvania, says of this species—"The Fox-Squirrel, as this species is called with us, will never, unless almost in the very jaws of a dog, ascend any other tree than that which contains its nest, differing very greatly in this respect from our gray squirrel."

The nest, which we have only seen on two occasions, was constructed of sticks and leaves, in the crotch of a tree about twenty feet from the ground, and in both cases the pair had a safer retreat in a hollow of the same tree above.

This species is said to have young but once a year. We have no positive evidence to the contrary, but suspect that it will hereafter be discovered that it produces a second litter in the summer, or toward autumn.

On taking some of them from the nest, we found on one occasion three, and on another four, young. These nests were placed in the hollows of oak trees.

GEOGRAPHICAL DISTRIBUTION.

The Cat-Squirrel is rather a rare species, but is not very uncommon in the oak and hickory woods of Pennsylvania; we have seen it near Easton and York; it is found occasionally in Maryland and Virginia, and is met with on Long Island and in some other portions of the State of New-York, but in the northern parts of that State is exceedingly rare, as we only saw two pair during fifteen years' close observation. At certain seasons we have found these squirrels tolerably abundant in the markets of the city of New-York, and have ascertained that persons who had them for sale were aware of their superior value, as we were frequently charged 37½ cents for one, whilst the common gray squirrel could easily be purchased for 12½ cents. The south-eastern portion of New-Jersey seems to be well suited to them. This species is rarely found in Massachusetts and one we received from the north-western part of that State was there regarded as a great curiosity.

GENERAL REMARKS.

This species has been sometimes confounded with the fox-squirrel, (*S. capistratus,*) and at other times with the northern gray squirrel, (*S. migratorius,*) and all three have by some been considered as forming but one species; it is however in size intermediate between the two former. and has some distinctive marks by which it may be known from either.

The northern gray squirrel has (as far as we have been able to ascertain from an examination of many specimens) permanently five molars on each side in the upper jaw, and the present species has but four. The Cat-Squirrel, however, like the young fox-squirrel, has no doubt a small deciduous tooth, which drops out in the very young state, and at so early a period that we have not succeeded in detecting it.

Sciurus capistratus is in all its varieties, as far as we have observed, invariably and permanently distinguished by its having white ears and a white nose, which is not the case with *S. cinereus.* The former is a southern species, the latter is found in the middle and northern States, but not in the colder portions of New England or in Canada.

S. capistratus is a longer, thinner and more active species, running with almost the speed of a hare, and ascending the tallest pines to so great a height that nothing but a rifle-ball can bring it down; the present species is heavy, clumsy, and prefers clinging to the body of a tree, not generally ascending to its extreme branches. The hair of *S. capistratus* is more rigid and smoother than that of *S. cinereus*, which is rather soft and woolly.

We have instituted this comparison in order to prove the inaccuracy of a statement contained in one of the last works published in our country on the American quadrupeds. The author says, "We suspect that GODMAN's fox-squirrel (*S. vulpinus*) as well as his Cat, (*S. cinereus*) are varieties only of the hooded squirrel." Under the above names GODMAN published only one and the same species, but the hooded squirrel, (*S. capistratus,*) with white ears and nose, is a very different species, and is not given by GODMAN.

The Cat-Squirrel was the first of the genus described from America. RAY characterizes it as *S. virginianus cinereus major.* CATESBY gives a tolerable description of it, and a figure, which although rather extravagant in the size of its tail, cannot from its short ears, which as well as the nose are destitute of the white marks of *S. capistratus*, be mistaken for the gray variety of the latter species.

He says—"These squirrels are as large as a half-grown rabbit; the whole structure of their bodies and limbs thicker in proportion and of a grosser and more clumsy make than our common squirrels." From this time it became for many years either lost or confounded with other species by naturalists. DESMAREST, under the name of *cinereus*, entirely mistook the species, and applied it to two others, the Carolina gray, and the northern gray squirrel. HARLAN copied the article, adopting and perpetuating the error. GODMAN, by the aid of LE CONTE as it appears to us, (see a reference to his letter—Amer. Nat. Hist., vol. ii., p. 129,) was enabled to correct this error, but fell into another, describing one species under two names, and omitting the southern fox-squirrel (*S. capistratus*) altogether, assigning its habits to his *S. vulpinus*. In our monograph of this genus, 1838, we endeavoured to correct the errors into which authors had fallen in regard to this species; time and further experience have only strengthened us in the views we then expressed.

No. 4 Plate XVIII

Drawn on Stone by R. Trembly

Marsh Hare

Drawn from Nature by J.J. Audubon F.R.S. F.L.S.

Printed by Nagel & Weingærtner N.Y.

LEPUS PALUSTRIS.—Bachman.

Marsh-Hare.

PLATE XVIII.—Male and Female.

L. corpore supra flavo-fuscente, subtus griseo, L. sylvatico minore auribus capite in multum brevioribus, oculis aliquantulum parvis, cauda brevissima, cruribus curtis varipilis.

CHARACTERS.

Smaller than the gray rabbit; ears, much shorter than the head; eyes, rather small; tail, very short; legs, short; feet, thinly clothed with hair; upper parts of body, yellowish-brown; beneath, gray.

SYNONYMES.

Lepus Palustris, Bach., Jour. Acad. of Natural Sciences, Philadelphia, vol. vii., pp. 194, 366, read May 10, 1836.

Lepus Douglassii, Gray, read, Zoological Society, London, Nov. 1837.

Lepus Palustris, Audubon—Birds of America, first edition—pounced upon by the common buzzard, (*Buteo vulgaris.*) Ornithological Biography, vol. iv., p. 510.

DESCRIPTION.

Upper incisors, longer and broader than those of the gray rabbit, marked like all the rest of the genus with a deep longitudinal furrow; the small accessory incisors are smaller and less flattened than those of the gray rabbit, the molars are narrower, and a little shorter. The transverse measurement of the cranium is much smaller, the vertical, about equal. Orbits of the eyes one-third smaller.

This last is a striking peculiarity, giving this a smaller and less prominent eye than that of any other American hare of equal size with which we are acquainted.

The zygomatic processes of the temporal bone run downwards nearly in a vertical line, whilst those of the gray rabbit are almost horizontal. Head, rather large; forehead, slightly arched; whiskers, numerous, rigid; nose, blunt; eyes, rather small; ears, short, rounded, broad, clothed on both surfaces with short hairs. Neck, moderately long; body, short, thick, and of rather a clumsy shape; hairs, rather long and much coarser than those of the gray rabbit. Legs, short, and rather small; feet so thinly clothed with hair, that the nails in most of the specimens are not covered, but project beyond the hair; the feet leave a distinct impression of the toes and claws on the mud or in moist places where their tracks can be seen. Heel, short, thinly covered with hair; nails, long, stout, and very acute; tail, short; scarcely visible whilst the animal is running.

COLOUR.

Teeth, yellowish-white; eyes, dark-brown, appearing in certain lights quite black. Upper part of the head, brown, and grayish-ash. Around the orbits of the eyes, slightly fawn-coloured; whiskers, black; ears, dark grayish-brown. Back, whole upper-parts, and upper-surface of the tail, yellowish-brown intermixed with many strong black hairs. The hairs, when examined singly, are bluish-gray at the roots, then light-brown, and are tipped with black. Throat, brownish-gray. Outer-surface of fore-legs, and upper-surface of thighs, reddish-yellow. The fur beneath, is light plumbeous; under the chin, gray; belly, and under-surface of tail, light-gray; the fur beneath, bluish, giving it a dark yellowish-brown appearance. Under-surface of the tail, ash-colour, edged with brown. During winter the upper surface becomes considerably darker than in summer, and the under-parts of the tail in a few specimens become nearly white.

DIMENSIONS.

A specimen in the flesh.

Length from point of nose to insertion of tail -	13 inches.
do. of tail, (vertebræ,) - - - - - -	1 "
do. do. do. including fur - - - - -	1½ "
Height from end of middle claw to top of shoulder	7 "
Length of head - - - - - - - -	3½ "
do. ears - - - - - - - -	2½ "
do. hind-foot - - - - - - -	3 "

Weight, 2½ lbs.

HABITS.

The Marsh-Hare chiefly confines itself to the maritime districts of the southern States, and is generally found in low marshy grounds that are sometimes partially inundated, near rivers subject to freshets that occasionally overflow their banks, or near the large ponds called in Carolina "reserves," which are dammed up or otherwise made to retain the water intended to flood the rice-fields at the proper season.

In these situations—to which few persons like to resort, on account of the muddy nature of the ground, and the many thorny and entangling vines and other obstructions that abound near them; and which, besides, continually exhale from their stagnant waters a noxious vapour, which rapidly generates disease—surrounded by frogs, water-snakes and alligators, this species resides throughout the year, rarely molested by man, and enabled by its aquatic habits to make up for any want of speed when eluding the pursuit of its enemies.

It winds with great facility through miry pools, and marshes overgrown with rank weeds and willow bushes, and is quite at its ease and at home in the most boggy and unsafe parts of the swamps.

We have met with this animal a few miles from Columbia, South Carolina, one hundred and twenty miles north of Charleston, along the muddy shores of the sluggish rivers and marshes; but on arriving at the high grounds beyond the middle country, where the marshes disappear, it is no longer to be found.

In its movements it is unlike most of our other hares; it runs low on the ground, and cannot leap with the same ease, strength and agility they display. From the shortness of its legs and ears, and its general clumsy appearance as we see it splashing through the mud and mire, or plunging into creeks or ponds, it somewhat reminds us of an over-grown Norway rat endeavouring to escape from its pursuers.

The Marsh-Hare is so slow of foot, that but for the protection afforded it by the miry tangled and thorny character of its usual haunts, it would soon be overtaken and caught by any dog of moderate speed. We have observed the negroes of a plantation on a holiday, killing a good many of them by first setting fire to the half-dried grasses and weeds in a marshy piece of ground during a continued drought, when the earth had absorbed nearly all the moisture from it, and then surrounding the place, with sticks in their hands, and waiting until the flames drove the hares from their retreats, when they were knocked down and secured as they attempted to pass. Several gray-rabbits ran out of this place, but the men did not attempt to stop them, knowing their superior speed, but every Marsh-Hare that appeared was headed, and with a loud whoop set upon on all sides and soon captured.

The feet of the Marsh-Hare are admirably adapted to its aquatic habits. A thick covering of hair on its feet, like that on the soles of other species, would be inconvenient; they would not only be kept wet for a considerable length of time, but would retard the animal in swimming.

Quadrupeds that frequent the water, such as the beaver, otter, musk-rat, mink, &c., and aquatic birds, have nearly naked palms; and it is this peculiar structure, together with the power of spreading out its feet, and thus increasing the space between each of its toes, that enables this animal to swim with great ease and rapidity. Its track when observed in moist or muddy situations differs very much from that of other species. Its toes are spread out, each leaving a distinct impression like those of the rat. Some of the habits of this Hare differ greatly from those of others of the genus; it seeks the water, not only in order the easier to escape from its pursuers, but when in sportive mood; and a stranger in Carolina should he accidentally see one amusing itself by swimming about, if unacquainted with the habits of the animal, would be puzzled by its manœuvres.

When the Marsh-Hare is startled by the approach of danger, instead of directing its flight toward high grounds like the gray rabbit, it hastens to the thickest part of the marsh, or plunges into some stream, mill-pond, or "reserve," and very often stops and conceals itself where the water is many feet deep, among the leaves of lilies or other aquatic plants.

After a heavy rain had produced a flood, which inundated some swamps and rice-fields near us, we sallied forth to see what had become of the Marsh-Hares: and on beating the bushes, we started many of them which ran from their hiding places, plunged into the water, and swam off with such rapidity that some escaped from an active Newfoundland dog that we had with us. Several of them, supposing they were unobserved, hid themselves in the water, about fifteen yards from the shore, protruding only their eyes and the point of their nose above the surface; when thus almost entirely under the muddy water, with their ears pressed back and flat against their neck, they could scarcely be discovered. On touching them with a stick, they seemed unwilling to move until they perceived that they were observed, when they swam off with great celerity.

A few evenings afterwards when the waters had subsided and returned to their ordinary channels, we saw a good many of these Hares swimming in places where the water was seven or eight feet deep, meeting, or pursuing each other, as if in sport, and evidently enjoying themselves.

When the gray-rabbit approaches the water, it generally goes around or leaps over it, but the Marsh-Hare enters it readily and swims across.

We have on a few occasions seen this Hare take to a hollow tree when hard pressed by dogs, but (as we have just remarked) it usually depends more for its safety on reaching marshy places, ponds, or impenetrable thickets.

This species possesses a strong marshy smell at all times, even when kept in confinement and fed on the choicest food. Its flesh, however, although dark, is fully equal if not superior to that of the gray rabbit.

The Marsh-Hare never, that we are aware of, visits gardens or cultivated fields, but confines itself throughout the year to the marshes. It is occasionally found in places overflowed by salt, or brackish, water, but seems to prefer fresh-water marshes, where its food can be most conveniently obtained. It feeds on various grasses, and gnaws off the twigs of the young sassafras, and of the pond-spice (*Laurus geniculata.*) We have seen many places in the low grounds dug up, the foot-prints indicating that it was the work of this species in search of roots. It frequently is found digging for the bulbs of the wild potatoe, (*Apios tuberosa,*) as also for those of a small species of amaryllis, (*Amaryllis atamasco.*)

We kept an individual of this species in confinement, which had been captured when full-grown. It became so gentle in a few days that it freely took its food from the hand. It was fed on turnips and cabbage-leaves, but preferred bread to any other food that was offered to it. In warm weather it was fond of lying for hours in a trough of water, and seemed restless and uneasy when it was removed: scratching at the sides of its cage until the trough was replaced, when it immediately plunged in, burying the greater part of its body in the water.

This species, like all others of the genus existing in this country, as well as the deer and squirrels, is infested with a troublesome larva of an œstrus in the summer and autumn; which penetrating into the flesh and continually enlarging, causes pain to the animal and renders it lean.

The Marsh-Hare deposits its young in a pretty large nest, frequently composed of a species of rush, (*Juncus effusus,*) growing in convenient situations. The rushes appear to be cut by it into pieces of about a foot in length. We have seen these nests nearly surrounded by, and almost floating on the water. They were generally arched by carefully bending the rushes or grasses over them, admitting the mother by a pretty large hole in the side. A considerable quantity of hair was found lining them, but whether plucked out by the parent, or the result of the natural shedding of their coat, (it being late in the spring when these animals shed their hair,) we were unable to ascertain.

The young number from five to seven. They evidently breed several times in the season, but we have observed that the females usually produce their young at least a month later than the gray rabbit. Twenty-one specimens were obtained from the 9th to the 14th day of April; none of the females had produced young that season, although some of them would have done so in a very few days. On one occasion only, have we seen the young in March. They bear a strong resemblance to the adult, and may almost at a glance be distinguished from those of the gray rabbit.

GEOGRAPHICAL DISTRIBUTION.

The Marsh-Hare has been seen as far north as the swamps of the southern parts of North Carolina. In South Carolina, it is in some localities quite numerous. Nearly all the muddy swamps and marshes abound with it. We have known two persons kill twenty in the course of a few hours.

In high grounds it is never seen; it continues to increase in numbers as we proceed southwardly. It is abundant in the swamps of Georgia, Alabama, and Louisiana. We received a living specimen from Key West, the southern point of Florida. We have seen it in Texas, from whence the specimen described by GRAY was brought, and we are inclined to believe that it will be found to extend into the northern part of Mexico.

GENERAL REMARKS.

As a remarkable instance of a species continuing to exist in a thickly settled country without having found its way into scientific works, we may refer to this very common hare. We obtained specimens in Carolina in the spring of 1815. It was called by the inhabitants by the names of Swamp, and Marsh, Hare, and generally supposed to be only a variety of the gray rabbit. We did not publish a description of the species until 1836. In the following year, GRAY, who had not then seen the Transactions of the Acad of Natural Sciences of Philadelphia, in which our description was contained, described it under the name of *Lepus Douglassii.*

This species may always be distinguished from our other hares by its colour, its rather short and broad ears, its short tail, which is never pure white beneath, by its narrow hind-feet, and by its aquatic habits.

Nº 4 Plate XIX

Soft haired Squirrel

Drawn from Nature by J.J. Audubon F.R.S F.L.S.

Printed by Nagel & Weingaertner N.Y.

Drawn on Stone by R. Trembly.

SCIURUS MOLLIPILOSUS.—Aud. and Bach.

Soft-Haired Squirrel.

PLATE XIX.

S. cauda corpore curtiore; dorso fusco; iliis partibusque colli laterali-bus rufis; abdomine cinereo.

CHARACTERS.

Tail, shorter than the body; back, dark brown; sides of the neck, and flanks, rufous; under surface, cinereous.

SYNONYME.

Sciurus Mollipilosus, Aud. and Bach., Journal Acad. of Nat. Sciences, Philadel-phia, Oct. 1841, p. 102.

DESCRIPTION.

A little larger than the chickaree, (*S. Hudsonius;*) head, rather large, slightly arched; ears, round, broad, but not high, clothed on the outer and inner surfaces with short, smooth hairs; whiskers, longer than the head.

In form this species does not approach the Tamiæ, as *S. Hudsonius* does in some degree: it, on the contrary, very much resembles the Carolina gray-squirrel, *S. Carolinensis*, which is only an inch longer.

Legs, robust; toes, rather long; nails, compressed, arched; tail bushy, but apparently not distichous, as far as can be judged from the dried specimen; hairs of the tail about as long as those of the Carolina gray-squirrel. The hairs on the whole of the body are soft and very smooth.

COLOUR.

Teeth, light yellow; upper parts, inlcuding the nose, ears, and outer surface of the tail, dark-brown; this colour is produced by the hairs being plumbeous at the roots, tipped withlight-brown and black. On the sides of the neck, the shoulder, and near the thighs, it is of a reddish-brown colour. The tail is brown, twice annulated with black; a few of the

hairs are tipped with gray. On the under surface, the lips and chin are grayish-brown; inner surface of the fore-legs, throat, and abdomen, cinereous, lightly tinged in some places with rufous.

DIMENSIONS.

	Inches.	Lines.
Length of head and body - - - - - -	8	6
" of tail (vertebræ) - - - - - -	5	6
" " to end of hair - - - -	7	0
Height of ear - - - - - - - - -	0	5
From heel to end of nail - - - - - -	2	1

HABITS.

This species was procured in Upper California, near the Pacific ocean, and we are obliged to confess ourselves entirely unacquainted with its habits. From its form, however, we have no doubt of its having more the manners of the Carolina gray-squirrel than those of the chickaree. We may suppose that it lives on trees, and never burrows in the ground, as the chickaree sometimes does.

GEOGRAPHICAL DISTRIBUTION.

Our specimens were obtained in the northern part of California, near the Pacific ocean.

GENERAL REMARKS.

This species differs so widely in all its details from *S. Hudsonius*, that it is scarcely necessary to point out the distinctive marks by which it is separated from the latter. The space occupied by the lighter colours on the under surface is much narrower than in *S. Hudsonius*, and there is not, as in that species, any black line of separation between the colours of the back and under surface.

No. 4

Plate XX

Townsend's Ground Squirrel.

Drawn from Nature by J J Audubon F.R.S. F.L.S

Drawn on Stone by R Trembly

Printed by Nagel & Weingaertner N.Y.

Colored by J Lawrence

TAMIAS TOWNSENDII.—BACH.

TOWNSEND'S GROUND-SQUIRREL.

PLATE XX.

T. obscurus, supra flavo-fuscescens, striis quinque nigris longitudinalibus subequaliter distantibus dorsali usque ad caudam porrecta; subtus cinereus. T. Lysteri magnitudine superans.

CHARACTERS.

A little larger than Tamias Lysteri; tail much longer; upper surface, dusky yellowish-brown, with five nearly equidistant parallel black stripes on the back, the dorsal one extending to the root of the tail; under surface cinereous.

SYNONYME.

TAMIAS TOWNSENDII, Townsend's Ground Squirrel, Journal Acad. of Natural Sciences, Philadelphia, vol. viii., part 1, 1839.

DESCRIPTION.

Head, of moderate size; forehead, convex; nose, rather obtuse, clothed with very short hairs; nostrils, opening downward, their margins and septum naked; whiskers, as long as the head; eyes, large; ears, long, erect, obovate, clothed with short hair on the outer, and nearly naked on the inner surface; cheek-pouches, tolerably large. In form this species resembles *T. Lysteri;* it is, however, longer and stouter. Legs, of moderate size; toes, long; the fore-feet have four toes, with the rudiment of a thumb, protected by a short convex nail; the palms are naked, with five tubercles. Claws, curved, compressed, and sharp-pointed. On the hind-feet, five toes, the third and fourth nearly of equal length, the second a little shorter, and the first, or inner toe, shortest. Tail, long and subdistichous.

COLOUR.

Teeth, dark orange; whiskers, black; a line of fawn-colour, commencing at the nostrils, runs over the eyebrows, and terminates a little beyond them in a point of lighter colour; a patch of a similar colour commences under the eye-lids, and running along the cheeks, terminates at the ear.

A line of dark brown, commencing at the termination of the nose, where it forms a point, and bordering the fawn-colour above, is gradually blended with the colours of the head; fur on the outer surface of the ear, brown on the anterior parts, with a patch of white covering about one-fourth of the ear. On the posterior part of the ear there is a slight cinereous tint about six lines in length, terminating near the shoulder. A black stripe commences on the hind part of the head and runs over the centre of the back, where it spreads out to the width of four lines, terminating in a point at the insertion of the tail; a line of the same colour commences at the shoulders, and running parallel to the first, terminates a little beyond the hips; another, but narrower and shorter line of black runs parallel with this, low down on the sides, giving it five black stripes about equi-distant from each other. On the throat, belly, and inner parts of the legs and thighs, the colour is light cinereous; there is no line of separation between the colours of the back and belly. The tail is, on the upper surface, grayish-black, having a hoary appearance. Underneath, it is reddish-brown for two-thirds of its breadth, then a narrow line of black, tipped with light ash. Nails, brown.

DIMENSIONS.

	Inches.	Lines.
Length of head and body - - - - - -	6	9
" tail (vertebræ) - - - - - -	4	0
" " including fur - - - - - -	5	0
" head - - - - - - - - -	2	0
Height of ear - - - - - - - - -	0	6
Length from heel to end of nail - - - - -	1	6

HABITS.

No doubt the different species of this genus are as uniform in their habits as the true squirrels. They are usually found seated low, on stumps or rocks, at the roots of or near which they have their burrows. Their cheek-pouches enable them to carry to these hiding-places, nuts, grains, &c., to serve them for food in winter. Mr. Townsend, who procured the specimens from which we have drawn up our description, observes, "This pretty little fellow, so much resembling our common *T. striatus*, (*Lysteri*,) is quite common; it lives in holes in the ground; running over your foot as you traverse the woods. It frequently perches itself upon a log or stump, and keeps up a continual clucking, which is usually answered by another at some distance, for a considerable time. Their note so much resembles that of the dusky grouse, (*Tetrao obscurus*,) that I have more than once been deceived by it."

GEOGRAPHICAL DISTRIBUTION.

We have heard of this species as existing from the 37th to the 45th degree of latitude, on the Rocky Mountains. It probably does not extend to the eastward of that chain, as we saw nothing of it on our late expedition up the Missouri river, to the mouth of the Yellow-Stone, &c.

GENERAL REMARKS.

The markings of this Ground-Squirrel differ widely from those of any other known species. From *Tamias Lysteri* it differs considerably, being larger and having a much longer tail; it has a white patch behind the ear, and cinereous markings on the neck, of which the latter is destitute; the ears are a third longer than in *T. Lysteri*. The stripes on the back are also very differently arranged. In *Tamias Lysteri* there is first a black dorsal stripe, then a space of grayish-brown, half an inch wide, then two shorter stripes, within two lines of each other; which narrow intervening portion is yellowish-white. The stripes in the present species are at a uniform distance from each other, the dorsal one running to the tail; whereas, in the other it does not reach within an inch of it, and the intervening spaces are filled up by a uniform colour.

This species has not the whitish stripes on the sides, nor the rufous colour on the hips, which are so conspicuous in *T. Lysteri*.

VULPES VIRGINIANUS.—Schreber.

Gray Fox.

PLATE XXI.—male.

V. griseo nigroque variegatus, lateribus et partibus colli lateralibus fulvis, genis nigris.

CHARACTERS.

Gray, varied with black, sides of neck and flank, fulvous; black on the sides of the face between the eye and nose.

SYNONYMES.

Fox of Carolina, Lawson, Car., p. 125.
Gray Fox, Catesby, Car., vol. ii., p. 78, fig. C.
" " Pennant, Synop., p. 157, 114.
Canis Virginianus, Schreber, Säugethiere, p. 361, 10 to 92 B, 1775.
" " Erxleben, Syst., p. 567, 10, 1777.
" " Linn., Syst. Nat., ed. Gmel., vol. i., p. 74, 16, 1788.
" Cinereo-Argenteus, Erxleben, Syst., p. 576, 9.
" Cinereo-Argentatus, Say, Long's Expedition, vol. ii., p. 340.
" Virginianus, Desm., Mamm., p. 204.
" Cinereo-Argentatus, Godman, Nat. Hist., vol. i., p. 280, fig. 2.
" (Vulpes) Virginianus, Rich., F. Boreali A., p. 96.
Vulpes Virginianus, Dekay, Nat. Hist. of New-York, p. 45.

DESCRIPTION.

Head, considerably broader and shorter than that of the red fox, (*Vulpes fulvus;*) nose, also shorter, and a little more pointed; teeth, not so stout; ears, a little longer than in the latter animal, of an oval shape, and thickly clothed with hair on both surfaces; whiskers, half the length of the head. Body, rather thicker and more clumsy in appearance than that of either the swift fox, (*V. velox,*) or the red fox; fur, much coarser than that of the other species. Legs, rather long; nails, strong, slightly arched, visible beyond the fur; soles, with five stout tubercles, not clothed with hair; tail, large, bushy, clothed like the body with two kinds of hair; the fur, or inner hair, being soft and woolly, the outer hairs longer and coarser.

No. 5 Plate XXI

Gray Fox.

Male.

Drawn from Nature by J. J. Audubon F.R.S. F.L.S.

Printed by Nagel & Weingaertner N.Y.

COLOUR.

There are slight differences in the colour of different specimens; we will, however, give a description of one which is of the colour most common to this species in every part of the United States. Head, brownish-gray; muzzle, black; a broad patch of dark brown runs from the eye to the nose, on each side of the face; whiskers, black; inner surface of ears, dull white; outer surface of ears, sides of neck, outer surface of fore-legs and thighs, tawny; a yellowish wash under the throat, and along the sides; chin, and around the mouth, dark-brown; cheeks, throat, and under surface of body, dull white, occasionally tinged with a yellowish shade; under surface of hind and fore-feet, yellowish-brown; upper surface of feet and legs, grizzly black and white; nails, dark-brown. The soft inner fur on the back, which is about an inch and a half long, is for half its length from the roots, plumbeous, and pale yellowish-white at the tips. The long hairs which give the general colour to the body above, are white at their roots, then for more than a third of their length black, then white, and are broadly tipped with black, giving the animal a hoary or silver-gray appearance. It is darkest on the shoulder, along the back and posterior parts. The fur on the tail has a little more fulvous tinge than that of the back; the longer hairs are much more broadly tipped with black. When the fur lies smooth, there is a black line along the upper surface of the tail from the root to the extremity; end of brush, black. Some specimens are a little lighter coloured, having a silver-gray appearance. Specimens from the State of New-York are rather more fulvous on the neck, and darker on the back, than those of Carolina. In some specimens there is a dark spot on the sides of the throat about an inch from the ear.

We possessed for many years a beautiful specimen of a variety of the Gray Fox, which was barred on the tail like the racoon, and had a dark cross on the back like that of *Canis crucigera* of Gesner, which latter is regarded by Baron Cuvier as a mere variety of the European fox.

DIMENSIONS.

Length of head and body	28	inches.
" of tail (vertebræ)	12½	do.
" " to end of hair	14	do.
Height of ear	2½	do.
From heel to end of nail	5	do.

HABITS.

Throughout the whole of our Atlantic States, from Maine to Florida, and westwardly to Louisiana and Texas, there are but two *species* of fox known, viz., the red fox, (*V. fulvus,*) and the present species, (*V. Virginianus,*) although there are several permanent *varieties.* The former may be regarded as a Northern, the latter as a Southern species. Whilst the Northern farmer looks upon the red fox as a great annoyance, and detests him as a robber who is lying in wait for his lambs, his turkeys, and his geese, the Gray Fox, in the eyes of the Southern planter, is the object of equal aversion. To ourselves, however, who have witnessed the predatory dispositions of each in different portions of our country, it appears that the red fox is far more to be dreaded than the gray; the latter is a pilfering thief, the former a more daring and cunning plunderer. When they have whelps, the females of both species, urged by the powerful pleadings of their young, become more bold and destructive than at any other time; the red fox produces its young very early in the season, sometimes indeed whilst the snow is still remaining here and there in large banks unthawed on the ground, and becomes more daring in consequence of being stinted for food; whilst the present species, having its young later when breeding in the Northern States, and finding a more abundant supply of food when inhabiting the Middle or Southern States, is less urged by necessity to depredate on the poultry of the planter.

We have never, indeed, heard any well authenticated account of this species having entered the poultry-yard of the farmer; it is true, it will seize on a goose, or a turkey hen, that happens to stray into the woods or fields and make its nest at some distance from the house; but we have not heard of its having attempted to kill pigs, or like the red fox, visited the sheep pasture in spring, and laid a contribution, from day to day, on the young lambs of the flock.

The Gray Fox is shy and cowardly, and the snap of a stick or the barking of a dog will set him off on a full run. Although timid and suspicious to this degree, his cunning and voracity place him in a conspicuous rank among the animals that prey upon other species weaker than themselves. The wild turkey hen often makes an excavation in which she deposits her eggs, at a considerable distance from the low grounds, or makes her nest on some elevated ridge, or under a pile of fallen logs covered over with scrub oaks, ferns, tall weeds and grasses; we have often seen traces of a violent struggle at such places; bunches of feathers scattered about, and broken egg-shells, giving sufficient evidence that the Fox has been there, and that there will be one brood of wild turkeys less that season. Coveys of partridges, which generally at the dusk of the evening fly into some sheltered place and hide in the tall grass, arrange themselves for the night in a circle, with their tails touching each other and their heads turned outward; the Gray Fox possessing a considerable power of scent, winds them like a pointer dog, and often discovers where they are thus snugly nestled, and pounces on them, invariably carrying off at least one of the covey.

On a cold, drizzly, sleety, rainy day, while travelling in Carolina, we observed a Gray Fox in a field of broom-grass, coursing against the wind, and hunting in the manner of the pointer dog. We stopped to witness his manœuvres: suddenly he stood still and squatted low on his haunches; a moment after he proceeded on once more, but with slow and cautious steps; at times his nose was raised high in the air, moving about from side to side. At length he seemed to be sure of his game and went straight forward, although very slowly, at times crawling on the earth; he was occasionally hidden by the grass, so that we could not see him very distinctly; however, at length we observed him make a dead halt. There was no twisting or horizontal movement of the tail, like that made by the common house-cat when ready to make a spring, but his tail seemed resting on the side, whilst his ears were drawn back and his head raised only a few inches from the earth; he remained in this attitude nearly half a minute and then made a sudden pounce upon his prey; at the same instant the whirring of the distracted covey was heard as the affrighted birds took wing; two or three sharp screams succeeded, and the successful prowler immediately passed out of the field with an unfortunate partridge in his mouth, evidently with the intention of seeking a more retired spot to make a dainty meal. We had a gun with us, and he passed within long gun-shot of us. But why wound or destroy him? He has enabled us for the first time to bear witness that he is not only a dog, but a good pointer in the bargain; he has obeyed an impulse of nature, and obtained a meal in the manner in which it was intended by the wise Creator that he should be supplied. He seized only a single bird, whilst man, who would wreak his vengeance on this poacher among the game, is not satisfied till he has killed half the covey with the murderous gun, or caught the whole brood in a trap and wrung off their necks in triumph. Condemn not the Fox too hastily; he has a more strikingly carnivorous tooth than yourself, indicating the kind of food he is required to seek; he takes no wanton pleasure in destroying the bird, he exhibits to his companions no trophies of his skill, and is contented with a meal whilst you are perhaps not satisfied when your capacious bird-bag is filled.

That this Fox occasionally gives chase to the gray rabbit, pursuing him in the manner of the dog, we have strong reason to suspect. We on one occasion observed a half-grown rabbit dashing by us with great rapidity, and running as if under the influence of fear; an instant afterwards a Fox followed, seeming to keep the object of his pursuit fairly in sight; scarcely had they entered the woods when we heard the repeated cry of the rabbit, resembling somewhat that of a young child in pain, and although we were not eye witnesses of his having captured it by sheer speed, we have no doubt of the fact. We do not believe, however, that the Fox is an enemy half as much to be dreaded by the family of hares as either the Bay lynx, or the great horned owl, (*Strix Virginianus.*)

In the Southern States this species is able to supply itself with a great variety and abundance of food, and is consequently generally in good condition and often quite fat. We have followed the track of the Gray Fox in moist ground until it led us to the scattered remains of a marsh hare, which no doubt the Fox had killed; many nests of the fresh water marsh hen (*Rallus elegans*) are torn to pieces and the eggs devoured by this prowler. In Pennsylvania and New-Jersey, the meadow-mouse (*Arvicola Pennsylvanica*) is often eaten by this species; and in the Southern States, the cotton-rat, and Florida rat, constitute no inconsiderable portion of its food. We have seen places where the Gray Fox had been scratching the decayed logs and the bark of trees in order to obtain insects.

This species is not confined exclusively to animal food; a farmer of the State of New-York called our attention to a field of corn, (maize,) which had sustained no inconsiderable injury from some unknown animals that had been feeding on the unripe ears. The tracks in the field convinced us that the depredation had been committed by Foxes, which was found to be the case, and they were afterwards chased several successive mornings, and three of them, apparently a brood of the previous spring, were captured.

Although this Fox is nocturnal in his habits we have frequently observed him in search of food at all hours of the day; in general, however, he lies concealed in some thicket, or in a large tuft of tall broom-grass, till twilight invites him to renew his travels and adventures.

On a cold starlight night in winter, we have frequently heard the hoarse querulous bark of this species; sometimes two of them, some distance apart, were answering each other in the manner of the dog.

Although we have often seen this Fox fairly run down and killed by hounds, without his having attempted to climb a tree, yet it not unfrequently occurs that when his strength begins to fail he ascends one that is small or sloping, and standing on some horizontal branch 20 or 30 feet from the ground, looks down on the fierce and clamorous pack which soon comes up and surrounds the foot of the tree. We were on one occasion, in company with a friend, seeking for partridges in an old field partially overgrown with high grass and bushes, when his large and active pointer dog suddenly started a Gray Fox, which instantly took to its heels, pursued by the dog: after a race of a minute, the latter was so close upon the Fox that it ascended a small tree, and our friend soon came up, and shot it. We were unable to obtain any information in regard to the manner in which the Fox climbs trees, as he does not possess the retractile nails of the cat or the sharp claws of the squirrel, until we saw the animal in the act. At one time when we thus observed the Fox, he first leaped on to a low branch four or five feet from the ground, from whence he made his way upwards by leaping cautiously and rather awkwardly from branch to branch, till he attained a secure position in the largest fork of the tree, where he stopped. On another occasion, he ascended in the manner of a bear, but with far greater celerity, by clasping the stem of a small pine. We have since been informed that the Fox also climbs trees occasionally by the aid of his claws, in the manner of a racoon or a cat. During winter only about one-fifth of the Foxes chased by hounds will take to a tree before they suffer themselves to be run down; but in summer, either from the warmth of the weather causing them to be soon fatigued, or from the greater number being young animals, they seldom continue on foot beyond thirty or forty minutes before they fly for protection to a tree. It may here be observed, that as long as the Fox can wind through the thick underbrush, he will seldom resort to a tree, a retreat to which he is forced by open woods and a hard chase.

In general, it may be said that the Gray Fox digs no burrow, and does not seek concealment in the earth; we have, however, seen one instance to the contrary, in a high, sandy, pine-ridge west of Albany, in the State of New-York. We there observed a burrow from which a female Gray Fox and four young were taken. It differed widely from the burrows of the red fox, having only a single entrance. At about eight feet from the mouth of the burrow there was an excavation containing a nest composed of leaves, in which the young had been deposited. We have on several occasions seen the kennel of the Gray Fox—it is usually in a prostrate hollow log; we once, however, discovered one under the roots of a tree. In the State of New-York we were shown a hollow tree, leaning on another at an angle of about forty-five degrees, from a large hole in which two Gray Foxes had been taken; they were traced to this retreat by their footsteps in the deep snow, and from the appearance of the nest it seemed to have been their resort for a long time.

This species, in many parts of the country where caves, fissures, or holes in the rocks, offer it a safe retreat from danger, makes its home in such places. Some little distance above the city of New-York, in the wild and rocky woods on the Jersey side of the Hudson river, a good many Gray Foxes abide, the number of large fissures and holes in the rocks thereabouts furnishing them secure dwelling places, or safe resorts in case they are pursued. In this neighbourhood they are most easily killed by finding the paths to their hole, and, after starting the animal, making the best of your way to near the entrance of it, while he doubles about a little before the dogs; you can thus generally secure a shot at him as he approaches his home, which if the dogs are near he will do without looking to see if he be watched. The Gray Fox is frequently caught in steel-traps, and seems to possess far less cunning than the red species; we have never, however, seen it taken in box-traps, into which the Bay lynx readily enters; and it is not often caught in dead-falls, which are very successful in capturing the racoon and opossum.

The Gray Fox does not possess the rank smell of the red fox or the European fox; as a pet, however, we have not found him particularly interesting. It is difficult to subdue the snappish disposition of this species, and we have never seen one that was more than half tamed. It does not at any time become as playful as the red fox, and continually attempts to escape.

This species affords good sport when chased, winding and doubling when in favourable ground, so that when the hunter is on foot even, he can occasionally obtain a "view," and can hear the cry of the pack almost all the while. When started in an open part of the country the Gray Fox, however, generally speeds toward some thickly grown and tangled retreat, and prefers the shelter and concealment of a heavy growth of young pines along some elevated sandy ridge; having gained which, he threads along the by-paths and dashes through the thickets, some of which are so dense that the dogs can hardly follow him. He does not, like the red fox, run far ahead of the pack, but generally courses along from seventy to a hundred yards in advance of his pursuers.

We have been told that the Gray Fox has been run down and caught in the winter season, by a remarkably fleet pack of hounds, in forty minutes; but a two hours' chase is generally necessary, with tolerably good dogs, to tire out and capture him. As many as two or three Foxes have been occasionally caught on the same day by one pack of hounds; but in most cases both hunters and dogs are quite willing to give over for the day, after they have captured one.

From Maryland to Florida, and farther west, through Alabama to Mississippi and Louisiana, fox-hunting, next to deer-hunting, is the favourite amusement of sportsmen, and the *chase* of that animal may in fact be regarded exclusively as a Southern sport in the United States, as we believe the fox is never followed on horseback in the Northern portions of our country, where the rocky and precipitous character of the surface in many districts prevents the best riders from attempting it; whilst in others, our sturdy independent farmers would not much like to see a dozen or more horsemen leaping their fences, and with break-neck speed galloping through the wheat-fields or other "fall" crops. Besides, the red fox, which is more generally found in the Northern States than the Gray species, runs so far before the dogs that he is seldom seen, although the huntsmen keep up with the pack, and after a chase of ten miles, during which he may not have been once in view, he perhaps takes refuge in some deep fissure of a rock or in an impenetrable burrow, which of course ends the sport very much to the satisfaction of—the Fox!

In the Southern States on the contrary, the ground is in many cases favourable for this amusement, and the planter sustains but little injury from the passing hunt, as the Gray Fox usually courses through woods, or worn-out old fields, keeping on high dry grounds, and seldom during the chase running across a cultivated plantation.

Fox-hunting, as generally practised in our Southern States, is regarded as a healthful manly exercise, as well as an exhilarating sport, which in many instances would be likely to preserve young men from habits of idleness and dissipation. The *music* of the hounds, whilst you breathe the fresh sweet morning air, seated on a high-mettled steed, your friends and neighbours at hand with light hearts and joyous expectations, awaiting the first break from cover, is, if you delight in nature and the recreation we are speaking of, most enlivening; and although we ourselves have not been fox-hunters, we cannot wholly condemn the young man of leisure who occasionally joins in this sport; at the same time let him not forget that whilst exercise and amusement are essential to health and cheerfulness of mind; the latter especially was not intended to interfere with the duties of an active and useful life, and should never be more than a relaxation, to enable him to return the more energetically to the higher and nobler pursuits which are fitted for an intelligent and immortal mind.

In fox-hunting, the horse sometimes becomes as much excited as his rider, and at the cry of the hounds we have known an old steed which had been turned loose in the woods to pick up a subsistence, prick up his ears, and in an instant start off full gallop until he overtook the pack, keeping in the van until the chase was ended. Although exercise and amusement are the principal inducements to hunt the Fox, we may mention that it is also a desirable object in many parts of our country, to get rid of this thievish animal, which exists in considerable numbers in some neighbourhoods.

We will now return to our subject, and try to make you familiar with the mode of hunting the Gray Fox generally adopted in Carolina and Louisiana. The hounds are taken to some spot where the animal is likely to be found, and are kept as much as possible out of the "drives" frequented by deer. Thickets on the edges of old plantations, briar patches, and deserted fields covered with broom-grass, are places in which the Fox is most likely to lie down to rest. The trail he has left behind him during his nocturnal rambles is struck, the hounds are encouraged by the voices of their masters, and follow it as fast as the devious course it leads them will permit. Now they scent the Fox along the field, probably when in search of partridges, meadow-larks, rabbits, or field-mice; presently they trace his footsteps to a large log, from whence he has jumped on to a worm-fence, and after walking a little way on it, has leaped a ditch and skulked toward the borders of a marsh. Through all his crooked ways the sagacious hounds follow his path, until he is suddenly aroused, perchance from a sweet, dreamy vision of fat hens, geese, or turkeys, and with a general cry the whole pack, led on by the staunchest and best dogs, open-mouthed and eager, join in the chase. The startled Fox makes two or three rapid doublings, and then suddenly flies to a cover perhaps a quarter of a mile off, and sometimes thus puts the hounds off the scent for a few minutes, as when cool and at first starting, his scent is not so strong as that of the red fox; after the chase has continued for a quarter of an hour or so, however, and the animal is somewhat heated, his track is followed with greater ease and quickness and the scene becomes animating and exciting. Where the woods are free from underbrush, which is often the case in Carolina, the grass and bushes being burnt almost annually, many of the sportsmen keep up with the dogs, and the Fox is very frequently in sight and is dashed after at the horses' greatest speed. He now resorts to some of the manœuvres for which he is famous; he plunges into a thicket, doubles, runs into the water, if any be at hand, leaps on to a log, or perhaps gets upon a worm-fence and runs along the top of it for a hundred yards, leaping from it with a desperate bound and continuing his flight instantly, with the hope of escape from the relentless pack. At length he becomes fatigued, he is once more concealed in a thicket where he doubles hurriedly; uncertain in what direction to retreat, he hears, and perhaps sees, the dogs almost upon him, and as a last resort climbs a small tree. The hounds and hunters are almost instantly at the foot of it, and whilst the former are barking fiercely at the terrified animal, the latter determine to give him another chance for his life. The dogs are taken off to a little distance, and the Fox is then forced to leap to the ground by reaching with a long pole, or throwing a billet of wood at him. He is allowed a quarter of an hour before the hounds are permitted to pursue him, but he is now less able to escape than before; he has become stiff and chill, is soon overtaken, and falls an easy prey, turning however upon his pursuers with a growl of despair, and snapping at his foes until he bites the dust and the chase is ended.

The following anecdotes of the sagacity of this animal, we hope, may interest our readers. Shortly after the railroad from Charleston to Hamburgh, South Carolina, had been constructed, the rails for a portion of the distance having been laid upon timbers at a considerable height from the ground, supported by strong posts, we observed a Fox which was hard pressed by a pack of hounds, mounting the rails, upon which he ran several hundred yards; the dogs were unable to pursue him, and he thus crossed a deep cypress swamp over which the railroad was in this singular manner carried, and made his escape on the opposite side. The late BENJAMIN C. YANCEY, ESQ., an eminent lawyer, who in his youth was very fond of fox-hunting, related the following: A Fox had been pursued, near his residence at Edgefield, several times, but the hounds always lost the track at a place where there was a foot-path leading down a steep hill. He, therefore, determined to conceal himself near this declivity the next time the Fox was started, in order to discover his mode of baffling the dogs at this place. The animal was accordingly put up and chased, and at first led the hounds through many bayous and ponds in the woods, and at length came running over the brow of the hill along the path, stopped suddenly and spread himself out flat and motionless on the ground; the hounds came down the hill in pursuit at a dashing pace, and the whole pack passed and did not stop until they were at the bottom of the hill. As soon as the immediate danger was over, the Fox, casting a furtive glance around him, started up, and ran off at his greatest speed on his "back track."

The Gray Fox produces from three to five young at a time. In Carolina this occurs from the middle of March to the middle of April; in the State of New-York they bring forth somewhat later. Gestation continues for about three months.

GEOGRAPHICAL DISTRIBUTION.

The Gray Fox is scarce in New-England, and we have not heard of it to the north of the State of Maine; in Canada we have heard of its occasional, but rare appearance. In the vicinity of Albany, N. Y., it is not an uncommon species; south of this, through Pennsylvania and New Jersey, it is about as abundant as the red fox. In the Southern States, except in the mountains of Virginia, it is the only species and is abundant. It exists plentifully in Florida, Mississippi, and Louisiana; it is found on the prairies of the West, and we have received a specimen from California, scarcely differing in any of its markings from those of Carolina.

GENERAL REMARKS.

This species was noticed by LAWSON, CATESBY, and PENNANT. SCHREBER, in 1775, gave it a specific name; he was followed two years afterwards by ERXLEBEN, and in 1788 by GMELIN. In the meantime ERXLEBEN, SCHREBER, and GMELIN published a variety of the Gray Fox, which was a little more cinereous in colour, as a new species, under the name of *Canis cinereo-argenteus.* RICHARDSON was correct in having applied the specific name of *Virginianus* to the Gray Fox, but he erred in referring the Western kit-fox or swift-fox, (*V. velox*,) to *C. cinereo-argentatus.* To us, the short description of these authors, of *C. cinereo-argentatus*, appears to apply more strictly to the Gray Fox than to their accounts of *C. Virginianus*, the latter, we know, is intended for the present species, as it is the only fox in Virginia, with the exception of the red fox, which exists sparingly in the mountains. The views of DESMAREST in regard to our American foxes are very confused, and the translation by HARLAN partakes of all the errors of the original. RICHARDSON did not meet with this species in the Northern regions he visited, and on the whole, very little has been said of its habits by any author.

N°. 5 Plate XXII

Drawn on Stone by H. Trembly.

Grey Rabbit.

Old & Young.

Drawn from Nature by J. J. Audubon, F.R.S., F.L.S.

Printed by Nagel & Weingaertner, N.Y.

LEPUS SYLVATICUS.—Bachman.

Gray Rabbit.

PLATE XXII. Old Male, Female, and Young.

L. auribus capite curtioribus, aurium apice et margine aut nigro; corpore L. Americano minore, supra cinereo-fulva, fusco mixto, subtus subalbido.

CHARACTERS.

Smaller than the Northern hare; ears, shorter than the head, not tipped or margined with black; colour, grayish-fawn, varied with brown above; whitish beneath.

SYNONYMES.

Cony, Third Voyage of the English to Virginia, 1586, by Thomas Herriott. From Pinkerton's Voy., vol. xii., p. 600.
Hare, Hedge Coney, Lawson, p. 122, Catesby, Appendix 28.
American Hare, Kalm's Travels, vol. i., p. 105.
Lepus Americanus, Desmarest, Mam., p. 351.
" " Harlan, Fauna, p. 193.
" " Godman, Nat. Hist., vol. ii., p. 157.
" " Audubon, Birds of America, vol. ii., p. 51, in the talons of Falco Borealis; Ornithological Biography, vol. i., p. 272.
Lepus Americanus, Bach., Jour. Ac. Sc. Phil., vol. vii., p. 326.
" Sylvaticus, Bach., Jour. Ac. Sc. Phil., vol. vii., p. 403, & vol. viii., p. 78 & 326.
" Americanus, Emmons, Mass. Report, 1840, p. 56.
" Nanus, Dekay, Nat. Hist. of New-York, 1842.

DESCRIPTION.

This species bears some resemblance to the European burrowing rabbit, (*L. cuniculus,*) in the gray colour which is natural to the latter in a wild state, but does not change to the different colours the European rabbit presents in a state of domestication. It is a little smaller, and is of a more slender form than *L. cuniculus.* Head, short; eyes, large; ears, well clothed with short hairs on the outer surface; within, the hairs are a little longer, but less dense, the outer border for the fourth of an inch pretty well covered, but nearer the orifice the skin visible through the thinly scattered hairs; legs, of moderate size; claws, strong, sharp, and nearly straight, concealed by the hair; tail, longer in proportion than that of the Northern hare. Fur, compact and soft, about an inch and a quarter in length in winter.

COLOUR.

Summer dress.—Fur on the back, yellowish-brown; soft fur, from the roots to the surface, plumbeous; the long hairs which extend beyond the fur, and give the general colour to the animal, are for three-fourths of their length lead coloured, then yellowish, and are tipped with black. Ears, dark-brown on the outer surface, destitute of the distinct black border seen in the Northern hare, and not tipped with black like those of the Polar and the variable hare; whiskers, nearly all black; iris, light brownish-yellow; a circle of fawn colour around the eye, more conspicuous nearest the forehead. Cheeks, grayish; chin, under surface of body, and inner surface of legs, light grayish-white; tail, upper surface grayish-brown, beneath, white. Breast, light yellowish-gray; behind the ears, a broad patch of fawn colour; outer surface of fore-legs and thighs, yellowish-brown.

Winter colour.—Very similar to the above; in a few specimens, the hairs are whitest at the tips; in others, black tips prevail. This Hare never becomes white in any part of our country, and so far as our researches have extended, we have scarcely found any variety in its colouring.

DIMENSIONS.

Adult Male.

	Inches.	Lines.
Length of head and body	15	0
" head	3	5
" ears	3	0
" tail (vertebræ)	1	2
" tail, including fur	2	2
From heel to end of middle claw	3	7

Weight, 2lbs. 7oz.

HABITS.

This species abounds in our woods and forests, even in their densest coverts; it is fond of places overgrown with young pines thickly crowded together, or thickets of the high bush-blackberry, (*Rubus villosus;*) and is also fond of frequenting farms and plantations, and occupying the coppices and grassy spots in the neighbourhood of cultivation, remaining in its form by day, concealed by a brush-heap, a tuft of grass, or some hedge-row on the side of an old fence; from which retreat it issues at night, to regale itself on the clover, turnips, or corn-fields of the farmer. It not unfrequently divests the young trees in the nursery of their bark; it often makes inroads upon the kitchen-garden, feasting on the young green peas, lettuces, cabbages, &c., and doing a great deal of mischief; and when it has once had an opportunity of tasting these dainties, it becomes difficult to prevent its making a nightly visit to them. Although the place at which it entered may be carefully closed, the Rabbit is sure to dig a fresh hole every night in its immediate vicinity; and snares, traps, or guns, are the best auxiliaries in such cases, soon putting an end to farther depredations.

This animal, when first started, runs with greater swiftness, and makes fewer doublings than the Northern hare, (*L. Americanus;*) having advanced a hundred yards or more, it stops to listen; finding itself pursued by dogs, should the woods be open and free from swamps or thickets, it runs directly toward some hole in the root of a tree or hollow log. In the lower parts of Carolina, where it finds protection in briar patches, and places thickly overgrown with smilax and other vines, it continues much longer on foot, and by winding and turning in places inaccessible to larger animals, frequently makes its escape from its pursuers, without the necessity of resorting for shelter to a hollow tree.

The Gray Rabbit possesses the habit of all the other species of this genus with which we are acquainted, of stamping with its hind feet on the earth when alarmed at night, and when the males are engaged in combat. It is also seen during the spring season, in wood-paths and along the edges of fields, seeking food late in the mornings and early in the afternoons, and during the breeding season even at mid-day: on such occasions it may be approached and shot with great ease. This species, like all the true hares, has no note of recognition, and its voice is never heard except when wounded or at the moment of its capture, when it utters a shrill, plaintive cry, like that of a young child in pain; in the Northern hare this cry is louder, shriller, and of longer continuance. The common domesticated European rabbit seems more easily made to cry out in this way than any other of the genus.

Dr. Richardson, in his work on the American quadrupeds, expresses an opinion from a careful examination of many specimens in different States, that the change to the winter dress in the Northern hare is effected not by a shedding of its hair, but by a lengthening and blanching of the summer fur. Having watched the progress of this change in the present species in a state of confinement, and having also examined many specimens at all seasons of the year, we have arrived at the opposite conclusion as far as regards the Gray Rabbit. In autumn, the greater portion, if not all, the summer fur drops off in spots, and is gradually replaced by the winter coat. In this state, as there are shades of difference between the summer and winter colours, the animal presents a somewhat singular appearance, exhibiting at the same time, like the Northern hare, (although far less conspicuously,) patches of different colours. The Gray Rabbit, although it breeds freely in enclosed warrens, seldom becomes tame, and will probably never be domesticated. When captive, it seems to be constantly engaged in trying to find some means of escape; and though it digs no burrows in a state of nature, yet, when confined, it is capable of digging to the depth of a foot or more under a wall, in order to effect its object. We, however, at the house of Dr. De Benneville at Milestown, near Philadelphia, saw five or six that were taken from the nest when very young and brought up by hand, so completely tamed that they came at the call and leapt upon the lap of their feeder; they lived sociably and without restraint in the yard, among the dogs and poultry. The former, although accustomed to chase the wild rabbit, never molesting those which had, in this manner, grown up with them, and now made a part of the motley tenants of the poultry-yard. We have not only observed dogs peacefully associating with the hare, when thus tamed, but have seen hounds accustomed to the chase of the deer, eating from the same platter with one of those animals that was domesticated and loose in the yard, refraining from molesting it, and even defending it from the attacks of strangers of their own species that happened to come into the premises; and when this tame deer, which occasionally visited the woods, was started by the pack of hounds here referred to, they refused to pursue it.

The Gray Rabbit is one of the most prolific of all our species of this genus; in the Northern States it produces young about three times in the season, from five to seven at a litter; whilst in Carolina its young are frequently brought forth as early as the twentieth of February, as late as the middle of October, and in all the intermediate months. Nature seems thus to have made a wise provision for the preservation of the species, since no animal is more defenceless or possesses more numerous enemies. Although it can run with considerable swiftness for some distance, its strength in a short time is exhausted, and an active dog would soon overtake it if it did not take shelter in some hole in the earth, heap of logs, or stones, or in a tree with a hollow near its root; in these retreats it is often captured by young hunters.

In the Northern and Middle States, where the burrows of the Maryland marmot (*Arctomys monax*) and the holes resorted to by the common skunk, (*Mephitis chinga,*) are numerous, the Gray Rabbit in order to effect its escape when pursued betakes itself to them; and as they are generally deep, or placed among rocks or roots, it would require more labour to unearth it when it has taken possession of either of these animals' retreats than it is worth, and it is generally left unmolested. It is not always safe in these cases, however, for the skunk occasionally is "at home" when the Rabbit runs into his hole, and often catches and devours the astonished fugitive before it can retrace its steps and reach the mouth of the burrow.

This species is also captured occasionally by the skunk and other carnivorous animals when in its form. Its most formidable enemy, however, is the ermine, which follows its tracks until it retires to a hole in the earth or to a hollow tree, which the little but ferocious creature, although not one-fourth as large as the timid Rabbit, quickly enters and kills it—eating off the head, and leaving the body until a want of food compels it to return for more.

Whilst residing in the State of New-York many years ago, we were desirous of preserving a number of Rabbits during the winter from the excessive cold and from the hands of the hunters, who killed so many that we feared the race would be nearly extirpated in our neighbourhood; our design being to set them at liberty in the spring. At this period we had in confinement several weasels of two species existing in that part of the country, (*Putorius erminea* and *P. fusca,*) in order to ascertain in what manner their change of colour from brown in summer to white in winter, and *vice versâ*, was effected.

We bethought ourselves of using one of each species of these weasels instead of a ferret, to aid in taking the Rabbits we wanted, and having provided ourselves with a man and a dog to hunt the Rabbits to their holes, we took the weasels in a little tin box with us, having first tied a small cord around their necks in such a manner as to prevent them from escaping, or remaining in the holes to eat the Rabbits, whilst it could not slip and choke them.

We soon raced a Rabbit to its hole, and our first experiment was made with the little brown weasel, (*P. fusca;*) it appeared to be frightened, and refused to enter the hole; the common species, (*P. erminea,*) although we had captured the individual but a few days before, entered readily; but having its jaws at liberty, it killed the Rabbit. Relinquishing the weasel to our man, he afterwards filed its teeth down, to prevent it from destroying the Rabbits; and when thus rendered harmless, the ermine pursued the Rabbits to the bottom of their holes, and terrified them so that they instantly fled to the entrance and were taken alive in the hand; and although they sometimes scrambled up some distance in a hollow tree, their active and perservering little foe followed them and instantly forced them down. In this manner the man procured twelve Rabbits alive in the course of one morning, and more than fifty in about three weeks, when we requested him to desist.

On more than one occasion we have seen the tracks of this species on the snow, giving evidence by their distance from each other that the animal had passed rapidly, running under the influence of fear. Examining the surface of the snow carefully, we observed the foot-prints of the weasel, as if in pursuit, and following up the double trail, we found at the mouth of a hole a short distance beyond, the mutilated remains of the luckless Rabbit.

The Canada lynx, the Bay lynx, (wild cat,) the red and the gray fox, &c., capture this species by stratagem or stealth; various species of hawks and owls prey upon them, and the rattle-snake, chicken-snake, and other serpents, have been killed with the Gray Rabbit in their stomach. These reptiles probably caught their victims by stratagem, or by stealing upon them when in their form, and enclosing them in their twining folds, as the boa constrictor captures larger animals.

In order to catch or kill the Gray Rabbit, different means are resorted to according to the fancy of the hunter or the nature of the locality in which the animal may be. In the northern parts of the United States it is pursued with dogs, and either shot or taken from the hole or other retreat to which it may have been driven. It is also frequently captured in box-traps, or snares, placed in the gaps of some brush-fence made in the woods for the purpose. In the Southern States it is generally hunted with pointer dogs and shot at the moment when it leaps from its form.

GEOGRAPHICAL DISTRIBUTION.

We have not heard of the existence of this species farther north than the southern counties of the State of New Hampshire, beyond which it is replaced by other and larger species. It cannot be said to be abundant in the New England States, except in a few localities, and it does not seem to prefer high mountainous regions. In occasional botanical excursions among the Catskill mountains and those of Vermont and New Hampshire, where we saw considerable numbers of the Northern hare, we found scarcely any traces of the present species, especially in the mountains east of the Hudson river. It exists in the chain of the Alleganies running through Virginia to the upper parts of Carolina, but is there far from being abundant. It was exceedingly scarce north-east of Albany thirty-five years ago, where it has now become far more numerous than the Northern hare, which was then the only species usually met with. It abounds in the sandy regions covered with pine trees west of that city. From Dutchess county to the southern limits of New-York it is found in considerable numbers. In Pennsylvania, New-Jersey, Maryland, and all the Southern States, hunting the Gray Rabbit affords more amusement to young sportsmen than the pursuit of any other quadruped in the country. We have traced this species through all the higher portions of Florida. To the west we have seen it in all the Southern States, and it is very abundant on the upper Missouri River to nearly 1000 miles above Saint Louis.

GENERAL REMARKS.

This being the most common hare in the Atlantic States of America, it has been longest and most familiarly known. HERRIOTT, who gave an account of the third voyage of the English to Virginia in 1586, in enumerating the natural productions of that country, under the head of Conies, says, "Those that we have seen, and all that we can hear of, are of a gray colour like unto hares; in some places there are such plenty that all the people, of some towns, make them mantles of the fur, or fleece of the skins of those which they usually take." It is subsequently mentioned by the intrepid Governor SMITH of Virginia, by LAWSON and by CATESBY. KALM, in the 1st vol. of his Travels in America, gave a correct description, not only of the animal, but of its habits. The following is an extract from his Journal; the entry was made either at Philadelphia or his favourite retreat "Racoon," in the vicinity of that city, on the 6th Jan. 1749. "There are a great number of hares in this country, but they differ from our Swedish ones in their size, which is very small, and but little bigger than that of a rabbit; they keep almost the same gray colour both in summer and winter, which our Northern hares have in summer only; the tip of their ears is always gray, and not black; the tail is likewise gray on the upper side, at all seasons; they breed several times a year. In spring they lodge their young ones in hollow trees, and in summer, in the months of June and July, they breed in the grass. When they are surprised they commonly take refuge in hollow trees, out of which they are taken by means of a crooked stick, or by cutting a hole into the tree opposite to the place where they lie; or by smoke which is occasioned by making a fire on the outside of the tree. On all these occasions the grayhounds must be at hand. These hares never bite, and can be touched without any danger. In the day-time they usually lie in hollow trees, and hardly ever stir from thence unless they be disturbed by men or dogs; but in the night they come out and seek their food. In bad weather, or when it snows, they lie close for a day or two, and do not venture to leave their retreats. They do a great deal of mischief in the cabbage-fields, but apple-trees suffer infinitely more from them, for they peel off all the bark next to the ground. The people here are agreed that the hares are fatter in a cold and severe winter than in a mild and wet one, for which they could give me several reasons from their own conjectures. The skin is useless, because it is so loose that it can be drawn off; for when you would separate it from the flesh, you need only pull at the fur and the skin follows. These hares cannot be tamed. They were at all times, even in the midst of winter, plagued with a number of common fleas."

In 1820 (as we have observed in our article on *L. Americanus*) Desmarest mistaking the species, gave a pretty good description of the Gray Rabbit, and unfortunately referred it to *L. Americanus*. He had evidently been misled by Forster, Schœpff, Pennant, Erxleben and Bodd, who having confounded these two species, induced him to believe that as he was describing an American hare, only one American species at that time being known, it must be the one referred to by previous authors. Hence he quoted Gmelin, Schœpff, Erxleben, Pallas and Bodd, and gave to the species the extravagant geographical range, from Churchill, Hudson's Bay, to California, and assigned it a habitation in New-Albion, Louisiana, Florida, the two Carolinas, &c. Harlan, in giving an account of the American quadrupeds in 1825, finding the Gray Rabbit described by Desmarest, translated the article very literally, even to its faults, from the French of that author, (See Encyclopédie de Mammalogie, p. 351.) Harlan's translation represents the fur as "becoming whiter during winter, but the ears and tail remaining always of the same gray." In the following year Godman (Amer. Nat. Hist., vol. ii., p. 157) once more described this species under the (wrong) name of *Lepus Americanus*. In speaking of its colour, he says, "in winter the pelage is nearly or altogether white," and he gives it the extraordinary weight of seven pounds. This is rather surprising, as we know no city in the union where the market in winter is better supplied with this species of hare than Philadelphia.

In this singular manner the Gray Rabbit, the most common and best known of all the species of quadrupeds in America, had never received a specific name that was not pre-occupied. In 1827, we proposed the name of *Lepus sylvaticus*, and assigned our reasons for so doing in a subsequent paper, (See Journ. Acad. Nat. Sc., vol. viii., part 1, p. 75.) In 1840, Dr. Emmons also, (Report on Quadrupeds of Massachusetts,) described it under the (wrong) name of *L. Americanus*, giving as synonymous, *L. Hudsonius*, PALLAS; American hare, FORSTER, PENNANT, Arct. Zool. HEARNE'S Journey, SABINE, PARRY and RICHARDSON; who each described the Northern hare, and not this species. He, however, quoted HARLAN and GODMAN correctly, with the exception of the name which they had misapplied.

In 1842 Dr. DEKAY (See Nat. Hist. N. York, part 1st, p. 93) refers this species to *Lepus nanus* of SCHREBER, supposing the description of that author, (which is contained in an old work that is so scarce in America that our naturalists have seldom had an opportunity of referring to it,) to have escaped the notice of modern authors. After giving a translation from SCHREBER, he remarks, "The whole history of the habits of this species, and its abundance, sufficiently confirm the fact that SCHREBER had our Rabbit in view, although he was misled by SCHŒPFF and PENNANT, and confounded two species."

We regret that we are obliged to differ from an author who is generally accurate, and who is always courteous in his language towards other naturalists, but in this case we must do so.

In order to save the student of natural history the labour of searching for SCHREBER'S work, to refer to his description, we have concluded to insert it here, together with our translation of the article, adding the references to authors, &c., which were omitted by DEKAY, and which we conceive very important in pursuing our inquiries.

EXTRACT FROM SCHREBER.

"DER WABUS, ODER AMERIKANISCHE HASE.

TAB. CCXXXIV. B.

Lepus nanus. Lepus auribus extrorsum nigro marginatis, cauda supra nigricante.

SYNONYMEN.

LEPUS HUDSONIUS.

LEPUS APICE AURIUM CAUDÆQUE CINEREO, Pall., Nov. Spec. Glis., p. 30, 45, Zimmerm., E. E. z. 336.

LEPUS AMERICANUS, Lepus cauda abbreviata pedibus postici corpore dimidio longioribus auricularum caudæque apicibus griseis, Erxleben, Mamm., p. 330.

AMERICAN HARE, Forster, Phil. Tr., lxxii., p. 376, Pennant, Hist., p. 372 u. 243.

HARE, HEDGE CONEY, Lawson, Car., p. 122, Catesby's App., p. xxviii.

HARAR, en art som är midt emellan hare ach canin, Kalm, Rese, vol. ii., p. 236, vol. iii., p. 8, 285.

DER AMERIKANISCHE HASE, Forster, von den Thieren in Hudson's Bay, in Sprenge's Beyt.

DER NORDAMERIKANISCHE HASE, Schœpff.

WABUS, (ALGONQUINISCH,) Jefferson's Notes, (Phil. 1788,) p. 51, 57.

BESCHREIBUNG.

Der Kopf hat nichts Unterscheidendes. Die Backen sind dickhärig. Die Ohren dünne, auswendig dünne behaart, inwendig kahl, und reichen, vorwärts gebogen, noch nicht bis an die Nasenspitze; nach hinten gelegt, bis an die Schulterblätter. Ueber den grossen schwarzen Augen vier bis fünf Börsten. Die Barthörsten grossentheils schwarz; einige weiss; die längsten scheinen länger als der Kopf zu sein.

Die Sommerfarbe ist folgende. Die Ohren bräunlich, mit einer sehr schmalen schwarzen Einfassung am äussern Rande, die an der Spitze eben die Breite behält, oder gegen die Spitze hin gar verschwindet. Stirne, Backen, Rücken und Seiten, Aerme und Schenkel auswendig leicht braun mit Schwarz überlaufen.

Der Umfang des Afters weiss. Die Füsse dicht und kurz behaart, von einem hellern leicht Braun, ohne alles Schwarz, an der innern Seite stärker in grau-weiss abfallend. Der Schwanz oben auf von der Farbe des Rückens, (vermuthlich stärker mit Schwarz überlaufen, denn Herr PENNANT beschreibt ihn oben schwarz,) unten weiss. Die Kehle weiss; der Untertheil des Halses leicht braun, mit Weiss überlaufen.

Brust, Bauch, innere Aerme und Schenkel, einem weichen Weiss. Die Winterfarbe, wo sie verschieden, ist weiss. Backenzähne oben und unten auf jeder Seite fünf. Die Länge des Körpers höchstens anderthalb englische Fuss, des Schwanzes nicht viel über zwei Zoll. Das Gewicht $2\frac{1}{4}$ bis 3 Pfund; nach Herrn PENNANT 3 bis $4\frac{1}{2}$ Pfund.

Die underscheidenden Merkmale dieser Art sind nach den Herren FORSTER, PENNANT und SCHŒPFF, 1. die Grösse; er kommt dem gemeinen und veränderlichen Hasen lange nicht bei, und ist kaum grösser als ein Kaninchen, daher er auch in Nord-Amerika nicht selten den Namen Rabbit oder Kaninchen bekommt. 2. Das Verhältniss der Füsse; die Vorterfüsse sind kürzer und die Hinterfüsse länger als an allen Dreien. 3. Die Farbe der Ohren; sie haben eine schwarze Einfassung auswendig, aber keinen schwarzen Fleck an der Spitze. Ihre geringere Länge unterscheidet von den Ohren des gemeinen Hasen. 4. Die Farbe des Schwanzes; diese ist oben auf nicht schwarz, oder doch nicht so sattschwarz als am Hasen. 5. Die Farbe des Körpers. 6. Die Lebensart und Eigenschaften. Er kann also unmöglich etwas anders als eine für sich bestehende Art sein. Sein Vaterland ist ganz Nord-Amerika, von Hudson's Bay an bis nach Florida hinab. Er schweift nicht herum, sondern schränkt sich auf kleine Räume ein.

In Hudson's Bay, Canada und Neu-England vertauscht er sein kurzes Sommerhaar im Herbste gegen ein langes seitenartiges und bis an die Wurzel silberweisses Haar, und nur der Rand der Ohren und der Schwanz behalten ihre Farbe, (PENNANT, KALM.) In den südlichen Ländern bleibt die Farbe, auch in den härtesten Wintern, unverändert, (KALM.)

Daher könnte man diesen Hasen füglich den *halb*-veränderlichen nennen."

In carefully reading the above description, the attentive reader can scarcely have failed to remark that if *Lepus Americanus* of ERXLEBEN, and *Lepus Hudsonius* of PALLAS, are the Northern hare, *Lepus nanus* must be the same species, as the descriptions agree in every particular; and where SCHREBER enters more into detail, he describes the Northern hare still more minutely, and only confirms us still farther in the conviction that he had never seen the Gray Rabbit, and was describing the very species he professed to describe, viz., the Hudson's Bay quadruped of DAINES BARRINGTON, (See vol. lxii. Phil. Trans., p. 11,) and the "American hare, called rabbit at Hudson's Bay," of FORSTER, (See the above vol., p. 376,) which, however, had already received from two of his countrymen, PALLAS and ERXLEBEN, the names of *L. Americanus* and *L. Hudsonius*.

The time when this description was made must not be overlooked. At the close of the year 1772, the Philosophical Transactions, containing the two accounts of this new American hare, were published. No specific *Latin name*, such as would according to the binary system which was then coming into use, entitle the first describer to the species, had as yet been given to it; and whilst the English naturalists were looking for decided characters by which it could be distinguished, (and we know from experience with how much difficulty these characteristics are found in the hares,) the German naturalists, with the example of LINNÆUS, their next door neighbour, before their eyes, went forward in hot haste to describe the species. Leaving the English philosophers to *cook* their animal, to ascertain by the colour of its flesh whether it was a hare or a rabbit, they sought for a Latin cognomen, desirous that their own names should be handed down to posterity along with it. Hence ERXLEBEN, PALLAS and SCHREBER, (the two former evidently without the knowledge of the latter,) named the species, very likely, as we are inclined to think, without having had any specimen before them, and simply attaching a name to the descriptions of the English naturalists. Be this as it may, in less than three years it had already received in Germany alone, the several names of *L. Americanus*, *nanus*, and *Hudsonius*. If SCHREBER, who had the Philosophical Transactions lying before him when he drew up his description, (for he quotes both the accounts,) and who also possessed the accounts of ERXLEBEN and PALLAS, had examined a different species, surely *he* would have made the discovery; but after a careful examination, and not a bad description, he gives the size, colour, and measurements of the Northern hare, and finally quotes FORSTER, PENNANT, SCHŒPFF, &c., as his authorities for the species.

The name *Lepus nanus*, given to it by SCHREBER, might at first lead us to conjecture that as he meant to designate the species as a small hare, and as the Northern hare is rather large, he could not have intended it for the latter, but had in view the Gray Rabbit—hence the name, *nanus*, dwarf. There can, however, be no difficulty in accounting for the choice of that name. On turning to the eleventh page of the Philosophical Transactions, vol. xlii., where the species was first announced, it will be perceived that BARRINGTON had been closely investigating the several species of hare with which the naturalists of Europe were acquainted at that early day; and he gives the following measurements:—

	Fore-leg.*	Hind-leg.*	Back and Head.
Rabbit - - - - -	$4\frac{1}{2}$ inches	$6\frac{3}{4}$ inches	$16\frac{1}{2}$ inches
Hare - - - - -	$7\frac{3}{4}$ "	11 "	22 "
Hudson's Bay quadruped -	$6\frac{3}{4}$ "	$10\frac{3}{4}$ "	18 "
Alpine hare - - -	$6\frac{1}{2}$ "	$16\frac{3}{4}$ "	22 "

* From uppermost joint to toe.

Here then we have the relative sizes of the several species. The first is the common wild rabbit of England, (*L. cuniculus*,) which is a little larger than our Gray Rabbit. The second is the common English hare, (*L. timidus*.) The third, the American hare from Hudson's Bay; and the fourth, the Alpine or variable hare, (*L. variabilis*.) The rabbit being a *burrowing* animal with *white* flesh, was not considered a *hare*, and the American animal was smaller than either the European or the Alpine hare, measuring only eighteen inches in length, whilst these last measured twenty-two inches each. We perceive, therefore, that it was called *Lepus nanus*, because it was the smallest of the species then known. For the same reason our American woodcock was called *scolopax minor*, because it was smaller than the English woodcock, although it finally proved to be the largest snipe in America.

Let us compare the description of SCHREBER's *L. nanus*, with the Northern hare, of which we have a number of specimens (including all its various changes of colour) before us, to refer to as we proceed.

TRANSLATION.	REMARKS.
Lepus nanus.	*Lepus Americanus.*
The head has nothing peculiar; cheeks, thickly haired; ears, thin, externally with few hairs, naked within, and when bent forward do not reach the point of the nose; when bent backward they reach the shoulder blades.	This description agrees with *L. Americanus;* the ears in our dried specimens are none of them more than 3½ inches long, whilst from nose to ear they measure 4 inches; the ears therefore could not reach the nose.
Eyes, large and black, with four or five bristles above them; whiskers, mostly black; some are white, the longest appear to be longer than the head.	Applies perfectly to our specimens of *L. Americanus*, except the colour of the eyes, which applies to neither the Northern hare nor the Gray Rabbit, and which he must have obtained from some other source than a dried skin.
The following is the colour in summer: ears, brownish, with a very narrow black border on the outer margin, being at the tips the same breadth, or it even disappears towards the tips.	The very narrow black border on the outer margin betrays the species; it belongs to the Northern hare, but not to the Gray Rabbit. They only become effaced when covered with white hair in winter; and it is evident this last expression was taken from KALM, who says of the Rabbit, "the tip of their ears is always gray, and not black, as is the case in the European, common, and Alpine hares."
Forehead, cheeks, back and sides, fore and hind-legs externally, light brown, mixed with black; around the breech, white.	All agreeing with the description of the Northern hare.

TRANSLATION.	REMARKS.
Feet, thickly covered with short hairs of a light brown, unmixed with black, changing on the inside to a grayish white.	Such is the colour of the feet of several of our specimens of the Northern hare in summer pelage.
Upper part of the tail the colour of the back, (perhaps mixed with black, as Pennant describes it black above,) beneath white.	The upper part of the tail is like the back in most specimens, but it is seen how anxious he was not to depart from the views of Pennant, who describes it as black, which is the case in some specimens.
Throat, white; lower part of the neck, bright brown, mixed with white; chest and belly, inside of fore and hind-legs, a dull white.	These distinctive marks all belong to the Northern hare.
Colour in winter, when it does change, white.	The Gray Rabbit does not become white in winter.
Molars above and beneath, on each side, five. The length of the body at farthest eighteen inches, the tail not over two inches.	This size applies to the Northern hare, and not to the Gray Rabbit. None of our dried specimens of the former reach quite eighteen inches, and none of the Gray Rabbit beyond fifteen. Tail of the Northern hare, including fur, two inches; that of the Gray Rabbit is longer.
The weight is from $2\frac{1}{4}$ to 3 lbs.; according to Pennant, from 3 to $4\frac{1}{2}$ lbs.	These weights were compiled from authors. Carver, who had reference to the Gray Rabbit, gave the lesser weight; and Pennant, who referred to the Northern hare, gave the greater.
The most striking distinctions in this species, according to Forster, Pennant, and Schœpff, are, 1st, its size; it is not near as large as the common hare or the changeable	Forster says in regard to the Northern hare—"The proper characteristics of this species seem to be, 1st, its size, which is somewhat bigger than a rabbit, but less

TRANSLATION.	REMARKS.
hare, and scarcely larger than a rabbit; hence in North America he is frequently called rabbit.	than that of the Alpine or lesser hare."
2d, The proportion of the legs. The hind-feet being longer and the fore-feet shorter than either of the three.	2d, FORSTER says, "The proportion of its limbs. Its hind-feet being longer in proportion to the body than those of the rabbit and the common hare."
3d, The colour of the ears; they have a black margin outside, but no black spot at the tip.	3d, "The tip of the ears and tail, which are constantly gray, not black," KALM's Travels, vol. ii., p. 45.
The ear being less in length separates it from the common hare.	The ears of the Northern hare, the species here referred to, are considerably less in length than those of the common European hare.
4th, The colour of the tail; this is on the upper surface not black, or as intensely black as that of the hare.	The upper side of the tail of the European hare, (*L. timidus,*) is black, that of the Northern hare generally dark brown.
5th, The colour of the body.	That of the European hare is not as dark.
6th, Its mode of living and habits.	In the description of these habits by FORSTER, two species had been blended.
It can therefore only be a distinct species.	He meant distinct from those of Europe.
It is a native of all North America, from Hudson's Bay to Florida. It does not migrate far, but confines itself to a narrow compass.	The Gray Rabbit is not found at Hudson's Bay, where the other abounds. In his views of the Southern range of the Northern hare, he was misled by FORSTER, and supposing KALM's rabbit referred to the

TRANSLATION.

REMARKS.

same species, he quoted KALM as authority for its existence as far south as Florida.

In Hudson's Bay, Canada, and New-England, it changes in autumn this short summer hair into a long silky fur, white from the roots, and only the border of the ears and the tail preserve their colour, (PENNANT, KALM.)

The Gray Rabbit does not change in this manner. He meant by this to show that whilst this species became white in winter, the border of the ear and upper part of the tail underwent no change.

In the Southern parts, his colour, even in the coldest winters, remains unchanged, (KALM.) He might, therefore, be properly called the half changing hare.

SCHREBER, never having been in America, had to compile his account of its habits from others. It is easily seen that in this he was misled by FORSTER, who misunderstood KALM; the latter having here referred to the Gray Rabbit, which never changes its colour.

DEKAY conceives SCHREBER to have described the Gray Rabbit, from the abundance of the species; but the Northern hare, where it does exist, is not less abundant. In particular localities in the Northern States, it is more frequently met with than the Gray Rabbit in the Middle or Southern States.

HEARNE says that on the south side of Anawed Lake they were so plentiful, that several of the Indians caught twenty or thirty of a night with snares; and at Hudson's Bay, where all the specimens first brought to Europe were procured, it is represented as very abundant.

We think we have now shown that SCHREBER'S account of *L. nanus*—its size, length of legs, the black margin around the ear, its change of colour, and his references to authors, all prove explicitly that he had no reference to the Gray Rabbit, but described the Northern hare.

His name must therefore stand as a synonyme of *L. Americanus*, which is to be somewhat regretted, as although the name itself is very objectionable, his description of that species appears to us the best that was given, from its first describer, FORSTER, down to the time of RICHARDSON whose description is so accurate that nothing need be added to it.

Drawn on Stone by R Trembly.

Black Rat

Old & Young.

Drawn from Nature by J.J. Audubon, F.R.S. F.L.S.

Printed by Nagel & Weingærtner, N.Y.

GENUS MUS.—LINN.

Incisive $\frac{2}{2}$; *Canine* $\frac{0-0}{0-0}$; *Molar* $\frac{3-3}{3-3} = 16$.

Cheek-teeth, furnished with tubercles; ears, oblong or round, nearly naked; without cheek-pouches; fore-feet, with four toes, and a wart covered with an obtuse nail in place of a thumb; hind-feet, pendactylous; tail, long, usually naked and scaly; fur, with a few long, scattered hairs, extending beyond the rest.

The generic name MUS is derived from the Latin *mus*, a mouse, from the Greek μυς, (*mus*,) a mouse.

There are upwards of two hundred species of this genus described as existing in various quarters of the globe, of which about nine well-determined species are found in North America, three of which have been introduced.

MUS RATTUS.—LINN.

BLACK RAT.

PLATE XXIII.—OLD AND YOUNG, OF VARIOUS COLOURS.

M. cauda corpore longiore; pedibus anterioribus ungue pro pollice instructis; corpore atro, subtus cinereo.

CHARACTERS.

Tail, longer than the body; fore-feet, with a claw in place of a thumb, bluish-black above, dark ash-coloured beneath.

SYNONYMES.

MUS RATTUS, Linn., 12th ed., p. 83.
" " Schreber, Säugethiere, p. 647.
" " Desmar., in Nouv. Dict., 29, p. 48.
RAT, Buffon, Hist. Nat., vol. vii., p. 278, t. 36.
RAT ORDINAIRE, Cuv., Règne Anim., p. 197.
BLACK RAT, Penn., Arc. Zool., vol. i., p. 129.
ROLLER PONTOPP., Dan. i., p. 611.

Mus Rattus, Griffith's Animal Kingdom, vol. v., 578, 5.
" " Harlan, p. 148.
" " Godman, vol. ii., p. 83.
" " Richardson, p. 140.
" " Emmons, Report on Quadrupeds of Massachusetts, p. 63.
" " Dekay, Natural History of New-York, vol. i., p. 80.

DESCRIPTION.

Head, long; nose, sharp pointed; lower jaw, short; ears, large, oval, broad and naked. Whiskers, reaching beyond the ear.

Body, smaller and more delicately formed than that of the brown rat; thickly clothed with rigid, smooth, adpressed hairs.

Fore-feet, with four toes, and a claw in place of a thumb. Feet, plantigrade, covered on the outer surface with short hairs. Tail, scaly, slightly and very imperfectly clothed with short coarse hairs. The tail becomes square when dried, but in its natural state is nearly round. Mammæ, 12.

COLOUR.

Whiskers, head, and all the upper surface, deep bluish-black; a few white hairs interspersed along the back, giving it in some lights a shade of cinereous; on the under surface it is a shade lighter, usually cinereous. Tail, dusky; a few light-coloured hairs reaching beyond the toes, and covering the nails.

DIMENSIONS.

Length of head and body - - - - - -	8 inches.
" tail - - - - - - - - - -	$8\frac{1}{4}$ do.

HABITS.

The character of this species is so notoriously bad, that were we to write a volume in its defence we would fail to remove those prejudices which are every where entertained against this thieving cosmopolite. Possessing scarcely one redeeming quality, it has by its mischievous propensities caused the world to unite in a wish for its extermination.

The Black Rat is omnivorous, nothing seeming to come amiss to its voracious jaws—flesh, fowl or fish, and grain, fruit, nuts, vegetables, &c., whether raw or cooked, being indiscriminately devoured by it. It is very fond of plants that contain much saccharine or oleaginous matter.

The favourite abodes of this species are barns or granaries, holes under out-houses or cellars, and such like places; but it does not confine itself to any particular locality. We have seen its burrows under cellars used for keeping the winter's supply of sweet potatoes in Carolina, in dykes surrounding rice-fields sometimes more than a mile from any dwelling, and it makes a home in clefts of the rocks on parts of the Alleghany mountains, where it is very abundant.

In the neighbourhood of the small streams which are the sources of the Edisto river, we found a light-coloured variety, in far greater numbers than the Black, and we have given three figures of them in our Plate. They were sent to us alive, having been caught in the woods, not far from a mill-pond. We have also observed the same variety in Charleston, and received specimens from Major LECONTE, who obtained them in Georgia.

During the summer season, and in the autumn, many of these rats, as well as the common or Norway rat, (*Mus decumanus,*) and the common mouse, (*Mus musculus,*) leave their hiding places near or in the farmer's barns or hen-houses, and retire to the woods and fields, to feed on various wild grasses, seeds, and plants. We have observed Norway rats burrowing in banks and on the borders of fields, far from any inhabited building; but when the winter season approaches they again resort to their former haunts, and possibly invite an additional party to join them. The Black Rat, however, lives in certain parts of the country permanently in localities where there are no human habitations, keeping in crevices and fissures in the rocks, under stones, or in hollow logs.

This species is by no means so great a pest, or so destructive, as the brown or Norway rat, which has in many parts of the country either driven off or exterminated it. The Black Rat, in consequence, has become quite rare, not only in America but in Europe.

Like the Norway rat this species is fond of eggs, young chickens, ducks, &c., although its exploits in the poultry house are surpassed by the audacity and voraciousness of the other.

We have occasionally observed barns and hen-houses that were infested by the Black Rat, in which the eggs or young chickens remained unmolested for months together; when, however, the Rats once had a taste of these delicacies, they became as destructive as usual, and nothing could save the eggs or young fowls but making the buildings rat-proof, or killing the plunderers.

The following information respecting this species has been politely communicated to us by S. W. ROBERTS, ESQ., civil engineer:—

"In April, 1831, when leading the exploring party which located the portage railroad over the Alleghany mountains, in Pennsylvania, I found a multitude of these animals living in the crevices of the silicious limestone rocks on the Upper Conemaugh river, in Cambria county, where the large viaduct over that stream now stands. The county was then a wilderness, and as soon as buildings were put up the rats deserted the rocks, and established themselves in the shanties, to our great annoyance; so that one of my assistants amused himself shooting at them as he lay in bed early in the morning. They ate all our shoes, whip-lashes, &c., &c., and we never got rid of them until we left the place."

We presume that in this locality there is some favourite food, the seeds of wild plants and grasses, as well as insects, lizards, (*Salamandra*,) &c., on which these Rats generally feed. We are induced to believe that their range on the Alleghanies is somewhat limited, as we have on various botanical excursions explored these mountains at different points to an extent of seven hundred miles, and although we saw them in the houses of the settlers, we never observed any locality where they existed permanently in the woods, as they did according to the above account.

The habits of this species do not differ very widely from those of the brown or Norway rat. When it obtains possession of premises that remain unoccupied for a few years, it becomes a nuisance by its rapid multiplication and its voracious habits. We many years ago spent a few days with a Carolina planter, who had not resided at his country seat for nearly a year. On our arrival, we found the house infested by several hundreds of this species; they kept up a constant squeaking during the whole night, and the smell from their urine was exceedingly offensive.

The Black Rat, although capable of swimming, seems less fond of frequenting the water than the brown rat. It is a more lively, and we think a more active, species than the other; it runs with rapidity, and makes longer leaps; when attacked, it shrieks and defends itself with its teeth, but we consider it more helpless and less courageous than the brown or Norway rat.

It is generally believed that the Black Rat has to a considerable extent been supplanted both in Europe and America by the Norway rat, which it is asserted kills or devours it. We possess no positive facts to prove that this is the case, but it is very probably true.

We have occasionally found both species existing on the same premises, and have caught them on successive nights in the same traps; but we have invariably found that where the Norway rat exists in any considerable numbers the present species does not long remain. The Norway rat is not only a gross feeder, but is bold and successful in its attacks on other animals and birds. We have known it to destroy the domesticated rabbit by dozens; we have seen it dragging a living frog from the banks of a pond; we were once witnesses to its devouring the young of its own species, and we see no reason why it should not pursue the Black

Rat to the extremity of its burrow, and there seize and devour it. Be this as it may, the latter is diminishing in number in proportion to the multiplication of the other species, and as they are equally prolific and equally cunning, we cannot account for its decrease on any other supposition than that it becomes the prey of the more powerful and more voracious Norway rat.

The Black Rat brings forth young four or five times in a year; we have seen from six to nine young in a nest, which was large and composed of leaves, hay, decayed grasses, loose cotton, and rags of various kinds, picked up in the vicinity.

GEOGRAPHICAL DISTRIBUTION.

This species is constantly carried about in ships, and is found, although very sparingly, in all our maritime cities. We have met with it occasionally in nearly all the States of the Union. On some plantations in Carolina, particularly in the upper country, it is the only species, and is very abundant. We have, however, observed that in some places where it was very common a few years ago, it has altogether disappeared, and has been succeeded by the Norway rat. The Black Rat has been transported to every part of the world where men carry on commerce by means of ships, as just mentioned.

GENERAL REMARKS.

PENNANT, KALM, LINNÆUS, PALLAS, DESMAREST, and other European writers, seem disposed to consider America the Fatherland of this pest of the civilized world. HARLAN adopted the same opinion, but BARTRAM, (if he was not misunderstood by KALM,) did more than any other to perpetuate the error.

In the course of a mutual interchange of commodities, the inhabitants of the Eastern and Western Continents have presented each other with several unpleasant additions to their respective productions, especially among the insect tribe.

We are willing to admit that the Hessian fly was not brought to America in straw from Hanover, as we sought in vain for the insect in Germany; but we contend that the Black Rat and the Norway rat, which are in the aggregate greater nuisances, perhaps, than any other animals now found in our country, were brought to America from the old world. There are strong evidences of the existence of the Black Rat in Persia, long before the discovery of America, and we have no proof that it was known in this country till many years after its colonization. It is true, there were rats in our country which by the common people might have been regarded as similar to those of Europe, but these have now been proved to be of very different species. Besides, if the species existed in the East from time immemorial, is it not more probable that it should have been carried to Europe, and from thence to America, than that it should have been originally indigenous to both continents? As an evidence of the facility with which rats are transported from one country to another, we will relate the following occurrence: A vessel had arrived in Charleston from some English port, we believe Liverpool. She was freighted with a choice cargo of the finest breeds of horses, horned cattle, sheep, &c., imported by several planters of Carolina. A few pheasants (*Phasianus colchicus*) were also left on board, and we were informed that several of the latter had been killed by a singular looking set of rats that had become numerous on board of the ship. One of them was caught and presented to us, and proved to be the Black Rat. Months after the ship had left, we saw several of this species at the wharf where the vessel had discharged her cargo, proving that after a long sea voyage they had given the preference to terra firma, and like many other sailors, at the clearing out of the ship had preferred remaining on shore.

We have seen several descriptions of rats that we think will eventually be referred to some of the varieties of this species. The *Mus Americanus* of GMELIN, *Mus nigricans* of RAFINESQUE, and several others, do not even appear to be varieties; and we have little doubt that our light-coloured variety, if it has not already a name, will soon be described by some naturalist who will consider it *new*. To prevent any one from taking this unnecessary trouble, we subjoin a short description of this variety, as observed in Carolina and Georgia.

Whole upper surface, grayish-brown, tinged with yellow; light ash beneath; bearing so strong a resemblance to the Norway rat, that without a close examination it might be mistaken for it.

In shape, size, and character of the pelage, it does not differ from the ordinary black specimens.

N°5

Plate XXIV

Drawn on Stone by R. Trembly

Four striped Ground Squirrel.

1 Male, 2 Female, 3 & 4 Young.

Drawn from Nature by J. J. Audubon F.R.S., F.L.S.

Printed by Nagel & Weingærtner NY

TAMIAS QUADRIVITTATUS.—Say.

Four-Striped Ground-Squirrel.

PLATE XXV.—Male, Female, and Young.

T. striis quinque sub nigris longitudinalibus, cum quatuor sub albidis dorso alternatum distributis; corpore magnitudine T. Lysteri minore; lateribus rufo fuscis, ventre albo.

CHARACTERS.

Smaller than Tamias Lysteri; five dark brown stripes and four light-coloured stripes occupying the whole back; sides, reddish-brown; underneath, white.

SYNONYMES.

Sciurus Quadrivittatus, Say, Long's Expedition, vol. ii., p. 349.
" " Griffith, Animal Kingdom, vol. v., No. 665.
" " Harlan, Fauna, p. 180.
" " Godman, vol. ii., p. 137.
Sciurus (Tamias) Quadrivittatus, Rich., Zool. Journ., No. 12, p. 519, April, 1828; Fauna Boreali Americana, p. 184, pl. 16.
Tamias Minimus, Bach., Journ. Acad. Nat. Sc. Phila., vol. viii., part 1, Young.

DESCRIPTION.

Head, of moderate size; nose, tapering, but not very sharp. The mouth recedes very much, (as in all the other species of Tamias;) cheek-pouches, of moderate size; whiskers, about the length of the head; eye, small; ears, erect, of moderate length, clothed on both surfaces with very short hairs; body, rather slender; fore-feet, with four toes and a small thumb, armed with an obtuse nail; palms, naked; claws, compressed, and curved like those of *Tamias Lysteri*. Hind-feet, with five slender toes; soles, covered with short hairs for three-fourths of their length; tail, long, narrow and sub-distichous.

COLOUR.

Forehead, dark-brown, with a few whitish hairs inter.

row black line from the nostril to the corner of the eye; above and beneath the eye, a line of white, which continues downward to the point of the nose.

A dark-brown dorsal line, commencing behind the ears, continues along the back to the insertion of the tail; another line, which is not quite so dark, begins at each shoulder and ends on the buttocks, near the tail; on each flank there is another shorter and broader line, which runs along the sides to near the haunches; on each side of the dorsal line there is a light-coloured stripe running down to near the insertion of the tail. The outer brown stripes are also separated by a line of yellowish-white; thus the whole back is covered by five dark and four pale lines. From the neck a broad line of reddish-brown extends along the sides, terminating at the hips; feet, light yellowish-brown; under surface of the body, and inner surface of the legs, grayish-white.

The tail, which is slightly distichous, is composed of hairs yellowish-brown at the roots, then dark-brown, and tipped with reddish-brown; on its under surface they are reddish-brown, then black for a narrow space, and reddish-brown at the tips.

DIMENSIONS.

A fine Male (killed Aug. 19th, 1843, on the Upper Missouri river.)

Nose to anterior canthus - - - - -	$\frac{1}{2}$ inch.
Nose to opening of ear - - - - -	$1\frac{1}{8}$ do.
Height of ear - - - - - - - -	$\frac{1}{2}$ do.
Width of ear - - - - - - - -	$\frac{7}{16}$ do.
Between centre of eyes - - - - -	$\frac{5}{8}$ do.
Length of head and body - - - - -	$4\frac{3}{8}$ do.
Tail (vertebræ) - - - - - - - -	$3\frac{1}{4}$ do.
Tail to end of hair - - - - - - -	$4\frac{1}{4}$ do.
Heel to end of hind-claws - - - - -	$1\frac{1}{16}$ do.
Palm and fore-feet to claws - - - - -	$1\frac{1}{16}$ do.

Weight 4 oz.

HABITS.

This pretty little species was discovered by Mr. Say, during Colonel Long's expedition. Mr. Say does not however appear to have seen much of its habits, and gives us but the following short account of them:—

"It does not seem to ascend trees by choice, but nestles in holes, and on the edges of rocks. We did not observe it to have cheek-pouches. Its nest is composed of a most extraordinary quantity of the burrs of the cactus, and their branches, and other portions of the large upright cactus, and small branches of pine trees and other vegetable productions, sufficient in some instances to fill an ordinary cart. What the object of so great and apparently so superfluous an assemblage of rubbish may be we are at a loss to conjecture; nor do we know what peculiarly dangerous enemy it may be intended to exclude by so much labour. Their principal food, at least at this season, is the seeds of the pine, which they readily extract from the cones."

We met with this species as we were descending the Upper Missouri river in 1843; we saw it first on a tree; afterwards we procured both old and young, among the sandy gulleys and clay cliffs on the sides of the ravines near one of our encampments.

These Ground Squirrels ascend trees when at hand and offering them either shelter or food, and seem to be quite as agile as the common species *Tamias Lysteri*.

Dr. Richardson, who found this Ground Squirrel during his long and laborious journeyings across our great continent, says of it—"It is an exceedingly active little animal, and very industrious in storing up provisions, being very generally observed with its pouches full of the seeds of leguminous plants, bents and grasses. It is most common in dry sandy spots, where there is much underwood, and is often seen in the summer, among the branches of willows and low bushes. It is a lively restless animal, troublesome to the hunter, and often provoking him to destroy it, by the angry chirruping noise that it makes on his approach, and which is a signal of alarm to the other inhabitants of the forest. During winter it resides in a burrow with several openings, made at the roots of a tree; and is even seen on the surface of the snow. At this season, when the snow disappears, many small collections of hazel-nut shells, from which the kernel has been extracted by a minute hole gnawed in the side, are to be seen on the ground near its holes."

Dr. Richardson further informs us that on the banks of the Saskatchawan, the mouths of the burrows of this species are not protected with heaps of vegetable substances, as described by Mr. Say, and we have no doubt the animal adapts its nest (as many of our birds do) to the locality and circumstances that surround it.

These animals bite severely when captured, and probably resemble *Tamias Lysteri* in their general habits and mode of living.

GEOGRAPHICAL DISTRIBUTION.

This species was originally discovered by Say, who procured it on the Rocky Mountains, near the sources of the Arkansas and Platte rivers.

We obtained it on the Upper Missouri, and Mr. DRUMMOND brought specimens from the sources of the Pearl river. It is found as far north as Lake Winnipeg, in lat. 50°.

GENERAL REMARKS.

When we published *Tamias minimus*, we had some misgivings lest it might prove the young of the present species. The discoverer however assured us that the two species did not exist within many hundred miles of each other, and that the specimens he sent us were those of full grown animals; we consequently ventured on their publication. Having, however, since procured young specimens of *T. quadrivittatus*, we are satisfied of the error we committed, and hasten to correct it. In the investigation of species existing in distant and little known portions of country, it always requires a length of time to settle them beyond the danger of error. The traveller who makes these investigations very hastily, and seizes on a specimen wherever there is a moment's pause in the journey, is often himself deceived, and the describer, having perhaps only a single specimen, is very apt to fall into some mistake. The investigation of described species in every branch of natural history, both in Europe and America, occupied much of the time of the naturalists of our generation, who corrected many of the errors of a former age; most fortunate are they who are permitted to live to correct their own.

No. 5 Plate XXV

Downy Squirrel.

Drawn from Nature by J.J. Audubon F.R.S. F.L.S.

Printed by Nagel & Weingaertner

SCIURUS LANUGINOSUS.—Bachman.

Downy Squirrel.

PLATE XXV.

S. auribus brevibus, cauda subdisticha; S. Hudsonico paullo robustior, supra castaneo-fuscus, subtus albus, naso concolori; lateribus argenteis; occipite maculo distincto.

CHARACTERS.

Ears, short; tail, sub-distichous; light chesnut-brown on the upper surface; sides, silver-gray. A spot on the hind part of the head, nose, and under surface of body, pure white. A little stouter than S. Hudsonius.

SYNONYME.

Sciurus Lanuginosus, Bach., Jour. Acad. Nat. Sc. of Phila., vol. viii., pt. 1, p. 67, 1838.

DESCRIPTION.

Head, broader than in *S. Hudsonius;* forehead, much arched; ears, short and oval; whiskers, longer than the head; feet and toes, short; thumb, armed with a broad flat nail. Nails, compressed and acute; the third, on the fore-feet, longest.

The tail, (which bears some resemblance to that of the flying squirrel, *P. volucella,*) is clothed with hairs a little coarser than those on the back, and is much shorter than the body. On the fore-feet the palms are nearly naked, the under surface of the toes being only partially covered with hair; but on the hind-feet, the under surface from the heel to the extremity of the nails is thickly covered with soft short hairs. Fur, softer and more downy than that of any other of our species. The fur indicates that the animal is an inhabitant of a cold region.

COLOUR.

Teeth, dark orange; whiskers, brown; fur on the back from the roots to near the tip of the hair, light plumbeous, tipped with light chesnut-brown; on the sides tipped with silver-gray. A broad line of white around the eyes, a spot of white on the hind part of the head, a little in advance of the anterior portion of the ears; nose, white, which colour extends along the forehead over the eyes, where it is gradually blended with the colour of the back; the whole under surface, feet, and inner surface of the legs, pure white. Tail, irregularly covered with markings of black, light brown, and white, scarcely two hairs being uniform in colour.

In general it may be said that the tail, when examined without reference to its separate hairs, is light-ash at the roots of the hairs, a broad but not well defined line of light rufous succeeding, then a dark brown space in the hairs, which are tipped with rufous and gray.

DIMENSIONS.

	Inches.	Lines.
Length of head and body - - - - - -	7	11
" tail (vertebræ) - - - - - -	4	8
" tail, including fur - - - - -	6	0
Palm, and middle fore-claw - - - - -	1	0
Sole and middle hind-claw - - - - -	1	9
Length of fur on the back - - - - -	0	7
Height of ear, measured posteriorly - - -	0	5
Distance between the orbits - - - - -	0	6

HABITS.

This downy and beautifully furred squirrel exists in the north-western portions of our continent. The specimen from which our drawing was made, is the only one which we have seen, and was brought from near Sitka, by Mr. J. K. Townsend, who kindly placed it in our hands, in order that we might describe it. As the animal was presented to Mr. Townsend by an officer attached to the Hudson's Bay Company, and was not observed by him, he could give us no account of its habits. We think, however, that from its close approximation to that group of squirrels of which the Hudson's Bay, or chickaree squirrel, is the type, and with which we are familiar, we can form a pretty correct judgment in regard to its general characteristics, and we will venture to say that it is less agile and less expert in climbing than the chickaree; it no doubt burrows in the earth in winter like the latter species, and as its tail is more like that of a spermophile than the tail of a squirrel, although the rest of its specific characters are those of the true squirrels, we are disposed to consider it a closely connecting link between these two genera, and it very probably, according to circumstances, adopts the mode of life commonly observed in each.

GEOGRAPHICAL DISTRIBUTION.

This species is found several degrees to the north of the Columbia river, and is said to extend through the country adjoining the sea-coast as far as into the Russian settlements. Mr. TOWNSEND says, "It was killed on the coast near Sitka, and given me by my friend, W. F. TOLMIE, Esq., Surgeon of the Honourable Hudson's Bay Company."

GENUS GULO.—STORR.

DENTAL FORMULA.

$$Incisive\ \frac{6}{6};\ Canine\ \frac{1-1}{1-1};\ Molar\ \frac{5-5}{6-6} = 38.$$

The three first molars in the upper, and the four first in the lower jaw, small; succeeded by a larger carnivorous or trenchant tooth, and a small tuberculous tooth at the back.

In the upper jaw the three first molars are uni-cuspidateous, and may be called false-carnivorous teeth, increasing successively in size; the following or carnivorous tooth is large and strong, furnished with two points on the inner side, and a trenchant edge in front; the last tooth is small, and tuberculous or flattish.

In the lower jaw the first four molars are false, each presenting only one point or edge; the fifth is long and large, with two trenchant points; the last molar is nearly flat. All the teeth touch each other successively. (Cuv.)

Head, of moderate length; body, long; legs, short; tail, bushy; feet, with five deeply divided toes, terminated by long curved nails.

No glandular pouch in some of the species, but a simple fold beneath the tail.

Habits, carnivorous and nocturnal.

The generic name is derived from the Latin *gulo*, a glutton.

Four species of this genus have been described; one existing in the Arctic regions of both continents, two in South America, and one in Africa.

GULO LUSCUS.—LINN.

THE WOLVERENE, OR GLUTTON.

PLATE XXVI.

G. subniger; fasciâ subalbida utrinque a humero per ilia producta, fasciis supra coxas se jungentibus; caudâ pilis longis hirsutâ.

Nº 6

Plate XXVI

Drawn on stone by R. Trembly

Wolverene.

Drawn from Nature by J. J. Audubon, F.R.S. F.L.S.

Printed & Col^d by J. T. Bowen, Philad^a

CHARACTERS.

Dark-brown, passing into black, above; a pale band on each side, running from the shoulders around the flanks, and uniting on the hips; tail, with long bushy hairs.

SYNONYMES.

MUSTELA GULO, Linn., Syst. Nat., 12th edit.
URSUS LUSCUS, Linn., Syst., Nat., 12th edit.
URSUS GULO, Pallas, do., Schreber, Säugeth., p. 525.
" " F. Cuv., in Dict. des Sc. Nat., 19th edit., p. 79, c. fig.
QUICKHATCH or WOLVERINE, Ellis, Voy. Hudson's Bay, p. 42.
URSUS FRETI HUDSONIS, Briss, Quad., p. 188.
WOLVERING, Cartwright's Journal, vol. ii., p. 407.
WOLVERINE, Pennant's Hist. Quad., vol. ii., p. 8, t. 8, Hearne's Journey, p. 372.
GULO ARCTICUS, var. A. GLOUTON WOLVERINE., Desm., Mamm., p. 174.
GULO LUSCUS, (Capt.) Sabine, supp. Parry's 1st Voyage, p. 184.
" " Sabine, (Mr.) Franklin's 1st Journey, p. 650.
" " Richardson's Appendix Parry's 2d Voyage, p. 292.
" " Fischer's Mammalium, p. 154.
THE GLUTTON, Buffon, vol. vii., p. 274, pl. 243.
URSUS GULO, Shaw's Gen. Zool., vol. i., p. 46.
GULO VULGARIS, Griffith's Animal Kingdom, sp. 331.
GULO WOLVERINE, Griffith's Animal Kingdom, sp. 332.
GULO LUSCUS, Rich., F. B. A., p. 41.
" " Capt. Ross, Expedition, p. 8.
CARCAJOU, French Canadians; QUICKHATCH, English residents.

DESCRIPTION.

Head, of moderate size, broad on the hinder part, much arched, rounded on all sides; nose, obtuse, naked; eyes small; ears, short, broad, rounded, and partially hidden by the surrounding fur. The whole head bears a strong resemblance to that of some of the varieties of the dog.

Body, very long, stout, and compactly made; back, arched; the whole form indicating strength without much activity. The Wolverene is covered with a very thick coat of two distinct kinds of hair. The inner fur, soft and short, scarcely an inch long; the intermixed hairs, numerous, rigid, smooth, and four inches long; giving the animal the appearance of some shaggy dog.

Legs, short and stout; feet, broad, clothed on the under surface with a compact mass of woolly hair. Toes, distinct, and armed with five strong, rounded, and pretty sharp claws. The tracks made in the snow by this species are large, and not very unlike those of the bear. There are five tubercles on the soles of the fore-feet, and four on the hind-feet; no tubercle on the heel.

The tail is rather short, hangs low, and is covered with pendulous hairs. "There are two secretory organs about the size of a walnut, from which it discharges a fluid of a yellowish-brown colour and of the consistence of honey, by the rectum, when hard pressed by its enemies."—Ross.

COLOUR.

Under fur, deep chesnut-brown, a shade lighter near the roots; the longer hairs are blackish-brown throughout their whole length, the hair having very much the appearance of that of the bear. Eyes, nose, and whiskers, black; a pale reddish-brown band commences behind the shoulder, and running along the flanks, turns up on the hip, and unites on the rump with similar markings on the opposite side. There is a brownish-white band across the forehead running from ear to ear. On the sides of the neck there are tufts of white hair extending nearly in a circle from the inside of the legs around the chest. Legs and tail, brownish-black; claws, dark-brown. The colour varies greatly in different specimens, and although there is a strong general resemblance among all we have examined, we are not surprised that attempts have been made from these varieties to multiply the species. There are however no permanent varieties among the many specimens we have examined. The peculiar lateral band, although it exists in all, differs a few shades in colour. In some specimens it is of a chesnut colour, in others light ferruginous, and in a few cases ash-coloured. We find these differences of colour existing in both continents, and not confined to either. We have never seen a specimen of a Wolverene as light in colour as that to which Linnæus gave the specific name of *luscus*, and we regard it as a mere accidental variety. We have found American specimens obtained in the Polar regions fully as black as those from Russia.

DIMENSIONS.

Recent specimen, obtained in Rensselaer county, N. Y.

	Feet.	Inches.	Lines.
From point of nose to root of tail	2	9	0
Tail (vertebræ)	0	8	0
Height to shoulder	1	0	0
" of ear, posteriorly	0	1	6
Length of hair on body	0	4	0
From heels to point of nails	0	5	0
Breadth of hind-foot	0	4	7

Specimen from which our figure was made.

	Feet.	Inches.	Lines.
From point of nose to root of tail - -	2	6	0
Tail (vertebræ) - - - - - -	0	6	0
Tail, including fur - - - - - -	0	10	0
Height of ear - - - - - - -	0	1	4

HABITS.

The Wolverene, or Glutton as he is generally called, is one of the animals whose history comes down to us blended with the superstitions of the old writers. Errors when once received and published, especially if they possess the charm of great singularity or are connected with tales of wonder, become fastened on the mind by early reading and the impressions formed in youth, until we are familiarized with their extravagance, and we at length regret to find ideas (however incorrect) adopted in early life, not realized by the sober inquiries and investigations of maturer years.

The Wolverene, confined almost exclusively to Polar regions, where men have enjoyed few advantages of education and hence have imbibed without much reflection the errors, extravagances and inventions of hunters and trappers, has been represented as an animal possessing extraordinary strength, agility, and cunning, and as being proverbially one of the greatest gormandizers among the "brutes." Olaus Magnus tells us that "it is wont when it has found the carcass of some large beast to eat until its belly is distended like a drum, when it rids itself of its load by squeezing its body betwixt two trees growing near together, and again returning to its repast, soon requires to have recourse to the same means of relief." It is even said to throw down the moss which the reindeer is fond of, and that the Arctic fox is its jackal or provider. Buffon, in his first description of this animal, seems to have adopted the errors and superstitions of Olaus Magnus, Schoeffer, Gesner, and the early travellers into Sweden and Lapland. He says of this animal, (vol. vii., p. 277,) "the defect of nimbleness he supplies with cunning: he lies in wait for animals as they pass, he climbs upon trees in order to dart upon his prey and seize it with advantage; he throws himself down upon elks and reindeer, and fixes so firmly on their bodies with his claws and teeth that nothing can remove him. In vain do the poor victims fly and rub themselves against trees; the enemy, attached to the crupper or neck, continues to suck their blood, to enlarge the wound, and to devour them gradually and with equal voracity, till they fall down."

"More insatiable and rapacious than the wolf, if endowed with equal agility the Glutton would destroy all the other animals; but he moves so heavily that the only animal he is able to overtake in the course is the beaver, whose cabins he sometimes attacks, and devours the whole unless they quickly take to the water, for the beaver outstrips him in swimming. When he perceives that his prey has escaped, he seizes the fishes; and when he can find no living creature to destroy, he goes in quest of the dead, whom he digs up from their graves and devours with avidity."

Even the intelligent GMELIN, who revised and made considerable additions to the great work of LINNÆUS, on a visit to the North of Europe imbibed many of the notions of the Siberian hunters, and informs us, without however giving full credence to the account, that the Wolverene "watches large animals like a robber, or surprises them when asleep," that "he prefers the reindeer," and that "after having darted down from a tree like an arrow upon the animal, he sinks his teeth into its body and gnaws the flesh till it expires; after which he devours it at his ease, and swallows both the hair and skin."

However, although BUFFON in his earlier history of the species adopted and published the errors of previous writers, he subsequently corrected them and gave in a supplementary chapter not only a tolerable figure but a true history. He received a Wolverene alive from the northern part of Russia, and preserved it for more than eighteen months at Paris. And when the Count was thus enabled to examine into its habits, as they were developed from day to day, he found them of a very ordinary character, and it was discovered to be an animal possessing no very striking peculiarities. He informs us, "He was so tame that he discovered no ferocity and did not injure any person. His voracity has been as much exaggerated as his cruelty: he indeed ate a great deal, but when deprived of food he was not importunate."

"The animal is pretty mild; he avoids water, and dreads horses and men dressed in black. He moves by a kind of leap, and eats pretty voraciously. After taking a full meal he covers himself in the cage with straw. When drinking he laps like a dog. He utters no cry. After drinking, he throws the remainder of the water on his belly with his paws. He is almost perpetually in motion. If allowed, he would devour more than four pounds of flesh in a day; he eats no bread, and devours his food so voraciously, and almost without chewing, that he is apt to choke himself."

We have seen this species in a state of confinement in Europe; the specimens came, we were informed, from the north of that continent. In Denmark, a keeper of a small caravan of animals allowed us the privilege of examining a Wolverene which he had exhibited for two years.

We took him out of his cage; he was very gentle, opened his mouth to enable us to examine his teeth, and buried his head in our lap whilst we admired his long claws and felt his woolly feet; he seemed pleased to escape from the confinement of the cage, ran round us in short circles, and made awkward attempts to play with and caress us, which reminded us very much of the habit of the American black bear. He had been taught to sit on his haunches and hold in his mouth a German pipe. We observed he was somewhat averse to the light of the sun, keeping his eyes half closed when exposed to its rays. The keeper informed us that he suffered a good deal from the heat in warm weather, that he drank water freely and ate meat voraciously, but consumed more in winter than in summer. There was in the same cage a marmot from the Alps, (*Arctomys marmota,*) to which the Wolverene seemed much attached. When returned to his cage, he rolled himself up like a ball, his long shaggy hairs so completely covering his limbs that he presented the appearance of a bear-skin rolled up into a bundle.

In the United States the Wolverene has always existed very sparingly, and only in the Northern districts. About thirty-five years ago, we saw in the possession of a country merchant in Lansingburg, New-York, three skins of this species, that had as we were informed been obtained on the Green Mountains of Vermont; about the same time we obtained a specimen in Rensselaer county, near the banks of the Hoosack river. While hunting the Northern hare, immediately after a heavy fall of snow, we unexpectedly came upon the track of an animal which at the time we supposed to be that of a bear, a species which even then was scarcely known in that portion of the country, (which was already pretty thickly settled.) We followed the broad trail over the hills and through the devious windings of the forest for about five miles, till within sight of a ledge of rocks on the banks of the Hoosack river, when, as we found the night approaching, we were reluctantly compelled to give up the pursuit for that day, intending to resume it on the following morning. It snowed incessantly for two days afterwards, and believing that the bear had retired to his winter retreat, we concluded that the chance of adding it to our collection had passed by. Some weeks afterwards a favourite servant who was always anxious to aid us in our pursuits, and who not only knew many quadrupeds and birds, but was acquainted with many of their habits, informed us that he had on a previous day seen several tracks similar to those we had described, crossing a new road cut through the forest. As early on the following morning as we could see a track in the snow, we were fully accoutred, and with a gun and a pair of choice hounds, started on what we conceived our second bear-hunt. Before reaching the spot where the tracks had been observed, however, we met a fresh trail of the previous night, and pursued it without loss of time. The animal had joined some foxes which were feeding on a dead horse not a hundred yards from a log cabin in the forest, and after having satiated itself with this delicate food, made directly for the Hoosack river, pursuing the same course along which we had formerly traced it. To our surprise it did not cross the river, now firmly bound with ice, but retired to its burrow, which was not far from the place where we had a few weeks before abandoned the pursuit of it. The hounds had not once broke into full cry upon the track, but no sooner had they arrived at the mouth of the burrow than they rushed into the large opening between the rocks, and commenced a furious attack on the animal within. This lasted but for a few moments, and they came out as quickly as they had entered. They showed some evidence of having been exposed to sharp claws and teeth, and although they had been only a moment engaged in battle, had no disposition to renew it. No effort of ours could induce them to re-enter the cavern, whilst their furious barking at the mouth of the hole was answered by a growl from within. The animal, although not ten feet from the entrance, could not be easily reached with a stick on account of his having retreated behind an angle in the chasm. As we felt no particular disposition to imitate the exploits of Colonel Putnam in his rencontre with the wolf, we reluctantly concluded to trudge homeward through the snow, a distance of five miles, to obtain assistance. On taking another survey of the place, however, we conceived it possible to effect an opening on one of its sides. This was after great labour accomplished by prying away some heavy fragments of the rock. The animal could now be reached with a pole, and seemed very much irritated, growling and snapping at the stick, which he once succeeded in tearing from our hand, all the while emitting a strong and very offensive musky smell. He was finally shot. What was our surprise and pleasure on discovering that we had, not a bear, but what was more valuable to us, a new species of quadruped, as we believed it to be. It was six months before we were enabled, by consulting a copy of Buffon, to discover our mistake and ascertain that our highly prized specimen was the Glutton, of which we had read such marvellous tales in the school-books.

In some of the figures that we have seen of the Wolverene, or Glutton, he is represented as touching the ground to the full extent of his heel, and in several of the descriptions this habit is also assigned to him. Our notes in reference to this point were made in early life, and it is possible that we may have laboured under a mistake; but we are confident, from our own observation, that the animal treads upon its hind-feet in the manner of the dog, that the impression of the tarsus or heel can only be observed in deep snow, and that in its ordinary walk on the ground the heel seldom touches the earth. We made no note in regard to the living Wolverene we saw in Europe, but are under an impression that its method of walking was similar to that stated above. There is another peculiarity in the tracks of the animal: in walking, the feet do not cross or approach each other in the manner of the feet of a fox or wolf, but make a double track in the snow, similar in this respect to that of the skunk.

There was a large nest of dried leaves in the cavern, which had evidently been a place of resort for the Wolverene we have been speaking of, during the whole winter, as its tracks from every direction led to the spot. It had laid up no winter store, and evidently depended on its nightly excursions for a supply of food. It had however fared well, for it was very fat.

It has been asserted that the Wolverene is a great destroyer of beavers: but we are inclined to think that this can scarcely be the case, unless it be in summer, when the beaver is often found some distance from the water. In such cases we presume that the Wolverene, although not swift of foot, could easily overtake that aquatic animal. But, should he in winter attempt to break open the frozen mud-walls of the beaver-huts, which would be a very difficult task, this would only have the effect of driving the occupants into their natural element, the water, where their hungry pursuer could not follow them. The statement of his expertness in swimming, diving, and catching fish, we believe to be apocryphal.

We are inclined to adopt the views of RICHARDSON in regard to the Wolverene, that it feeds chiefly on the carcasses of beasts that have been killed by accident. "It also devours meadow-mice, marmots, and other rodentia, and occasionally destroys disabled quadrupeds of a larger size."

That it seizes on deer or large game by pouncing on them, is incredible; it neither possesses the agility nor the strength to accomplish this feat. This habit has also been ascribed to the Canada lynx as well as to the Bay lynx; we do not think it applies to either. That the Wolverene occasionally captures the grouse that have plunged into the fresh snow as a protection from the cold, is probable.

RICHARDSON observes that he saw one chasing an American hare, which was at the same time harassed by a snowy owl. The speed of the hare however is such that it has not much to fear from the persevering but slow progress of the Wolverene; and the one seen by RICHARDSON, in his efforts to catch the tempting game must have been prompted by a longing desire after hare's flesh, rather than by any confidence in his ability to overtake the animal.

All Northern travellers and writers on the natural history of the Arctic regions, ELLIS, PENNANT, HEARNE, PARRY, FRANKLIN, RICHARDSON, &c., speak of the indomitable perseverance of the Wolverene in following the footsteps of the trappers, in order to obtain the bait, or take from the traps the Arctic fox, the marten, beaver, or any other animal that may be caught in them. They demolish the houses built around the dead-falls, in order to obtain the bait, and tear up the captured animals apparently from a spirit of wanton destructiveness. HEARNE (p. 373) gives an account of their amazing strength, one of them having overset the greatest part of a large pile of wood, measuring upwards of seventy yards round, to get at some provisions that had been hid there. He saw another take possession of a deer that an Indian had killed, and though the Indian advanced within twenty yards he would not relinquish his claims to it, but suffered himself to be shot, standing on the deer. HEARNE farther states, "they commit vast depredations on the foxes during the summer, while the young ones are small; their quick scent directs them to their den, and if the entrance be too small, their strength enables them to widen it, and go in and kill the mother and all her cubs; in fact they are the most destructive animals in this country."

Capt. J. C. ROSS, R. N., F. R. S., who gave an interesting account of the animals seen in the memorable expedition of Sir JOHN ROSS, relates the following anecdote of this species:—"In the middle of winter, two or three months before we abandoned the ship, we were one day surprised by a visit from a Wolverene, which, hard pressed by hunger, had climbed the snow wall that surrounded our vessel, and came boldly on deck where our crew were walking for exercise. Undismayed at the presence of twelve or fourteen men, he seized upon a canister that had some meat in it, and was in so ravenous a state that whilst busily engaged at his feast he suffered me to pass a noose over his head, by which he was immediately secured and strangled."

The Wolverene is at all times very suspicious of traps, and is seldom taken in the log-traps set for the marten and Arctic fox; the usual mode in which it is obtained is by steel-traps, which must be set with great caution and concealed with much art.

Captain CARTWRIGHT in his journal speaks of having caught all he obtained at Labrador in this manner, and we have seen several skins giving evidence that the animals had been taken by the foot.

Captain Cartwright (see Journal, vol. ii., p. 407) records an instance of strength and cunning in this species that we cannot pass by in giving its history; we will use his own words. "In coming to the foot of Table Hill I crossed the track of a Wolvering with one of Mr. Callingham's traps on his foot; the foxes had followed his bleeding track. As this beast went through the thick of the woods, under the north side of the hill, where the snow was so deep and light that it was with the greatest difficulty I could follow him even in Indian rackets, I was quite puzzled to know how he had contrived to prevent the trap from catching hold of the branches of the trees or sinking in the snow. But on coming up with him I discovered how he had managed; for after making an attempt to fly at me, he took the trap in his mouth and ran upon three legs. These creatures are surprisingly strong in proportion to their size; this weighed only twenty-six pounds and the trap eight; yet including all the turns he had taken he had carried it six miles."

The Wolverene produces young but once a year, from two to four at a litter. Richardson says the cubs are covered with a downy fur of a pale or cream colour. The fur of the Wolverene resembling that of the bear, is much used for muffs, and when several skins are sewed together makes a beautiful sleigh-robe.

GEOGRAPHICAL DISTRIBUTION.

The Wolverene exists in the north of both continents. On the Eastern continent it inhabits the most northern parts of Europe and Asia, occurring in Sweden, Norway, Lapland, and Siberia, as well as in some of the Alpine regions, and in the forests of Poland and Courland. In North America it is found throughout the whole of the Arctic circle. They were caught to the number of ten or twelve every winter by Capt. Cartwright in Labrador. It exists at Davis' Straits, and has been traced across the continent to the shores of the Pacific. It is found on the Russian islands of Alaska. Richardson remarks, "It even visits the islands of the Polar sea, its bones having been found in Melville Island, nearly in latitude 75°. It occurs in Canada, although diminishing in numbers the farther we proceed southerly. We have seen specimens procured at Newfoundland, and have heard of its existence, although very sparingly, in Maine. Professor Emmons, (Dekay, Nat. Hist. of New-York,) states that it still exists in the Hoosack Mountains of Massachusetts. We examined a specimen obtained in Jefferson county, near Sackett's Harbour, N. Y., in 1827, and in 1810 we obtained a specimen in Rensselaer county, latitude 42° 46′; we have never heard of its existence farther south.

GENERAL REMARKS.

This species has been arranged by different authors under several genera. LINNÆUS placed it under both MUSTELLA and URSUS. STORR established for it the genus GULO, which was formed from the specific name, as it had been called *Ursus Gulo*, by LINNÆUS. STORR's generic name has been since adopted by CUVIER and other modern naturalists. GRAY named it GRISONIA. LINNÆUS is notwithstanding entitled to the specific name, although this is the result of an error into which he was led in this manner: EDWARDS had made a figure from a living specimen imported from America. It was a strongly marked variety, with much white on its forehead, sides, and neck. LINNÆUS regarding it as a new species, described it as such. In seeking for some name by which to designate it, he observed that it had lost one eye, and it is supposed applied the trivial name "luscus," one-eyed, to the animal, merely on account of the above accidental blemish.

The vulgar names Glutton, Carcajou, &c., have given rise to much confusion in regard to the habits of the species.

The name Glutton induced many ancient authors to ascribe to it an appetite of extravagant voraciousness.

Carcajou appears to be some Indian name adopted by the French, and this name has evidently been applied to different species of animals. CHARLEVOIX, in his Voyage to America, vol. i., p. 201, speaks of the "carcajou or quincajou, a kind of cat, with a tail so long that he twists it several times round his body, and with a skin of a brownish red." He then refers to his climbing a tree, where after two foxes have driven the elk under the tree, the cat being on the watch pounces on it in the manner ascribed to the Wolverene. Here he evidently alludes to the cougar, as his long tail and colour apply to no other animal in our country. LAWSON refers the same singular habit to the wild cat of Carolina; he says, (p. 118,) "the wild cat takes most of his prey by surprise, getting up the trees which they pass by or under, and thence leaping directly upon them. Thus he takes deer, which he cannot catch by running, and fastens his teeth into their shoulders. They run with him till they fall down for want of strength and become a prey to the enemy."

In the last work published on American Quadrupeds, LAWSON is quoted as authority for the former existence of the Wolverene in Carolina, and a reference is also made to a plate of that species. On looking over the work of LAWSON, (London, 1709,) we find that no mention is made of the Wolverene, and no plate of the animal is given. We have supposed it possible that the author of the "Natural History of New-York" might have intended to refer to CATESBY; but the latter gave no plate of the species, and only noticed it as existing in the very northern parts of America. We feel confident that the geographical range of the Wolverene has never extended to Carolina, that it existed only as a straggler in the northern portion of the Middle States, and that it is now, and ever has been, almost entirely confined to the Northern regions.

SCIURUS LANIGERUS.—Aud. and Bach.

WOOLLY SQUIRREL.

PLATE XXVII.

Sc. migratorii magnitudine; pilis longis et lanosis; cauda ampla, villosa vixque disticha; naso, auriculis, pedibusque pene nigris; vellere supra ex cinereo fusco; subtus dilute fusco.

CHARACTERS.

Size of Sciurus migratorius; hair, long and woolly; tail, large and bushy; nose, ears, and feet, nearly black; upper surface, grizzly dark gray and brown; under parts, pale brown.

SYNONYME.

SCIURUS LANIGERUS, Aud. and Bach., Journal of the Acad. Nat. Sc., Philad., 1841, p. 100

DESCRIPTION.

Head, short; forehead, arched; nose, blunt; clothed with soft hair; whiskers, longer than the head; eyes, large; ears, large, broad at base, ovate.

Body, stout, covered with long and woolly hairs, which are much longer and a little coarser than those of the Northern gray squirrel.

Legs, stout; feet, of moderate size; claws, strong, compressed, arched and sharp. The third toe, longest; a blunt nail in place of a thumb. Palms, naked; toes, hairy to the extremity of the nails.

Tail, long and bushy, and the hairs long and coarse.

COLOUR.

Incisors, dark orange on the outer surface; the head, both on the upper and lower surface, as far as the neck, the ears, whiskers, fore-legs to the shoulder, feet, and inner surface of hind-legs, black; with a few yellowish-brown hairs intermixed. The long fur on the back is for half its

Drawn on stone by R. Trembly

Long Haired Squirrel.

Drawn from Nature by J. J. Audubon F.R.S. F.L.S. Printed & Col^d by J. T. Bowen, Philad.

length from the roots, light plumbeous, then has a line of light-brown, and is tipped with reddish-brown and black.

The hairs on the tail, in which the annulations are very obscure, are for one-third of their length brownish-black, then light-brown, then brownish-black, and are tipped with ashy-white. On the under surface the hairs, which are short, are at the base light-plumbeous, tipped with light-brown and black; the throat is light grayish-brown.

Of two specimens received from the same locality, the head of one is lighter-coloured than that of the other, having a shade of yellowish-brown; in other respects they are precisely similar; a figure of each is given on the plate.

DIMENSIONS.

Length of head and body - - - -	$11\frac{7}{8}$	inches.
Tail (vertebræ) - - - - - - -	10	do.
Tail, to end of fur - - - - -	12	do.
Height of ear posteriorly - - - -	$0\frac{3}{4}$	do.
Breadth of ear - - - - - - -	$0\frac{3}{4}$	do.
From heel to end of middle claw - - -	$2\frac{1}{2}$	do.
Hairs on the back - - - - - - -	$1\frac{1}{8}$	do.

HABITS.

We have been unable to obtain any information in regard to the habits of this species. Its form, however, indicates that it is a climber, like all the species of the genus, living in forests, feeding on nuts and seeds. Its long woolly coat proves its adaptation to cold regions.

GEOGRAPHICAL DISTRIBUTION.

Our specimens were procured from the northern and mountainous portions of California.

GENERAL REMARKS.

The difficulty in finding characters by which the various species of this genus can be distinguished, is very great. There is, however, no variety of any other species of squirrel that can be compared with that here described. Its black head and legs, brown back and belly, its broad ears and long woolly hair, are markings by which it may be easily distinguished from all others.

PTEROMYS VOLUCELLA.—GMEL.

COMMON FLYING-SQUIRREL.

PLATE XXVIII.—MALES, FEMALES, AND YOUNG.

Pt. Tamias Lysteri magnitudine, supra ex fusco-cinereo et albido, infra ex albo.

CHARACTERS.

Size of Tamias Lysteri; above, brownish-ash tinged with cream colour; beneath, white.

SYNONYMES.

ASSAPANICK, Smith's Virginia, p. 27, 1624.
SCIURUS AMERICANUS VOLANS, Ray, Syn. Quad.
FLYING SQUIRREL, Lawson's Carolina, p. 124.
LA PALATOUCHE, Buff., X., pl. 21.
SCIURUS VOLUCELLA, Pallas, Glires, p. 353, 359.
" " Schreber, Säugethiere, p. 808, 23, t. 222.
" " Gmelin, Linn., Syst. Nat., p. 155, 26.
SCIURUS VIRGINIANUS, Gmelin, Syst. Nat.
" " Shaw's Gen. Zool., vol. ii., p. 155. t. 150.
FLYING SQUIRREL, Catesby's Carolina, vol. ii., p. 76.
" " Pennant's Quadrupeds, p. 418, 283.
PTEROMYS VOLUCELLA, Desm., Mamm., p. 345, 554.
" " Harlan, p. 187.
" " Godman, vol. ii., p. 146.
" " Emmons, Report, p. 69.
" " Dekay, p. 65.

DESCRIPTION.

Head, short and rounded; nose, blunt; eyes, large and prominent; ears, broad and nearly naked; whiskers, numerous, longer than the head; neck, short; body, rather thicker than that of the chipping squirrel. The flying membrane is distended by an additional small bone of about half an inch in length, articulated with the wrist. The fur on the whole body is very fine, soft and silky; legs, rather slender; claws, feeble, compressed, acute, and covered with hair; tail, flat, distichous, rounded at the tip, and very thickly clothed with fine soft fur. Ten mammæ.

No. 6 Plate XXVIII

Drawn on Stone by R. Trembly

Common Flying Squirrel.

1. 2. Males, 3. 4 Females, 5 Young

Drawn from Nature by J.J. Audubon F.R.S. F.L.S. Printed by Nagel & Weingærtner N.Y.

COLOUR.

A line of black around the orbits of the eye; whiskers, nearly all black, a few are whitish toward their extremities. Ears, light-brown. In most specimens there is a light-coloured spot above the eyes; sides of the face and neck, light cream-colour; fur on the back, dark slate-colour, tipped with yellowish-brown. On the upper side of the flying membrane the colour gradually becomes browner till it reaches the lower edge, where it is of a light cream-colour; throat, neck, inner surface of legs, and all beneath, white; with occasionally a tint of cream-colour. The upper surface of the tail is of the colour of the back; tail, beneath, light fawn.

DIMENSIONS.

Length of head and body - - - - - -	$5\frac{1}{4}$	inches.
" head - - - - - - -	1	do.
" tail (vertebræ) - - - - -	4	do.
" tail, including fur - - - - -	5	do.

Of a specimen from which one of our figures was drawn.

From nose to eye - - - - - - -	$\frac{5}{8}$	inches.
" " opening of ear - - - - -	$1\frac{3}{4}$	do.
" " root of tail - - - - - -	$5\frac{1}{4}$	do.
Tail (vertebræ) - - - - - - -	$3\frac{3}{4}$	do.
Tail, to end of hair - - - - - - -	$4\frac{1}{2}$	do.
Breadth of tail, hair extended - - - - -	$1\frac{3}{8}$	do.
Spread of fore-legs to extremity of claws - -	$6\frac{7}{8}$	do.
Spread of hind-legs - - - - - - -	7	do.

HABITS.

It has sometimes been questioned whether the investigation of objects of natural history was calculated to improve the moral nature of man, and whether by an examination into the peculiar habits of the inferior animals he would derive information adapted to the wants of an immortal being, leading him from the contemplation of nature up to nature's God.

Leaving others to their own judgment on this subject, we can say for ourselves that on many occasions when studying the varied characters of the inferior creatures, we have felt that we were reading lessons taught us by nature, that were calculated to make us wiser and better. Often, whilst straying in the fields and woods with a book under our arm, have we been tempted to leave HOMER or ARISTOTLE unopened, and attend to the teachings of the quadrupeds and birds that people the solitudes of the wilderness. Even the gentle little Flying-Squirrel has more than once diverted our attention from the pages of GRIESBACH and MICHAELIS, and taught us lessons of contentment, of innocence, and of parental and filial affection, more impressive than the theological disquisitions of learned commentators.

We recollect a locality not many miles from Philadelphia, where, in order to study the habits of this interesting species, we occasionally strayed into a meadow containing here and there immense oak and beech trees. One afternoon we took our seat on a log in the vicinity to watch their lively motions. It was during the calm warm weather peculiar to the beginning of autumn. During the half hour before sunset nature seemed to be in a state of silence and repose. The birds had retired to the shelter of the forest. The night-hawk had already commenced his low evening flight, and here and there the common red bat was on the wing; still for some time not a Flying-Squirrel made its appearance. Suddenly, however, one emerged from its hole and ran up to the top of a tree; another soon followed, and ere long dozens came forth, and commenced their graceful flights from some upper branch to a lower bough. At times one would be seen darting from the topmost branches of a tall oak, and with wide-extended membranes and outspread tail gliding diagonally through the air, till it reached the foot of a tree about fifty yards off, when at the moment we expected to see it strike the earth, it suddenly turned upwards and alighted on the body of the tree. It would then run to the top and once more precipitate itself from the upper branches, and sail back again to the tree it had just left. Crowds of these little creatures joined in these sportive gambols; there could not have been less than two hundred. Scores of them would leave each tree at the same moment, and cross each other, gliding like spirits through the air, seeming to have no other object in view than to indulge a playful propensity. We watched and mused till the last shadows of day had disappeared, and darkness admonished us to leave the little triflers to their nocturnal enjoyments.

During the day this species avoids the light, its large eyes like those of the owl cannot encounter the glare of the sun; hence it appears to be a dull and uninteresting pet, crawling into your sleeve or pocket, and seeking any dark place of concealment. But twilight and darkness are its season for activity and pleasure. At such times, in walking through the woods you hear a rattling among the leaves and branches, and the falling acorns, chesnuts, and beech-nuts, give evidence that this little creature is supplying itself with its food above you.

This is a harmless and very gentle species, becoming tolerably tame in a few hours. After a few days it will take up its residence in some crevice in the chamber, or under the eaves of the house, and it or its progeny may be seen in the vicinity years afterwards. On one occasion we took from a hollow tree four young with their dam; she seemed quite willing to remain with them, and was conveyed home in the crown of a hat. We had no cage immediately at hand, and placed them in a drawer in our library, leaving a narrow space open to enable them to breathe; next morning we ascertained that the parent had escaped through the crevice, and as the window was open, we presumed that she had abandoned her young rather than be subject to confinement in such a narrow and uncomfortable prison. We made efforts for several days to preserve the young alive by feeding them on milk; they appeared indifferent about eating, and yet seemed to thrive and were in good order. A few evenings afterwards we were surprised and delighted to see the mother glide through the window and enter the still open drawer; in a moment she was nestled with her young. She had not forsaken them, but visited them nightly and preserved them alive by her attentions. We now placed the young in a box near the window, which was left partly open. In a short time she had gained more confidence and remained with them during the whole day. They became very gentle, and they and their descendants continued to reside on the premises for several years.

During the first winter they were confined to the room, boxes were placed in different parts of it containing Indian meal, acorns, nuts, &c. As soon as it was dark they were in the habit of hurrying from one part of the room to the other, and continued to be full of activity during the whole night. We had in the room a wheel that had formerly been attached to the cage of a Northern gray squirrel. To this they found an entrance, and they often continued during half the night turning the wheel; at times we saw the whole group in it at once. This squirrel, we may conclude, resorts to the wheel not from compulsion but for pleasure.

In an interesting communication which we have received from Gideon B. Smith, Esq., M. D., of Baltimore, he has given us the following details of the singular habits of this species:—

"After having arrived at the top of a tree from which they intend to make their airy leap, they spring or jump, stretch their fore-legs forward and outward and their hind-legs backward and outward, by this means expanding the loose skin with which they are clothed, and which forms a sort of gliding elevator. In this way they pass from tree to tree, or to any other object, not by flying as their name imports, but by descending from a high position by a gliding course; as they reach the vicinity of tne earth, their impetus, aided by their expanded skin, enables them to ascend in a curved line and alight upon the tree aimed at, about one-third as high from the ground as they were on the tree they left. On reaching a tree in this manner they run briskly up its trunk as high as they wish to give them a start for another; in this way they will travel in a few minutes, from tree to tree or object to object, a quarter of a mile or more. There is nothing resembling flying in their movements.

"They are gregarious, living together in considerable communities, and do not object to the company of other and even quite different animals. For example, I once assisted in taking down an old martin-box, which had been for a great number of years on the top of a venerable locust tree near my house, and which had some eight or ten apartments. As the box fell to the ground we were surprised to see great numbers of Flying-Squirrels, screech-owls, and leather-winged bats running from it. We caught several of each, and one of the Flying-Squirrels was kept as a pet in a cage for six months. The various apartments of the box were stored with hickory-nuts, chesnuts, acorns, corn, &c., intended for the winter supply of food. There must have been as many as twenty Flying-Squirrels in the box, as many bats, and we know there were six screech owls. The crevices of the house were always inhabited by the Squirrels. The docility of the one we kept as a pet was remarkable; although he was never lively and playful in the day-time, he would permit himself to be handled and spread out at the pleasure of any one. We frequently took him from the cage, laid him on the table or on one hand, and exposed the extension of his skin, smoothed his fur, put him in our pocket or bosom, &c., he pretending all the time to be asleep.

"It was a common occurrence that these Squirrels flew into the house on a summer's evening when the windows were open, and at such times we caught them. They were always perfectly harmless. Although I frequently seized them in my hand I was never bitten. We caught so many of them one season that the young girls bordered their winter capes with their tails, which are very pretty. It was a curious circumstance that the Flying-Squirrels never descended to the lower parts of the house, and we never knew of any rats in the upper rooms. Whether the Squirrels or the rats were the repulsive agents I do not know; certain it is they never inhabited the lower location in common."

The Flying-Squirrel, as is shown above, is gregarious. In Carolina, we have generally found six or seven in one nest; it is difficult, however. to count them, as on cutting down a tree which they inhabit, several escape without being noticed. In New Jersey, Pennsylvania, and Virginia, they appear to be more numerous, and the families are larger.

The Flying-Squirrels never build their nest of leaves on the trees during summer like the true squirrels, but confine themselves to a hollow, or some natural cavity in the branches or trunk. We have very frequently found them inhabiting the eaves and roofs of houses, and we discovered a considerable number of them in the crevices of a rock in the vicinity of the Red Sulphur Springs in Virginia.

Although the diet of this species generally consists of nuts and seeds of various kinds, together with the buds of trees in winter, yet we have known many instances in which it manifested a strong desire for animal food. On several occasions we found it caught in box-traps set for the ermine, which had been baited only with meat. The bait, (usually a blue jay,) was frequently wholly consumed by the little prisoner. In a room in which several Flying-Squirrels had been suffered to go at large, we one evening left a pine grosbeak, (*Corythus enucleator*,) a rare specimen, which we intended to preserve on the following morning. On searching for it however next day it was missing; we discovered its feet and feathers at last in the box of the Flying-Squirrels, they having consumed the whole body.

This species has from three to six young at a time. We have been assured by several persons that they produce young but once a year in the Northern and Middle States. In Carolina, however, we think they have two litters in a season, as we have on several occasions seen young in May and in September.

A writer in LOUDON's Magazine, under the signature of D. W. C., says at p. 571, vol. ix., in speaking of the habits of this animal in confinement in England, "I found that as soon as the female was pregnant she would not allow any one to approach her; and as the time went on, she became more savage and more tenacious of the part of the cage which she had fixed upon for her nest, which she made of leaves put in for that purpose. Two of the females produced young last spring. I think the period of their gestation is a month; but of this fact I am not certain. The young are blind for three weeks after their birth, and do not reach puberty till the next spring. I never obtained more than two young ones at a time, nor more than one kindle in a year from the same female. The young were generally born in March or April. The teats of the female appear through the fur some time before she brings forth. One of them produced two young ones without making a distinct nest, or separating herself from the rest, but the consequence was that they disappeared on the third day."

"If on any occasion we disturbed the young in their nest, the mother removed them to another part of the cage. The common squirrel of this country, (England,) is said to remove her young in the same manner, if disturbed. Finding this the case, we often took the young Squirrels out of their nest for the purpose of watching the mother carry them away, which she did by doubling the little one up under her body with her fore-feet and mouth till she could take hold of the thigh and the neck, when she would jump away so fast that it was difficult to see whether she was carrying her young one or not.

"As the young increased in size (which they soon do) and in weight, the undertaking became more difficult. We then saw the mother turn the young one on its back, and while she held the thigh in her mouth, the fore-legs of the young one were clasped round her neck. Sometimes when she was attempting to jump upon some earthen pots which I had placed in the cage, she was overbalanced and fell with her young to the ground, she would drop the young Squirrel, so as to prevent her own weight from crushing it, which would have been the case if they had fallen together. I have seen the young ones carried in this manner till they were half-grown."

GEOGRAPHICAL DISTRIBUTION.

This species is far more numerous than it is generally supposed to be; in traps set for the smaller rodentia in localities where we had never seen the Flying Squirrel, we frequently caught it. We have met with it in all the Atlantic States, and obtained specimens in Upper Canada, within a mile of the falls of the Niagara. In Lower Canada it is replaced by a larger species, (*P. sabrinus*,) and we have reason to believe that it does not exist much to the north of the great lakes; we obtained specimens in Florida and in Texas, and have seen it in Missouri, and according to LICHTENSTEIN it is found in Mexico.

GENERAL REMARKS.

This species was among the earliest of all our American quadrupeds noticed by travellers. Governor SMITH of Virginia, in 1624, speaks of it as "a small beaste they call Assapanick, but we call them Flying Squirrels, because spreading their legs, and so stretching the largeness of their skins, that they have been seen to fly thirty or forty yards." RAY and LINNÆUS supposed it to be only a variety of the European *P. volans*, from which it differs very widely. LINNÆUS arranged it under MUS; GMELIN, PALLAS, CUVIER, RAY, and BRISSON, under SCIURUS; F. CUVIER and DESMAREST under SCIUROPTERUS; FISCHER under PETAURISTUS; and GEOFFROY and more recent naturalists, under PTEROMYS.

N°. 6

Plate XXIX

Drawn on Stone by R. Trembly

Rocky Mountain Neotoma

Drawn from Nature by J.J. Audubon F.R.S.F.L.S.

Printed by Nagel & Weingaertner N.Y.

NEOTOMA DRUMMONDII.—Richardson.

Rocky Mountain Neotoma.

PLATE XXIX. Winter and Summer colours.

N. subtus albida; supra hyeme flavo-fuscescens, æstate saturate cinereus; cauda crassa, corpore longiore muse decumano robustior.

CHARACTERS.

Colour, above, yellowish-brown in winter and dark-ash in summer; whitish beneath; tail, bushy and longer than the body; larger than the Norway rat.

SYNONYMES.

Rat of the Rocky Mountains, Lewis and Clarke, vol. iii., p. 41.
Myoxus Drummondii, Rich., Zool. Jour., 1828, p. 5, 7.
Neotoma Drummondii, Rich., Fauna Boreali Americana, p. 137, pl. 7.

DESCRIPTION.

This species bears a striking resemblance to the Florida rat. It differs from the Norway rat by its longer and broader ears, and by its bushy tail and light active form. Fur, long and loose, bearing a considerable resemblance to that of the gray rabbit; nose, rather obtuse; the nostrils have a very narrow naked margin; the tip of the nose is covered with short hairs; ears, large, oval, and rounded, nearly naked within, except near the margins, where they are slightly clothed with short hairs. On the outer surface there are a few more hairs, but not enough to conceal the skin beneath; eyes, small, much concealed by the fur; whiskers, like hog's bristles, very strong, the longest reaching to the shoulders; neck, short, and fully as thick as the head.

Fore-legs, short; feet, of moderate size, with four toes; claws, small, compressed, and pointed. The third toe nearly equals the middle one, which is the longest, the first is a little shorter, and the outer one not more than half the length of the other two; there is also the rudiment of a thumb, which is armed with a minute nail. The toes of the hind-feet are longer than those of the fore-feet, and the claws less hooked; the middle toe is the longest, those on each side of it of nearly an equal length; the outer one a little shorter, and the inner shortest of all. The palms on the fore and hind-feet are naked; but the toes, even beyond the nails, are covered with short, adpressed hairs. The hairs of the tail (which are not capable of a distichous arrangement) are short near the root, and gradually lengthen toward the end, where it is large and bushy, the hairs being one inch in length.

COLOUR.

Incisors, yellow; on the whole of the back, the head, shoulders, and outsides of the thighs, a dusky darkish-brown, proceeding from a mixture of yellowish-brown and black hairs. From the roots to near the tips, the fur is of a dark lead-colour, tipped with light-brown and black. The sides of the face and the ventral aspect, are bluish-gray. Margin of the upper lip, chin, feet, and under surface, dull white; whiskers, black and white, the former colour predominating; tail, grayish-brown above, dull yellowish-white beneath.

The above is the colour of this species from the end of summer through the following winter to the time of shedding the hair in May; when in its new coat it has far less of yellowish-brown, and puts on a gray appearance on the back, this colour gradually assuming more of the yellowish hue as the autumn advances and the fur lengthens and thickens toward winter.

DIMENSIONS.

From point of nose to root of tail - - -	9 inches.
Tail (vertebræ) - - - - - - - -	7½ do.
Tail, including fur - - - - - - -	8½ do.
Height of ear, posteriorly - - - - -	1 do.
Length of whiskers - - - - - -	4 do.

HABITS.

We regret that from personal observation, we have no information to give in regard to the habits of this species, having never seen it in a living state. It was, however, seen by LEWIS and CLARKE, by DRUMMOND, DOUGLASS, NUTTALL, and TOWNSEND. According to the accounts given by these travellers, this Neotoma appears to have nearly the same general habits as the smaller species, (*N. Floridana,*) the Florida rat, but is much more destructive than the latter. It has a strong propensity to gnaw, cut to pieces, and carry to its nest every thing left in its way. The trappers dread its attacks on their furs more than they would the approach of a grizzly bear. These rats have been known to gnaw through whole packs of furs in a single night. The blankets of the sleeping travellers are sometimes cut to pieces by them, and they carry off small articles from the camp of the hunter.

"Mr. DRUMMOND," says RICHARDSON, "placed a pair of stout English shoes on the shelf of a rock, and as he thought, in perfect security; but on his return, after an absence of a few days, he found them gnawed into fragments as fine as saw-dust."

Mr. DOUGLASS, who unfortunately lost his life in ascending Mouna Roa, in the Sandwich Islands, by falling into a pit for catching wild bulls, where he was gored by one of those animals, was one of the most indefatigable explorers of the Western portions of our continent, and kept a journal of his travels and discoveries in natural history. It was never published, but a few copies were printed some time after his death, by his friend and patron, Sir WILLIAM HOOKER, who presented one of them to us. In it we found the following account of this animal:—

"During the night I was annoyed by the visit of a herd of rats, which devoured every particle of seed I had collected, ate clean through a bundle of dried plants, and carried off my soap, brush, and razor. As one was taking away my inkstand, which I had been using shortly before, and which lay close to my pillow, I raised my gun, which, with my faithful dog, always is placed under my blanket by my side with the muzzle to my feet, and hastily gave him the contents. When I saw how large and strong a creature this rat was, I ceased to wonder at the exploits of the herd in depriving me of my property. The body and tail together measured a foot and a half; the hair was brown, the belly white; it had enormous ears three quarters of an inch long, and whiskers three inches in length. Unfortunately the specimen was spoiled by the shot which in my haste to secure the animal and recover my inkstand, I did not take time to change; but a female of the same sort venturing to return some hours after, I handed it a smaller shot, which did not destroy the skin. It was in all respects like the former, except being a little smaller." This identical specimen is in the museum of the Zoological Society of London, where we examined it.

Mr. TOWNSEND has kindly furnished us with some remarks on this species, from which we make the following extracts:—"I never saw it in the Rocky Mountains, but it is very common near the Columbia river. It is found in the storehouses of the inhabitants, where it supplies the place of the common rat, which is not found here. It is a remarkably mischievous animal, destroying every thing which comes in its way—papers, books, goods, &c. It has been known not unfrequently to eat entirely through the middle of a bale of blankets, rendering the whole utterly useless; and like a pet crow carries away every thing it can lay its *hands* on. Even candlesticks, porter-bottles, and large iron axes, being sometimes found in its burrows."

The food of this species consists of seeds and herbage of various kinds; it devours also the small twigs and leaves of pine trees, and generally has a considerable store of these laid up in the vicinity of its residence.

It is said by Drummond to make its nest in the crevices of high rocks. The nest is large, and is composed of sticks, leaves, and grasses. The abode of this Rat may be discovered by the excrement of the animal, which has the colour and consistence of tar, and is always deposited in the vicinity. It is stated by those who have had the opportunity of observing, that this species produces from three to five young at a time.

GEOGRAPHICAL DISTRIBUTION.

We were informed by a gentleman who was formerly engaged as a clerk in the service of the Missouri fur company, that this Rat exists in the valleys, and along the sides, of the Rocky Mountains, through an extent of thirty degrees of latitude. Douglass states that it is very numerous near the Mackenzie and Peace rivers, latitude 69°. Townsend found it in Oregon. We have seen a specimen that was said to have been obtained in the Northern mountains of Texas, and have heard of its existence in North California.

GENUS SIGMODON.—SAY AND ORD.

DENTAL FORMULA.

$$Incisive\ \frac{2}{2};\quad Canine\ \frac{0-0}{0-0};\quad Molar\ \frac{3-3}{3-3} = 16.$$

As the present genus was instituted after a careful examination of the teeth of *Sigmodon hispidum*, by Messrs. SAY and ORD, who first described that species, we think it due to those distinguished naturalists, to give the dental formula in their own words, more especially as this species was named on the plate in our large edition, *Arvicola hispidus*, we having had some doubts whether it was sufficiently distinct from the arvicolæ in its generic characters, to warrant us in adopting the genus SIGMODON, to which we afterwards transfered it.

"Superior Jaw.—Incisor, slightly rounded on its anterior face, truncated at tip; first molar, equal to the second, composed of four very profound, alternate folds, two on each side, extending at least to the middle of the tooth; second molar, quadrate, somewhat wider, and a little shorter than the preceding, with three profound folds extending at least to the middle, two of which are on the exterior side; posterior molar, a little narrower, but not shorter than the preceding, with three profound folds, two of which are on the exterior side, extending at least to the middle; the inner fold, opposite to the anterior exterior fold, and not extending to the middle.

"Inferior Jaw.—Incisor obliquely truncate at tip, the acute angle being on the inner side; it originates in the ascending branch of the maxillary bone, passing beneath the molars; molars, subequal in breadth, inclining slightly forwards; first molar, a little narrower than the second, with five profound alternate folds, three of which are on the inner side; second molar, subquadrate, with two alternate profound folds, the inner one anterior; third molar, about equal in length and breadth to the anterior one, but rather larger and somewhat narrower than the second, with which it corresponds in the disposition of its folds, excepting that they are less compressed."

OBSERVATIONS.

"The enamel of the molars is thick, but on the anterior face of each fold excepting the first is obsolete. From the arrangement of the folds, as above described, it is obvious that the configuration of the triturating surface, (occasioned by the folds of enamel dipping deeply into the body of the tooth, in the second and third molar of the lower jaw,) accurately represents the letter S, which is reversed on the right side; that bearing considerable resemblance to the posterior tooth of the genus SPALAX, and to which also it has a slight affinity in the truncature of the inferior incisors. The configuration of the intermediate molar of the upper jaw may be compared to the form of the Greek letter Σ, whence our generic name."

"In respect to its generic affinities, it is very obvious that its system of dentition indicates a proximity to ARVICOLA; but the different arrangement of the folds, and the circumstance of the molars being divided into radicles, certainly exclude it from that genus. With respect to the radicles, it resembles the genus FIBER; but is allied to this genus in no other respect."

"We may further remark, that the teeth of our specimen are considerably worn, a condition that materially affects the depths of the folds."

Although the animal described below is the only species of SIGMODON at present admitted into this genus, there are several well known, and one undescribed, species, that we apprehend will yet be arranged under it.

SIGMODON HISPIDUM.—SAY AND ORD.

COTTON-RAT.

PLATE XXX.

S. flavo fuscescens, infra cinereum; cauda corpore breviore; auribus amplis rotundatisque; Tamiæ Lysteri magnitudine.

CHARACTERS.

Size of the chipping squirrel, (T. Lysteri;) tail, shorter than the body; ears, broad and rounded; above, dark yellowish-brown; cinereous beneath.

SYNONYMES.

MARSH-RAT, Lawson's Carolina, 1709, p. 125.
THE WOOD-RAT, Bartram's Travels in East Florida, 1791, p. 124.

N°. 6

Plate XXX

Drawn on Stone by R. Trembly.

Cotton Rat.

Drawn from Nature by J. J. Audubon F.R.S. F.L.S.

Printed by Nagel & Weingærtner N.Y.

SIGMODON HISPIDUM, Say and Ord, Journ. Ac. Nat. Sc., Phila., vol. iv., pt. 2, p. 354, read March 22d, 1825.
ARVICOLA HORTENSIS, Harlan, Fauna, 1825, p. 138.
" HISPIDUS, GODMAN, vol. ii., p. 68, 1826.
" HORTENSIS, Griffith, Cuvier, vol. v., sp. 547.

DESCRIPTION.

In its general external appearance this species approaches nearer to the genus ARVICOLA than to MUS. It has the thick short form of the former, and the broad and rather long ears of many species of the latter. The fur is long and coarse.

Head, of moderate size, rather long; nose, pointed; whiskers, few, weak, and shorter than the head; eyes, of moderate size and rather prominent; ears, broad, rounded, and slightly covered with hair.

Fore-legs, rather short and slender; four toes on each foot, the middle ones nearly of equal length, the inner one a size shorter, and the outer shortest; there is also a rudimentary thumb, protected by a strong conical nail. Hind-legs, stouter; five toes on each foot, much longer than those on the fore-feet; middle claw longest, the two on each side nearly equal, the outer, not one-third the length of the others, and the inner, which rises far back, shortest of all; nails, rather small, sharp, and slightly arched; toes, covered with hair extending to the roots of the nails; tail, clothed with short hairs.

COLOUR.

Hairs, on the whole upper surface of the body of a dark plumbeous colour from the roots to near the extremities, edged with brown, and irregularly tipped with black; giving it a rusty reddish-brown appearance. The ears, head and tail, are of the colour of the back; chin, throat, and under surface of body, dull-white, the hairs being ashy-gray at the roots, and whitish at the points.

DIMENSIONS.

From point of nose to root of tail - - -	6	inches.
Tail - - - - - - - - -	4	do.
Length of ear - - - - - - -	$\frac{1}{2}$	do.
Breadth of ear - - - - - - -	$\frac{1}{2}$	do.
From eye to point of nose - - - - -	$\frac{5}{8}$	do.
From point of nose to ear - - - - -	$1\frac{1}{2}$	do.
From heel to point of longest nail - - -	$1\frac{1}{2}$	do.

HABITS.

This is the most common wood-rat existing in the Southern States, being even more abundant than any of the species of meadow-mice in the Northern and Eastern States. It is however a resident rather of hedges, ditches, and deserted old fields, than of gardens or cultivated grounds; it occasions very little injury to the planter. Although its paths are everywhere seen through the fields, it does not seem to destroy many plants or vegetables. It feeds on the seeds of coarse grasses and leguminous plants, and devours a considerable quantity of animal food. In its habits it is gregarious. We have seen spots of half an acre covered over with tall weeds, (*Solidago* and *Eupatorium*,) which were traversed in every direction by the Cotton-Rat, and which must have contained several hundred individuals.

Although this species does not reject grains and grasses, it gives the preference in all cases to animal food, and we have never found any species of rat more decidedly carnivorous. Robins, partridges, or other birds that are wounded and drop among the long grass or weeds in the neighbourhood of their burrows are speedily devoured by them. They may sometimes be seen running about the ditches with crayfish (*Astacus Bartoni*) in their mouths, and have been known to subsist on Crustacea, especially the little crabs called fiddlers, (*Gelasimus vocans.*)

We have frequently kept Cotton-Rats in cages; they killed and devoured every other species placed with them, and afterwards attacked each other; the weakest were killed and eaten by the strongest. They fight fiercely, and one of them will overpower a Florida rat twice its own size.

The old males when in confinement almost invariably destroy their young.

This species delights in sucking eggs, and we have known a Virginian partridge nest as completely demolished by these animals as if it had been visited by the Norway rat. They will sometimes leave Indian-corn and other grain untouched, when placed as a bait for them in traps, but they are easily caught when the traps are baited with meat of any kind.

Although the Cotton-Rat is nocturnal in its habits, it may frequently be seen by day; and in places where it is seldom disturbed, it can generally be found at all hours.

The galleries of this species often run twenty or thirty yards under ground, but not far beneath the surface; and the ridges thrown up as the animals excavate their galleries, can often be traced along the surface of the earth for a considerable distance, like those formed by the common shrew-mole.

Each burrow or hole contains apparently only one family, a pair of old ones with their young; but their various galleries often intersect each other, and many nests may be found within the compass of a few yards; they are composed of withered grasses, are not very large, and may usually be found within a foot of the surface. In summer the nests are often seen in a cavity of the earth, on the surface in some meadow, or among rank weeds.

This is a very prolific species, producing young early in spring, and through all the summer months, till late in autumn. We have on several occasions known their young born and reared in cages. They produce from four to eight at a litter. The young are of a bright chesnut-brown colour, and at the age of five or six days begin to leave the nest, are very active and sprightly, and attain their full growth in about five months.

This species has no other note than a low squeak, a little hoarser than that of the common mouse; when captured it is far more savage than the Florida rat. On one occasion, while seizing one of them, we were bitten completely through a finger covered by a buckskin glove.

The Cotton-Rat is fond of burrowing in the old banks of abandoned rice-fields. In such situations we have, during freshets, observed that it could both swim and dive like the water-rat of Europe, and WILSON's meadow-mouse of the Middle States.

This species supplies a considerable number of animals and birds with food. Foxes and wild-cats especially, destroy thousands; we have observed minks coursing along the marshes in pursuit of them, and have frequently seen them with one of these Rats in their mouth. Marsh-hawks, and several other species, may be constantly seen in the autumn and winter months sailing over the fields, looking out for the Cotton-Rat. No animal in the Southern States becomes more regularly the food of several species of owls than this. The barred owl (*Syrnium nebulosum*) is seen as early as the setting of the sun, flitting along the edges of old fields, seeking to make its usual evening meal on it or carry it off as food for its young. We were invited some years since to examine the nest of the American barn-owl (*Strix Americana*) in the loft of a sugar refinery in Charleston. There were several young of different sizes, and we ascertained that the only food on which they were fed was this Rat, to obtain which the old birds must have gone several miles.

The Cotton-Rat has obtained its name from its supposed habit of making its nest with cotton, which it is said to collect for the purpose in large quantities. We have occasionally, although very seldom, seen cotton in its nest, but we have more frequently found it composed of leaves and withered grasses. Indeed, this species does not appear to be very choice in selecting materials for building its nest, using indiscriminately any suitable substance in the vicinity. We should have preferred a more characteristic English name for this Rat, but as it already has three names, Cotton-Rat, Hairy Campagnol, and Wood-Rat, the latter being in Carolina applied both to this and the Florida rat, we have concluded not to add another, although one more appropriate might be found.

GEOGRAPHICAL DISTRIBUTION.

We have traced the Cotton-Rat as far north as Virginia, and have seen it in North Carolina, near Weldon and Wilmington. It is exceedingly abundant in South Carolina, Georgia, and Florida; in Alabama, Mississippi, and Louisiana, traces of it are every where seen. We have received a specimen from Galveston, Texas, but have had no opportunity of ascertaining whether it exists farther south.

GENERAL REMARKS.

Although this species was noticed by Lawson a century and a half ago, it was not described until a comparatively recent period. Ord obtained specimens in Florida in 1818, and it was generally supposed that it was not found further to the north. In the spring of 1815, three years earlier than Mr. Ord, we procured a dozen specimens in Carolina, which we neglected to describe. Say and Ord, and Harlan, described it about the same time, (in 1825,) and Godman a year afterwards. We prefer adopting the name given to it by the individual who first brought it to the notice of naturalists. In its teeth it differs in a few particulars from Arvicola, and approaches nearer to Mus.

N°. 7

Plate XXXI

Drawn on Stone by R Trembly

Collared Peccary.

Drawn from Nature by J.J. Audubon F.R.S.F.L.S.

Printed by Nagel & Weingærtner NY

GENUS DYCOTYLES.—F. Cuvier.

DENTAL FORMULA.

$$Incisive\ \frac{4}{6};\quad Canine\ \frac{1-1}{1-1};\quad Molar\ \frac{6-6}{6-6} = 38.$$

Tusks or canine teeth, projecting slightly, not curved near the points as in the common hog, (Sus,) small, triangular, and very sharp; molars, with tubercular crowns; tubercles, rounded and irregularly disposed. Head, broad and long; snout, straight, terminated by a cartilage; ears, of moderate size and pointed; eyes, rather small, pupil round. Fore-feet, with four toes, the two middle toes largest, the lateral toes quite short, not reaching to the ground; hind-feet, with three toes, the external little toe of the hog wanting in this genus.

The metatarsal and metacarpal bones of the two largest toes on all the feet are united together like those of the ruminantia; all the toes are protected by hoofs. A gland situated on the back a few inches from the root of the tail, concealed by the hair, discharges an oily fœtid secretion. Body, covered with strong, stiff bristles; tail, a mere tubercle.

Only two species are known, both inhabiting the warmer climates of America; the generic name Dycotyles, is derived from the Greek words, δις, (*dis*,) *double*, and κοτυλη, (*kotule*,) a *cavity;* or double navel, from the opening on the back.

DYCOTYLES TORQUATUS.—F. Cuvier.

Collared Peccary.

PLATE XXXI.

D. pilis nigro alboque annulatis; vitta albida ab humeris in latere olli utroque decurrente.

CHARACTERS.

Hair, annulated with black and white; a light-coloured band extending from the sides of the neck around the shoulders, and meeting on the back.

SYNONYMES.

TAYTETOU, D'Azara, Quad. du Paraguay, vol. i., p. 31.
TAJACU, Buffon, vol. v., p. 272, pl. 135.
SUS TAJACU, Linn., 12th ed. vol. i., p. 103.
QUAVHTLA COYMATL, QUAHEROTL, Hern., Mex., 637.
TAJACU, Ray, Quad., p. 97.
SUS TAGASSA, Erxleben, Syst., p. 185.
SUS TAGASSA, Schreber, Säugethiere, t. 325.
APER AMERICANUS, Briss., Règne An., p. 3.
TAJACU CAAIGOANA MARCGR, Bras., p. 229.
MEXICAN HOG, Pennant, Quadr., p. 147.
PORCUS MOSCHIFERUS, Klein, Quadr., p. 25.
PECCARI, Shaw, Gen. Zool., vol. ii., p. 469, 224.
DYCOTYLES TORQUATUS, F. Cuvier, Dict. des Sciences Naturelles, tom. ix., p. 518.
" " Desm., Mamm., p. 393.
" " Cuv., Règne An., vol. i., p. 237.
" " Pr. Maxim. Beitr., vol. ii., p. 557.
" " Harlan, Fauna, p. 220.
" " Griffith's Animal Kingdom, sp. 740.

DESCRIPTION.

The form of the Collared Peccary bears a very striking resemblance to that of the common domesticated hog; it is however smaller in size, shorter, and more compact.

Head, rather large; snout, long; ears, upright, and of moderate size; eyes, rather small. The cartilage on the extremity of the nose is naked, with the exception of a few bristles on the upper lip. On the upper surface of the nose, near the cartilage, there is a spot half an inch in length that is naked; nostrils, large; the upper tusks, in the living animal, protrude downward below the lower lips half an inch; the ears are on both surfaces thinly clothed with hair that is softer than that on the remainder of the body. The hairs on the head are short. From the hind part of the head along the dorsal line on the back, there are long strong bristles, which are erected when the animal is irritated. Many of these bristles are five inches in length, whilst the hairs on the other parts of the body are generally about three.

On the lower part of the back, a slight distance from the rump, there is a naked glandular orifice surrounded by a few bristles in a somewhat radiated direction. From this orifice there exudes a strong scented fluid. This part of the animal has been vulgarly supposed to be its navel.

The legs, which strongly resemble those of the common hog, are rather short. There is not even a vestige of the small upper external hind-toe, which is always present in the common hog. There is a ruff under the throat, protruding about three inches beyond the surrounding hairs. The under surface of the body is rather thinly clothed with hair.

In place of a tail there is a mere protuberance about half an inch in length, which is rounded and like a knob.

COLOUR.

Eyes, dark-brown; nostrils, flesh-colour. The hairs are at their roots yellowish-white, are thrice annulated with dark-brown and yellowish-white, and are tipped with black. Head, cheeks, and sides of the neck, grayish; legs, dark-brown; a whitish band two inches broad runs from the top of the shoulder on each side toward the lower part of the neck. The long hairs on the dorsal line are so broadly tipped with black that the animal in those parts appears of a black colour; along the sides however the alternate annulations are so conspicuous that it has a deep gray or grizzled appearance. On the chest, outer surface of shoulders and thighs, it is of a darker colour than on the sides. Immediately behind the lightish collar on the shoulders the hairs are dark, rendering this collar or band more conspicuous.

The young have a uniform shade of red.

DIMENSIONS.

Living female.	Feet.	Inches.
Length of head and body	2	10
" head	0	11
" ear	0	3
Height to shoulder	1	8
Length of tail	0	0½

Adult male (recent) obtained in Texas.	Feet.	Inches.
From nose to anterior canthus	0	5½
From nose to beginning of ear	0	9¼
Length of ear	0	3½
Breadth of ear	0	2⅞
Length from snout to root of tail	3	4

	Feet.	Inches.
Tail - - - - - - - - - -	0	0¾
From knee to end of hoof - - - - - -	0	5¾
Hind-knee to end of hoof - - - - - -	0	7¼
Spread of fore-feet - - - - - - -	0	1¾
Girth across the centre of body - - - - -	2	5
Spread of mouth when fully extended - - -	0	5½
Breadth between the eyes - - - - - -	0	2¾

HABITS.

The accounts that have been handed down to us of the habits of this species by old travellers, ALDROVANDA, FERNANDEZ, Mons. DE LA BORDE, MARCGRAVE, ACOSTA, and others, who furnished the information from which BUFFON, BRISSON, RAY, and LINNÆUS, drew up their descriptions of the Mexican hog, are not to be fully relied on, inasmuch as their descriptions referred to two very distinct species, the white-lipped peccary, (*D. labiatus*,) and the subject of the present article. Neither LINNÆUS nor his contemporaries seem to have been aware of the difference which exists between the species; and although BUFFON was informed by M. DE LA BORDE that another and larger species existed at Cayenne, he does not appear to have drawn any line of distinction between it and our animal.

D'AZARA, who visited South America in 1783, (Essais sur l'Histoire Naturelle des Quadrupèdes de la Province du Paraguay, Paris, 1801,) endeavoured to correct the errors into which previous writers had fallen, and gave an account of the present species, which, although somewhat unmethodical, is nevertheless of such a character that it may on the whole be relied on. He commences his article on the "taytetou," as he designates this species, by first giving correct measurements; afterwards he describes the colour of the adult and young, points out the distinctive marks which separate this species from the white-lipped peccary, which he calls "tagnicate," and then gives a tolerable account of the habits of the species now under consideration. From the accounts which travellers have given us of the Collared Peccary it appears that this species is gregarious, and associates for mutual protection in pretty large families; it is however stated by D'AZARA that the white-lipped peccary is more disposed to congregate in very large herds than our animal.

Although they are usually found in the forests and prefer low and marshy grounds, like common hogs, Peccaries wander wherever they can find an abundance of food, often enter the enclosures of the planters, and commit great depredations on the products of their fields.

When attacked by the jaguar, the puma, the wolf, the dog, or the hunter, they form themselves into a circle, surrounding and protecting their young, repelling their opponents with their sharp teeth, and in this manner sometimes routing the larger predatory animals, or severely wounding the dogs and the hunters.

When angry, they gnash their teeth, raise their bristles, (which at such time resemble the quills of the porcupine,) and their sharp, shrill grunt can be heard at a great distance.

This species feeds on fruits, seeds, and roots; and like the domesticated hog is constantly rooting in the earth in quest of worms, insects, reptiles, or bulbous roots. It is said also to devour the eggs of alligators, turtles, and birds; and to be destructive to lizards, toads, and snakes. In fact, like the common hog it is omnivorous, feeds upon every thing that comes in its way, and is not particularly choice in the selection of its food.

Mons. De la Borde (D'Azara, Quad. du Paraguay, vol. i., p. 31,) relates that "they are easily shot; for instead of flying, they assemble together, and often give the hunters an opportunity of charging and discharging several times." He mentions "that he was one day employed, along with several others, in hunting these animals, accompanied by a single dog, which as soon as they appeared, took refuge between his master's legs. For greater safety he with the other hunters stood on a rock. They were nevertheless surrounded by the herd of hogs. A constant fire was kept up, but the creatures did not retire till a great number of them were slain." "These animals, however," he remarks, "fly after they have been several times hunted. The young, when taken in the chase, are easily tamed, but they will not associate or mix with the domestic species. In their natural state of liberty they frequent the marshes, and swim across large rivers. Their flesh," says he, "has an excellent taste, but is not so tender as that of the domestic hog; it resembles the flesh of the hare, and has neither lard nor grease."

The same author also states that "when pursued they take refuge in hollow trees, or in holes in the earth dug by the armadilloes. These holes they enter backwards and remain in as long as they can. But when highly irritated they instantly issue out in a body. In order to seize them as they come out, the hole is enclosed with branches of trees; one of the hunters, armed with a pitchfork, stands above the hole to fix them by the neck, while another forces them out, and kills them with a sabre."

"Where there is but one in a hole, and the hunter has not leisure to seize it, he shuts up the entrance, and is sure of his game next day."

All authors agree in stating that the dorsal glands of either the male or female should be cut off instantly after the animal is killed, for their retention for only a single hour gives the meat so strong an odour that it can scarcely be eaten.

The only recent account we have thus far received, that contains original and authentic information about this singular wild hog, was furnished us by Mr. William P. Smith. He had been sent to this country by our ever kind friend, the Right Honourable the Earl of Derby, for the purpose of procuring living animals to enrich his collection at Knowsley, near Liverpool. We engaged him also to obtain for us any rare species he could meet with in Texas, and to send descriptions of their habits, and any other information likely to be of interest to the readers of this work. Mr. Smith went to Texas in 1841, and shortly afterwards sent us the following account of the Peccary. He says,—

"The Mexican hogs previous to the overflowing of the bottom lands in 1833, struck terror into the hearts of the settlers in their vicinity, oftentimes pursuing the planter whilst hunting or in search of the lost track of his wandering cattle—at which time they frequently killed his dogs, or even at times forced him to ascend a tree for safety, where he would sometimes be obliged to wait until the hogs got tired of dancing attendance at the foot of his place of security, or left him to go and feed. These animals appeared quite savage, and would, after coming to the tree in which the planter had ensconsed himself, snap their teeth and run about and then lie down at the root of the tree to wait for their enemy to come down. At this early period of the settlement of Texas, (this refers to 1833,) they used to hunt this animal in company. From five to fifteen planters together, and occasionally a larger number of hunters, would join in the pursuit of these ravagers of their corn-fields, in order to diminish their number and prevent their farther depredations, as at times they would nearly destroy a farmer's crop. Since this time, however, their number has greatly decreased, and it is now a difficult matter to find them."

"On some parts of the Brazos they still exist, and in others are quite abundant."

Mr. Smith further says, "The two I send you are the only ones I have heard of since my arrival in this country. I happened, with the assistance of a person, to find out their lair, which is always in some hollow tree, although they have many sleeping places. Being late in the day I was determined not to disturb them until a more favourable time would present itself, as I was anxious, if possible, to procure them alive. Some time passed, and everything being ready, the dogs soon compelled them to make for home, when they having entered, we secured the entrance of their hole, and cut a large opening up the body of the tree, a few feet above them, from which "point of vantage" we were enabled easily to drop a noose round their necks, which we tightened until we thought they were nearly suffocated; we then drew them out, tied their legs and feet securely, and fastened their mouths by binding their jaws together with cords, and then left them lying on the ground for a time. On our return we found that they had got over the effect of the 'experimental hanging' they had gone through. We put them across a horse, and in trying to get loose they so tightened the ropes and entangled them about their necks, that they died before we observed this on our way home with them. This is the usual mode of taking these animals alive, although some are caught in pits. They have a large musk-bag upon the back, from which a very disagreeable odour is emitted whilst the animal is excited; but this is not observable after they are killed. The flesh of the female is good at some seasons of the year, but that of the male is strong, coarse and disagreeable at all times. Their principal food consists of nuts of every description (mast) during winter; but in summer they feed on succulent plants, with which the bottom lands in the Brazos abound. The male measured forty inches from the tip of its nose to that of its tail; the female is shorter by two inches. The eyes are very dark hazel colour."

"As soon as they get within their den, one of them, probably the oldest male, stands sentinel at the entrance. Should the hunter kill it, another immediately takes its place, and so in succession until all are killed. This animal, which in Texas is always called the wild hog, is considered the bravest animal of these forests, for it dreads neither man nor beast."

The Collared Peccary is easily domesticated, and breeds readily in confinement. We saw a pair on board of a ship that arrived in Charleston from South America, the female of which had produced two young whilst on the passage; they were then several weeks old, and seemed to be in a thriving condition.

Mons. M. L. E. Moreau Saint Mery, the translator of the work of D'Azara, from the Spanish into the French language, states that in 1787 he saw at the residence of the Governor General La Luzerne, a tame Collared Peccary, which he had procured from Carthagena, with the intention of multiplying the species in San Domingo, (Note du Traducteur D'Azara, tom. i., p. 42.) We observed at the Zoological Gardens in London, young Peccaries that had been born in the menagerie. This animal, however, is less prolific than the common domesticated hog, and its odorous glands being moreover offensive, the extensive domestication of it would not be attended with any profit to the agriculturist.

We have frequently seen the Collared Peccary in confinement. One that is at present (1846) in a menagerie in Charleston, is exceedingly gentle, taking its food from the hand, and allowing itself to be caressed even by strangers. It lies down in the manner of a pig, and next to giving it food, the greatest favour you can bestow on it is to scratch it either with the hand or a stick. It however is easily irritated. We noticed that it has a particular antipathy to the dog, and when approached by that animal immediately places itself in a defensive attitude, raising its bristles, showing its tusks, stamping its feet, and uttering a sharp cry which might be heard at the distance of seventy yards; when in a good humour, however, it occasionally utters a low grunt like a pig. It seems to suffer much from cold, and is always most lively and playful on warm days. It appears to prefer Indian-corn, potatoes, bread and fruits, but like the domestic hog evinces no unwillingness to take any kind of food that is presented to it. We remarked, however, that it is decidedly less carnivorous than the common hog.

It is stated by authors that this species produces but once a year, and brings forth only two at a litter.

GEOGRAPHICAL DISTRIBUTION.

The Collared Peccary has a most extensive geographical range. It was seen by NUTTAL at the Red River in Arkansas, north latitude 31°. Our specimens were obtained in Texas. It exists in all the lower portions of Mexico and Yucatan, and is found every where within the tropics. It is said by D'AZARA to be abundant at Paraguay, south latitude 37°, thus spreading itself through an extent of sixty-eight degrees of latitude.

GENERAL REMARKS.

This species has been noticed by all the early travellers in South America and Mexico. They however almost invariably confounded the habits of two species. D'AZARA pointed out the distinctive marks which separate these species. They differ so much from each other that they ought never to have been mistaken. LINNÆUS applied the name *Sus tajacu*, but as it is impossible to ascertain which species he had in view we cannot use his name for either. RAY, ERXLEBEN, and SCHREBER applied the same name, and committed the same error. BRISSON gave the name *Aper Americanus*, and KLEIN that of *Porcus muschiferus* in the same manner, without discriminating the species. Baron CUVIER established the genus DYCOTYLES, and F. CUVIER applied the specific name of *torquatus*. BUFFON, who had heard from M. DE LA BORDE that there were two distinct species in Cayenne, considered them as mere varieties produced by age, but gave, as he supposed, a figure of each; his figures, however, which are of no value, both refer to the present species, and bear no resemblance to the white-lipped Peccary, (*D. labiatus.*)

It is somewhat strange that GRIFFITH, in his "Animal Kingdom," which he states was arranged by Baron CUVIER, should have completely misunderstood D'AZARA, (Histoire Naturelle, tom. i., p. 31,) and reversed the habits of the two species, (CUVIER, Animal Kingdom, by GRIFFITH, vol. iii., p. 411,) giving D'AZARA as authority for applying the habits of the present species, Tajassu, (*Dycotyles torquatus,*) to those of his Tagnicati, (*D. labiatus,*) giving at the same time a pretty good figure of the latter. It may however be easily seen that the whole object of D'AZARA's article on this species was to correct the very error into which GRIFFITH has fallen.

LEPUS GLACIALIS.—Leach.

Polar Hare.

PLATE XXXII.—Male. In summer pelage.

L. æstate dilute cinereus, hyeme niveus, pilis apice ad radicem albis; aurium apicibus nigris; vulpes magnitudine.

CHARACTERS.

As large as a fox; colour, in summer, light gray above; in winter, white, the hairs at that season being white from the roots. Tips of ears, black.

SYNONYMES.

White Hares, Discoveries and Settlements of the English in America, from the reign of Henry VIII. to the close of that of Queen Elizabeth, quoted from Pinkerton's Voyages, vol. xii., p. 276.
Alpine Hare, Philosophical Transactions, London, vol. lxvi., p. 375, An. 1777.
Lepus Timidus, Fabri., Fauna Grœnlandica, p. 25.
Varying Hare, Pennant, Arc. Zool., vol. i., p. 94.
White Hare, Hearne's Journey, p. 382.
" " Cartwright's Journal, vol. ii., p. 75.
Lepus Glacialis, Leach, Zool. Miscellany, 1814.
" " Ross's Voyage.
" " Captain Sabine's Suppl. Parry's 1st Voyage, p. 188.
" " Franklin's Journal, p. 664.
" " Richardson, Appendix to Parry's 2d Voyage, p. 321.
Polar Hare, Harlan, Fauna, p. 194.
" " Godman, Nat. Hist., vol. ii., p. 162.
Lepus Glacialis, Richardson, Fauna Boreali Americana, p. 221.
" " Bachman, Acad. Nat. Sciences, Phila., vol. vii., part 2.

DESCRIPTION.

This fine species is considerably larger than the English hare, (*L. timidus.*) Head, larger and longer than that of the European hare; forehead, more arched; body, long; nose, blunt; eyes, large; ears, long; whiskers, composed of a few stiff long hairs; legs, long; soles of feet, broad, thickly covered with hair concealing the nails, which are long,

N° 7. Plate XXXII.

Drawn on stone by R. Trembly

Polar Hare

Drawn from Nature by J.J. Audubon, F.R.S.F.L.S. Printed & Col^d by J.T. Bowen, Philad^a.

moderately broad, and somewhat arched. Tail, of moderate length, woolly at the roots, intermixed with longer hairs. The fur on the back is remarkably close and fine; that on the under surface is longer, and not quite so close.

COLOUR.

In winter, the Polar Hare is entirely white on every part of the body except the tips of the ears; the hairs are of the same colour to the roots. The ears are tipped with hairs of a brownish-black colour. In its summer dress, this species is of a grayish-brown colour on the whole of the head extending to the ears; ears, black, bordered with white on their outer margins; under parts of the neck, and the breast, dark bluish-gray; the whole of the back, light brownish-gray. The fur under the long hairs of the back is soft and woolly, and of a grayish-ash; the hairs interspersed among the fur are dark blue near the roots, then black, tipped with grayish-fawn colour; a few black and white hairs are interspersed throughout. The wool on the under surface is bluish-white, interspersed with long hairs of a slate colour; the hairs forming the whiskers are white and black, the former predominating. The inner sides of the fore-legs, thighs, and under surface of the tail, pure white; the hairs on the soles are yellowish-brown; nails, nearly black. According to RICHARDSON, "the irides are of a honey-yellow colour." The skin of this species appears to be nearly as tender as that of the Northern hare.

DIMENSIONS.

Specimen, obtained at Labrador.

Length of head and body - - - - - -	26 inches.
" from point of nose to ear - - -	4½ do.
" of ear, measured posteriorly - - -	4¾ do.
" tail (vertebræ) - - - - - -	1¾ do.
" tail, including fur - - - -	3½ do.
" whiskers - - - - - - -	3 do.
" from wrist-joint to point of middle claw -	3¾ do.
" " heel to middle claw - - -	6½ do.

Weight, from 7 to 11 lbs.

These measurements were taken from the specimen after it had been stuffed. We are under the impression that it was a little longer in its recent state.

HABITS.

It is to the cold and inhospitable regions of the North, the rugged valleys of Labrador, and the wild mountain-sides of that desolate land, or to the yet wilder and more sterile countries that extend from thence toward the west, that we must resort, to find the large and beautiful Hare we have now to describe; and if we advance even to the highest latitude man has ever reached, we shall still find the Polar Hare, though the mercury fall below zero, and huge snow-drifts impede our progress through the trackless waste.

Both Indians and trappers are occasionally relieved from almost certain starvation by the existence of this Hare, which is found throughout the whole range of country extending from the Eastern to the Western shores of Northern America, and includes nearly thirty-five degrees of latitude, from the extreme North to Newfoundland.

In various parts of this thinly inhabited and unproductive region, the Polar Hare, perhaps the finest of all the American hares, takes up its residence. It is covered in the long dark winter with a coat of warm fur, so dense that it cannot be penetrated by the rain, and which is an effectual protection from the intense cold of the rigorous climate.

Its changes of colour help to conceal it from the observation of its enemies; in summer it is nearly of the colour of the earth and the surrounding rocks, and in winter it assumes a snow-white coat. The changes it thus undergoes, correspond with the shortness of the summers and the length of the Arctic winters. In the New England States the *Northern* hare continues white for about five months, that being the usual duration of the winters there; but in the Arctic regions, where the summer lasts for about three months only, whilst the earth during the remainder of the year is covered with snow, were the Polar Hare not to become white till November, (the time when the Northern hare changes,) it would for two months be exposed to the keen eyes of its greatest destroyers, the golden eagle and the snowy owl, as its dark fur would be conspicuous on the snow; or were it to become brown in April, it would wear its summer dress long before the earth had thrown off its mantle of white, or a single bud had peeped through the snow.

The eye of the Polar Hare is adapted to the twilight that reigns during a considerable part of the year within the Arctic circle; in summer it avoids the glare of the almost continual day-light, seeking the shade of the little thickets of dwarfish trees that are scattered over the barren grounds, the woods that skirt the streams, or the shelter of some overhanging rock.

In addition to the circumstance that the eye of this Hare is well fitted for seeing with a very moderate light, it may be remarked that in winter the frequent and long continued luminous appearance of the heavens caused by the aurora borealis, together with the brightness of the unsullied snow, afford a sufficient degree of light for it to proceed with its customary occupations.

During the summer this species is found on the borders of thickets, or in stony or rocky places. In winter it is often seen in the barren and open country, where only a few stunted shrubs and clumps of spruce fir (*Abies rubra*) afford it shelter, differing in this habit from the Northern hare, which confines itself to thick woods throughout the year, avoiding cleared fields and open ground.

Captain Ross says of the Polar Hare, "There is scarcely a spot in the Arctic regions, the most desolate and sterile that can be conceived, where this animal is not to be found, and that too, throughout the winter; nor does it seek to shelter itself from the inclemency of the weather by burrowing in the snow, but is found generally sitting solitarily under the lee of a large stone, where the snow drift as it passes along, seems in some measure to afford a protection from the bitterness of the blast that impels it, by collecting around and half burying the animal beneath it."

The food of this species varies with the season. Hearne tells us that "in winter it feeds on long rye-grass and the tops of dwarf willows, but in summer it eats berries and different sorts of small herbage."

According to Richardson, "it seeks the sides of the hills, where the wind prevents the snow from lodging deeply, and where even in the winter it can procure the berries of the Alpine arbutus, the bark of some dwarf willows, (*Salix*,) or the evergreen leaves of the Labrador tea-plant," (*Ledum latifolium*.) Captain Lyon, in his private journal, has noted that on the barren coast of Winter Island, the Hares went out on the ice to the ships, to feed on the tea-leaves thrown overboard by the sailors."

The Polar Hare is not a very shy or timid animal, but has on being approached much the same habits as the Northern hare. "It merely runs to a little distance, (says Richardson,) and sits down, repeating this manœuvre as often as its pursuer comes nearly within gun-shot, until it is thoroughly scared by his perseverance, when it makes off. It is not difficult to get within bow-shot of it by walking round it and gradually contracting the circle—a method much practised by the Indians." Hearne had previously made the same observations; he says also, "the middle of the day, if it be clear weather, is the best time to kill them in this manner, for before and after noon the sun's altitude being so small, makes a man's shadow so long on the snow as to frighten the Hare before he can approach near enough to kill it. The same may be said of deer when on open plains, which are frequently more frightened at the long shadow than at the man himself."

All travellers concur in stating the flesh of this animal to be of a finer flavour than that of any of our other hares. We obtained one while at St. George's Bay, in Newfoundland, and all our party made a meal of it; we pronounced it delicious food.

A lady residing at that place informed us that she had domesticated the Polar Hare, and had reared some of them for food. She said that the flesh was fine-flavoured, and the animals easily tamed, and that she had only been induced to discontinue keeping them in consequence of their becoming troublesome, and destructive in her garden.

The Polar Hare is stated by RICHARDSON, on the authority of Indian hunters, to bring forth once in a year, and only three young at a litter. That owing to the short summer of the Arctic regions, it does not produce more than once annually, is no doubt true; but the number of young brought forth at a time, we are inclined to believe, was not correctly given by the Indian hunters.

CARTWRIGHT (see Jour., vol. ii., p. 76) killed a female of this species at Labrador on the 11th June, from which he took five young.

Capt. Ross says, "a female killed by one of our party at Sheriff Harbour on the 7th of June, had four young in utero, perfectly mature, 5½ inches long, and of a dark gray colour. In one shot at Igloolik, on the 2d June, six young were found, not quite so far advanced."

An intelligent farmer who had resided some years in Newfoundland, informed us that he had on several occasions counted the young of the Polar Hare, and had never found less than five, and often had taken seven from one nest. He considered the average number of young to each litter as six. FABRICIUS, alluding to the habits of this species as existing in Greenland, says, "They pair in April, and in the month of June produce eight young at a birth."

Some idea may be formed of the very short period this species continues in its summer colours, from the following remarks of different observers. In BEACHY'S Narrative, (p. 447,) is the following notice:—"*May 5th.* The party killed a white Hare, it was getting its summer coat." CARTWRIGHT killed one on the 11th June, and remarks that it was yet white. We obtained a specimen on the 15th August, 1833, and ascertained that the change from summer to winter colours had already commenced. There was a large spot, nearly a hand's breadth, of pure white on the back, extending nearly to the insertion of the tail; three or four white spots about an inch in diameter were also found on the sides.

Captain Ross states—"One taken by us on the 28th of June, a few days after its birth, soon became sufficiently tame to eat from our hands, and was allowed to run loose about the cabin. During the summer we fed it on such plants as the country produced, and stored up a quantity of grass and astragali for its winter consumption; but it preferred to share with us whatever our table could afford, and would enjoy peas-soup, plum-pudding, bread, barley-soup, sugar, rice, and even cheese, with us. It could not endure to be caressed, but was exceedingly fond of company, and would sit for hours listening to a conversation, which was no sooner ended than he would retire to his cabin; he was a continual source of amusement by his sagacity and playfulness." * * * "The fur of the Polar Hare is so exceedingly soft, that an Esquimaux woman spun some of its wool into a thread, and knitted several pairs of gloves, one pair of which, beautifully white, came into my possession. It resembled the Angola wool, but was still softer."

The specimen we procured in Newfoundland weighed seven and a half pounds; it was obtained on the 15th August, in the midst of summer, when all hares are lean. It was at a period of the year also, when in that island they are incessantly harassed by the troublesome moose-fly. Deer, hares, &c., and even men, suffer very much in consequence of their attacks. The Indians we saw there, although tempted by a high reward, refused to go in search of these Hares, from a dread of this persecuting insect; and our party, who had gone on a moose-hunt, were obliged by the inflammation succeeding the bites inflicted on them to return on the same day they started.

Dr. Richardson sets down the weight of a full grown Polar Hare as varying according to its condition from seven to fourteen pounds.

In Beachy's Narrative there is an account of a Polar Hare, killed on the 15th May, that weighed nearly twelve pounds; and Hearne (see Journey, p. 383) says that, "in good condition many of them weigh from fourteen to fifteen pounds."

GEOGRAPHICAL DISTRIBUTION.

This species occupies a wide range in the northern portions of our continent; it extends from the shores of Baffin's Bay across the continent to Behring's Straits. It has been seen as far north as the North Georgian Islands, in latitude 75°. On the western portion of the American continent it has not been found further to the south than latitude 64°, but on the eastern coast it reaches much farther south. Richardson has stated that its most southerly known habitat is in the neighbourhood of Fort

Churchill, on Hudson's Bay, which is in the 58th parallel of latitude, but remarks, that it may perhaps extend farther to the southward on the elevated ridges of the Rocky Mountains, or on the Eastern coast, in Labrador. We have ascertained that on the eastern coast of America it exists at least ten and a half degrees south of the latitude assigned to it above; as we procured our specimen at Newfoundland, in latitude 47½°, where it was quite common; and we have been informed that it also exists in the northern portions of Nova Scotia. To the north-east, it has found its way across Baffin's Bay, and exists in Greenland.

GENERAL REMARKS.

Although the Polar Hare was noticed at a very early period in the history of America, until recently it was considered identical with other species that have since been ascertained to differ from it. The writer of the History of Discoveries and Settlements of the English in America, from the reign of Henry VII. to the close of that of Queen Elizabeth, speaking of the animals at Churchill and Hudson's Bay, (see PINKERTON, Voy., vol. vii., p. 276,) says, "the hares grow white in winter, and recover their colour in spring; they have very large ears which are always black; their skins in winter are very pretty, of fine long hair which does not fall; so that they make very fine muffs."

There can be no doubt that the Polar Hare was here alluded to. PENNANT remarked that its size was greater than that of the varying hare, with which it had so long been considered identical. HEARNE, who observed it on our continent, and FABRICIUS, who obtained it in Greenland, regarded it as the varying hare. LEACH, in 1814, (Zoological Miscellany,) characterized it as a new species. It was subsequently noticed by SABINE, FRANKLIN, and RICHARDSON. As an evidence of how little was known of our American hares until very recently, we would refer to the fact that in the last *general* work on American quadrupeds by an American author, published by Dr. GODMAN in 1826, only two hares were admitted into our Fauna—*Lepus Americanus,* by which he referred to our gray rabbit, and *Lepus glacialis,* which together with *Lepus Virginianus* of HARLAN, he felt disposed to refer to *Lepus variabilis* of Europe, leaving us but one native species, and even to that applying a wrong name. We hope in this work to be able to present our readers with at least fourteen species of true hares, that exist in America north of the tropic of Cancer, all peculiar to this country.

In 1829 Dr RICHARDSON gave an excellent description, (Fauna Boreali Americana, p. 221,) removing every doubt as to *Lepus glacialis* being a true species. In 1838, having obtained a specimen in summer pelage, the only one that as far as we have learned existed in any collection in our country, we were induced to describe it, (Journal Acad. Nat. Sciences, Philadelphia, vol. vii., p. 285.)

GENUS PUTORIUS.—Cuv

DENTAL FORMULA.

$$Incisive\ \frac{6}{6};\quad Canine\ \frac{1-1}{1-1};\quad Molar\ \frac{4-4}{5-5} = 34.$$

There are two false molars above, and three below; the great carnivorous tooth below, without an internal tubercle; the tuberculous tooth in the upper jaw, very long.

Head, small and oval; muzzle, short and blunt; ears, short and round; body, long and vermiform; neck, long; legs, short; five toes on each foot, armed with sharp crooked claws; tail, long and cylindrical. Animals of this genus emit a fetid odour, and are nocturnal in habit; they are separated from the martens in consequence of having one tooth less on each side of the upper jaw; their muzzle is also shorter and thicker than that of the marten. The species are generally small in size, and seldom climb trees like the true martens.

There are about fifteen well determined species of this genus, six of which belong to America, and the remainder to the Eastern continent.

The generic name *putorius* is derived from the Latin word *putor*—a fetid smell.

PUTORIUS VISON.—Linn.

Mink.

PLATE XXXIII. Male and Female.

P. fulvus, mente albo; auribus curtis; pedibus semi-palmatis; cauda corporis dimidiam longa. Mustela marte minor.

CHARACTERS.

Less than the pine marten; general colour, brown; chin white; ears short; feet semi-palmate; tail, half the length of the body.

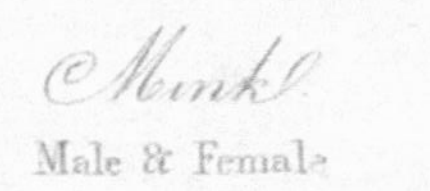

Drawn on stone by R. Trembly

Mink

Male & Female

Drawn from Nature by J. J. Audubon, F.R.S. F.L.S.

Printed & Col^{d} by J. T. Bowen, Philad^{a}

SYNONYMES.

THE MINK, Smith's Virginia, 1624. Quoted from Pinkerton's Voyages, vol. xiii., p. 31.
OTAY, Sagard Theodat, Hist. du Can., p. 749, A. D. 1636.
FOUTEREAU, La Hontan, Voy. 1., p. 81, A. D. 1703.
MINK, Kalm's Travels, Pinkerton's Voy., vol. xiii., p. 522.
LE VISON, Buffon, xiii., p. 308, t. 43.
MUSTELA VISON, Linn., Gmel., i., p. 94.
MINX, Lawson's Carolina, p. 121.
MUSTELA LUTREOLA, Forster, Phil. Trans., lxii., p. 371.
MINX OTTER, Pennant, Arct. Zool., i., p. 87.
VISON WEASEL, Ibid., i., p. 78.
JACKASH, Hearne's Journey, p. 376.
MUSTELA VISON, Cuv., Règne Anim., vol. i., p. 150, t. 1. fig. 2.
MUSTELA LUTREOLA, Sabine, Frank Journ., p. 652.
MUSTELA VISON, and M. LUTREOCEPHALA, Harlan, Fauna, p. 63, 65.
MINK, Godman, Nat. Hist., vol. i., p. 206.
PUTORIUS VISON, Dekay, Nat. Hist. New-York, p. 37, fig. 3, a. b. skull.

DESCRIPTION.

Body, long and slender; head, small and depressed; nose, short, flat, and thick; eyes, small, and placed far forward; whiskers, few, and reaching to the ears; ears broad, short, rounded, and covered with hair; neck, very long; legs, short and stout. The toes are connected by short hairy webs, and may be described as semi-palmated. There are short hairs on the webs above and below. Claws, very slightly arched, and acute. On the fore-feet, the third and fourth toes, counting from the inner side, are about of equal length; the second a line shorter, the fifth a little less, and the first, shortest. On the hind-feet, the third and fourth toes are equal, the second and fifth shorter and nearly equal, and the first very short. There are callosities on the toes resembling in miniature those on the toes of the Bay lynx. The feet and palms are covered with hair even to the extremity of the nails; tail, round, and thick at the roots, tapering gradually to the end; the longer hairs of the tail are inclined to stand out horizontally, giving it a bushy appearance. There are two brown-coloured glands situated on each side of the under surface of the tail, which have a small cavity lined by a thin white wrinkled membrane; they contain a strong musky fluid, the smell of which is rather disagreeable. Mammæ, six, ventral.

The coat is composed of two kinds of hair; a very downy fur beneath, with hairs of a longer and stronger kind interspersed. The hairs on the upper surface are longer than those on the lower. They are smooth and glossy both on the body and the tail, and to a considerable extent conceal the downy fur beneath.

COLOUR.

Under fur, light brownish-yellow; the longer hairs, and the surface of the fur, are of a uniform brown or tawny colour, except the ears, which are a little lighter, and the sides of the face, under surface, tail, and posterior part of the back, which are a little darker than the general tint, lower jaw white. In most specimens there is a white spot under the throat, and in all that we have seen, a longitudinal white stripe on the breast between the fore-legs, much wider in some specimens than in others; tail, darkest toward the end; for an inch or two from the tip it is often very dark-brown or black.

There are some striking and permanent varieties of the Mink, both in size and colour. We possess a specimen from Canada, which is considerably darker than those of the United States. Its tail is an inch longer than usual, and the white markings on its throat and chest are much narrower and less conspicuous than in most individuals of this species. In other respects we can see no difference.

In the Southern salt-water marshes this species is considerably larger in size, the white markings on the chin and under surface are broader, the hair is much coarser, the colour lighter, and the tail less bushy, than in Northern specimens. Those, however, which we obtained on the head waters of the Edisto river are as dark as specimens from Pennsylvania and New-York.

Along the mountain streams of the Northern and Middle States, we have often met with another species of Mink considerably smaller and darker than those found on large water-courses or around mill-ponds. This species was figured in the illustrations of our large edition, without distinguishing it from the lower figure on the plate, which represents the common species. We shall introduce a separate figure of it in the present work.

DIMENSIONS.

Length of head and body	13	inches.
" tail (vertebræ)	7	do.
" tail, to end of fur	8	do.

Another specimen.

Length from point of nose to root of tail	14	do.

Length of tail (vertebræ) - - - - - -	$7\frac{1}{2}$	inches.
" tail, to end of hair - - - - -	8	do.

Dimensions of the small species, (specimen from the Catskill mountains).

Length of head and body - - - - - -	11	inches.
" tail (vertebræ) - - - - - -	6	do.
" tail, to end of hair - - - - -	7	do.

HABITS.

Next to the ermine, the Mink is the most active and destructive little depredator that prowls around the farm-yard, or the farmer's duck-pond; where the presence of one or two of these animals will soon be made known by the sudden disappearance of sundry young ducks and chickens. The vigilant farmer may perhaps see a fine fowl moving in a singular and most involuntary manner, in the clutches of a Mink, towards a fissure in a rock or a hole in some pile of stones, in the gray of the morning, and should he rush to the spot to ascertain the fate of the unfortunate bird, he will see it suddenly twitched into a hole too deep for him to fathom, and wish he had carried with him his double-barrelled gun, to have ended at once the life of the voracious destroyer of his carefully tended poultry. Our friend, the farmer, is not, however, disposed to allow the Mink to carry on the sport long, and therefore straightway repairs to the house for his gun, and if it be loaded and ready for use, (as it always should be in every well-regulated farm-house,) he speedily returns with it to watch for the re-appearance of the Mink and shoot him ere he has the opportunity to depopulate his poultry-yard. The farmer now takes a stand facing the retreat into which the Mink has carried his property, and waits patiently until it may please him to show his head again. This, however, the cunning rogue will not always accommodate him by doing, and he may lose much time to no purpose. Let us introduce you to a scene on our own little place near New-York.

There is a small brook, fed by several springs of pure water, which we have caused to be stopped by a stone dam to make a pond for ducks in the summer and ice in the winter; above the pond is a rough bank of stones through which the water filters into the pond. There is a little space near this where the sand and gravel have formed a diminutive beach. The ducks descending to the water are compelled to pass near this stony bank. Here a Mink had fixed his quarters with certainly a degree of judgment and audacity worthy of high praise, for no settlement could promise to be more to his mind. At early dawn the crowing of several fine cocks, the cackling of many hens and chickens, and the paddling, splashing, and quacking of a hundred old and young ducks would please his ears; and by stealing to the edge of the bank of stones, with his body nearly concealed between two large pieces of broken granite, he could look around and see the unsuspecting ducks within a yard or two of his lurking place. When thus on the look-out, dodging his head backward and forward he waits until one of them has approached close to him, and then with a rush seizes the bird by the neck, and in a moment disappears with it between the rocks. He has not, however, escaped unobserved, and like other rogues deserves to be punished for having taken what did not belong to him. We draw near the spot, gun in hand, and after waiting some time in vain for the appearance of the Mink, we cause some young ducks to be gently driven down to the pond—diving for worms or food of various kinds while danger so imminent is near to them—intent only on the object they are pursuing, they turn not a glance toward the dark crevice where we can now see the bright eyes of the Mink as he lies concealed. The unsuspecting birds remind us of some of the young folks in that large pond we call the world, where, alas! they may be in greater danger than our poor ducks or chickens. Now we see a fine hen descend to the water; cautiously she steps on the sandy margin and dipping her bill in the clear stream, sips a few drops and raises her head as if in gratitude to the Giver of all good; she continues sipping and advancing gradually; she has now approached the fatal rocks, when with a sudden rush the Mink has seized her; ere he can regain his hole, however, our gun's sharp crack is heard and the marauder lies dead before us.

We acknowledge that we have little inclination to say anything in defence of the Mink. We must admit, however, that although he is a cunning and destructive rogue, his next door neighbour, the ermine or common weasel, goes infinitely beyond him in his mischievous propensities. Whilst the Mink is satisfied with destroying one or two fowls at a time, on which he makes a hearty meal; the weasel, in the very spirit of wanton destructiveness, sometimes in a single night puts to death every tenant of the poultry-house!

When residing at Henderson, on the banks of the Ohio river, we observed that Minks were quite abundant, and often saw them carrying off rats which they caught like the weasel or ferret, and conveyed away in their mouths, holding them by the neck in the manner of a cat.

Along the trout streams of our Eastern and Northern States, the Mink has been known to steal fish that, having been caught by some angler, had been left tied together with a string while the fisherman proceeded farther in quest of more. A person informed us that he had lost in this way thirty or forty fine trout, which a Mink dragged off the bank into the stream and devoured, and we have been told that by looking carefully after them, the Minks could be seen watching the fisherman and in readiness to take his fish, should he leave it at any distance behind him. Mr. HUTSON of Halifax informed us that he had a salmon weighing four pounds carried off by one of them.

We have observed that the Mink is a tolerably expert fisher. On one occasion, whilst seated near a trout-brook in the northern part of the State of New-York, we heard a sudden splashing in the stream and saw a large trout gliding through the shallow water and making for some long overhanging roots on the side of the bank. A Mink was in close pursuit, and dived after it; in a moment afterwards it re-appeared with the fish in its mouth. By a sudden rush we induced it to drop the trout, which was upwards of a foot in length.

We are disposed to believe, however, that fishes are not the principal food on which the Mink subsists. We have sometimes seen it feeding on frogs and cray-fish. In the Northern States we have often observed it with a WILSON's meadow-mouse in its mouth, and in Carolina the very common cotton-rat furnishes no small proportion of its food. We have frequently remarked it coursing along the edges of the marshes, and found that it was in search of this rat, which frequents such localities, and we discovered that it was not an unsuccessful mouser. We once saw a Mink issuing from a hole in the earth, dragging by the neck a large Florida rat.

This species has a good nose, and is able to pursue its prey like a hound following a deer. A friend of ours informed us that once while standing on the border of a swamp near the Ashley river, he perceived a marsh-hare dashing by him; a moment after came a Mink with its nose near the ground, following the frightened animal, apparently by the scent, through the marsh.

In the vicinity of Charleston, South Carolina, a hen-house was one season robbed several nights in succession, the owner counting a chicken less every morning. No idea could be formed, however, of the manner in which it was carried off. The building was erected on posts, and was securely locked, in addition to which precaution a very vigilant watch-dog was now put on guard, being chained underneath the chicken-house. Still, the number of fowls in it diminished nightly, and one was as before missed every morning.

We were at last requested to endeavour to ascertain the cause of the vexatious and singular abstraction of our friend's chickens, and on a careful examination we discovered a small hole in a corner of the building, leading to a cavity between the weather-boarding and the sill. On gently forcing outward a plank, we perceived the bright eyes of a Mink peering at us and shining like a pair of diamonds. He had long been thus snugly ensconced, and was enabled to supply himself with a regular feast without leaving the house, as the hole opened toward the inside on the floor. Summary justice was inflicted of course on the concealed robber, and peace and security once more were restored in the precincts of the chicken-yard.

This species is very numerous in the salt-marshes of the Southern States, where it subsists principally on the marsh-hen, (*Rallus crepitans,*) the sea-side finch, (*Ammodramus maritimus,*) and the sharp-tailed finch, (*A. caudacutus,*) which, during a considerable portion of the year, feed on the minute shell-fish and aquatic insects left on the mud and oyster-banks, on the subsiding of the waters. We have seen a Mink winding stealthily through the tall marsh-grass, pausing occasionally to take an observation, and sometimes lying for the space of a minute flat upon the mud: at length it draws its hind-feet far forwards under its body in the manner of a cat, its back is arched, its tail curled, and it makes a sudden spring. The screams of a captured marsh-hen succeed, and its upraised fluttering wing gives sufficient evidence that it is about to be transferred from its pleasant haunts in the marshes to the capacious maw of the hungry Mink.

It is at low tide that this animal usually captures the marsh-hen. We have often at high spring tide observed a dozen of those birds standing on a small field of floating sticks and matted grasses, gazing stupidly at a Mink seated not five feet from them. No attempt was made by the latter to capture the birds that were now within his reach. At first we supposed that he might have already been satiated with food and was disposed to leave the tempting marsh-hens till his appetite called for more; but we were after more mature reflection inclined to think that the high spring tides which occur, exposing the whole marsh to view and leaving no place of concealment, frighten the Mink as well as the marsh-hen; and as misery sometimes makes us familiar with strange associates, so the Mink and the marsh-hen, like neighbour and brother, hold on to their little floating islands till the waters subside, when each again follows the instincts of nature. An instance of a similar effect of fear on other animals was related to us by an old resident of Carolina: Some forty years ago, during a tremendous flood in the Santee river, he saw two or three deer on a small mound not twenty feet in diameter, surrounded by a wide sea of waters, with a cougar seated in the midst of them; both parties, having seemingly entered into a truce at a time when their lives seemed equally in jeopardy, were apparently disposed peaceably to await the falling of the waters that surrounded them.

The Minks which resort to the Southern marshes, being there furnished with an abundant supply of food, are always fat, and appear to us considerably larger than the same species in those localities where food is less abundant.

This species prefers taking up its residence on the borders of ponds and along the banks of small streams, rather than along large and broad rivers. It delights in frequenting the foot of rapids and waterfalls. When pursued it flies for shelter to the water, an element suited to its amphibious habits, or to some retreat beneath the banks of the stream. It runs tolerably well on high ground, and we have found it on several occasions no easy matter to overtake it, and when overtaken, we have learned to our cost that it was rather a troublesome customer about our feet and legs, where its sharp canine teeth made some uncomfortable indentations; neither was its odour as pleasant as we could have desired. It is generally supposed that the Mink never resorts to a tree to avoid pursuit; we have, however, witnessed one instance to the contrary. In hunting for the ruffed-grouse, (*T. umbellus,*) we observed a little dog that accompanied us, barking at the stem of a young tree, and on looking up, perceived a Mink seated in the first fork, about twelve feet from the ground. Our friend, the late Dr. WRIGHT, of Troy, informed us that whilst he was walking on the border of a wood, near a stream, a small animal which he supposed to be a black squirrel, rushed from a tuft of grass, and ascended a tree. After gaining a seat on a projecting branch, it peeped down at the intruder on its haunts, when he shot it, and picking it up, ascertained that it was a Mink.

We think, however, that this animal is not often seen to ascend a tree, and these are the only instances of its doing so which are known to us.

This species is a good swimmer, and like the musk-rat dives at the flash of a gun; we have observed, however, that the percussion-cap now in general use is too quick for its motions, and that this invention bids fair greatly to lessen its numbers. When shot in the water, the body of the Mink, as well as that of the otter, has so little buoyancy, and its bones are so heavy, that it almost invariably sinks.

The Mink, like the musk-rat and ermine, does not possess much cunning, and is easily captured in any kind of trap; it is taken in steel-traps and box-traps, but more generally in what are called dead-falls. It is attracted by any kind of flesh, but we have usually seen the traps baited with the head of a ruffed-grouse, wild duck, chicken, jay, or other bird. The Mink is exceedingly tenacious of life, and we have found it still alive under a dead-fall, with a pole lying across its body pressed down by a weight of 150 lbs., beneath which it had been struggling for nearly twenty-four hours.

This species, as well as the skunk and the ermine, emits an offensive odour when provoked by men or dogs, and this habit is exercised likewise in a moderate degree whenever it is engaged in any severe struggle with an animal or bird on which it has seized. We were once attracted by the peculiar and well known plaintive cry of a hare, in a marsh on the side of one of our southern rice-fields, and our olfactories were at the same time regaled with the strong fetid odour of the Mink; we found it in possession of a large marsh-hare, with which, from the appearance of the trampled grass and mud, it had been engaged in a fierce struggle for some time.

The latter end of February or the beginning of March, in the latitude of Albany, N. Y., is the rutting season of the Mink. At this period the ground is usually still covered with snow, but the male is notwithstanding very restless, and his tracks may every where be traced, along ponds, among the slabs around saw-mills, and along nearly every stream of water. He seems to keep on foot all day as well as through the whole night. Having for several days in succession observed a number of Minks on the ice hurrying up and down a mill-pond, where we had not observed any during a whole winter, we took a position near a place which we had seen them pass, in order to procure some of them.

We shot six in the course of the morning, and ascertained that they were all large and old males. As we did not find a single female in a week, whilst we obtained a great number of males, we came to the conclusion that the females, during this period, remain in their burrows. About the latter end of April the young are produced. We saw six young dug from a hole in the bank of a Carolina rice-field; on another occasion we found five enclosed in a large nest situated on a small island in the marshes of Ashley river. In the State of New-York, we saw five taken from a hollow log, and we are inclined to set down that as the average number of young this species brings forth at a time.

The Mink, when taken young, becomes very gentle, and forms a strong attachment to those who fondle it in a state of domestication. RICHARDSON saw one in the "possession of a Canadian woman, that passed the day in her pocket, looking out occasionally when its attention was roused by any unusual noise." We had in our possession a pet of this kind for eighteen months; it regularly made a visit to an adjoining fish-pond both morning and evening, and returned to the house of its own accord, where it continued during the remainder of the day. It waged war against the Norway rats which had their domicile in the dam that formed the fish-pond, and it caught the frogs which had taken possession of its banks. We did not perceive that it captured many fish, and it never attacked the poultry. It was on good terms with the dogs and cats, and molested no one unless its tail or foot was accidentally trod upon, when it invariably revenged itself by snapping at the foot of the offender.

It was rather dull at mid-day, but very active and playful in the morning and evening and at night. It never emitted its disagreeable odour except when it had received a sudden and severe hurt. It was fond of squatting in the chimney-corner, and formed a particular attachment to an arm-chair in our study.

The skins of the Mink were formerly an article of commerce, and were used for making muffs, tippets, &c.; they sold for about fifty cents each. RICHARDSON states that they at present are only taken by the traders of the fur company to accommodate the Indians, and that they are afterwards burnt, as they will not repay the expense of carriage. The fur, however although short, is even finer than that of the marten.

A short time since, we were kindly presented by CHARLES P. CHOUTEAU Esq., with a Mink skin of a beautiful silver-gray colour, the fur of which is quite different from the ordinary coat of the animal. These beautiful skins are exceedingly rare, and six of them, when they are united, will make a muff, worth at least a hundred dollars. A skin, slightly approaching the fine quality and colour of the one just mentioned, exists in the Academy of Natural Sciences at Philadelphia, but it is brownish, and the fur is not very good.

GEOGRAPHICAL DISTRIBUTION.

The Mink is a constant resident of nearly every part of the continent of North America. RICHARDSON saw it as far north as latitude 66°, on the banks of the Mackenzie river, and supposed that it ranged to the mouth of that river in latitude 69°; it exists in Canada, and we have seen it in every State of the Union. We observed it on the Upper Missouri and on the Yellow Stone river; it is said to exist also to the West of the Rocky Mountains and along the shores of the Pacific ocean.

GENERAL REMARKS.

This species appears, as far as we have been able to ascertain, to have been first noticed by Governor SMITH of Virginia, in 1624, and subsequently by SAGARD THEODAT and LA HONTAN. The latter calls it an amphibious sort of little pole-cat,—"Les fouteriaux, qui sont de petites fouines amphibies." KALM and LAWSON refer to it; the former stating that the English and the Swedes gave it the name of Mink, Moenk being the name applied to a closely allied species existing in Sweden.

The doubts respecting the identity of the American Mink (*P. vison*) and the *Mustela lutreola* of the north of Europe, have not as yet been satisfactorily solved. PENNANT in one place admits the American *vison* as a true species, and in another supposes the *M. lutreola* to exist on both continents. Baron CUVIER at one time regarded them as so distinct that he placed them under different genera; but subsequently in a note stated his opinion that they are both one species. Dr. GODMAN supposed that both the *Pekan* (*Mustela Canadensis*) and *vison* (*P. vison*) are nothing more than mere varieties of *Mustela lutreola;* in regard to the Pekan he was palpably in error. RICHARDSON considers them distinct species, although he does not seem to have had an opportunity of instituting a comparison. We have on two or three occasions compared specimens from both continents. The specimens, however, from either country differ so considerably among themselves, that it is somewhat difficult without a larger number than can generally be brought together, to institute a satisfactory comparison.

The fact that both species exist far to the northward, and consequently approach each other toward the Arctic circle, presents an argument favourable to their identity. In their semi-palmated feet, as well as in their general form and habits, they resemble each other.

The following reasons, however, have induced us, after some hesitation, and not without a strong desire for farther opportunities of comparison, especially of the skulls, to regard the American *P. vison* as distinct from the *lutreola* of the north of Europe.

P. lutreola, in the few specimens we have examined, is smaller than *P. vison*, the body of the latter frequently exceeding eighteen inches, (we have a large specimen that measures twenty-one inches,) but we have never found any specimen of the *lutreola* exceeding thirteen inches from nose to root of tail, and have generally found that specimens, even when their teeth were considerably worn, thereby indicating that the animals were adults, measured less than twelve inches.

P. lutreola is considerably darker in colour, resembling in this respect the small black species mentioned by us as existing along our mountain streams. The tail is less bushy, and might be termed sub-cylindrical. *P. lutreola* is, besides, more deficient in white markings on the under surface than the other species; the chin is generally, but not always, white; but there is seldom any white either on the throat or chest.

Black Squirrel.

Drawn from Nature by J.J.Audubon, F.R.S.F.L.S. Lith⁴, Printed & Col⁴ by J.T. Bowen, Philad^a.

SCIURUS NIGER.—Linn.

Black Squirrel.

PLATE XXXIV.—Male and Female.

S. corpore S. migratorio longiore; vellere molli nitidoque. auribus, naso et omni corporis parte nigerrimis, cirris albis dispersis.

CHARACTERS.

A little larger than the Northern gray squirrel; fur, soft and glossy; ears, nose, and all the body, black; a few white tufts of hair interspersed.

SYNONYMES.

Sciurus Niger, Godman, Nat. Hist., vol. ii., p. 133.
" " Bachman, Proceedings Zool. Society, 1838, p. 96.
" " Dekay, Nat. Hist. of New-York, part i., p. 60.

DESCRIPTION.

Head, a little shorter and more arched than that of the Northern gray squirrel, (in the latter species, however, it is often found that differences exist, in the shape of the head, in different individuals.) Incisors, compressed, strong, and of a deep orange colour anteriorly; ears, elliptical, and slightly rounded at the tip, thickly clothed with fur on both surfaces, the fur on the outer surface extending three lines beyond the margin; there are however no distinct tufts; whiskers, a little longer than the head; tail, long, not very distichous, thickly clothed with moderately coarse hair; the fur is softer than that of the Northern gray squirrel.

COLOUR.

The whole of the upper and lower surfaces, and the tail, glossy jet black; at the roots the hairs are a little lighter. Specimens procured in summer do not differ materially in colour from those obtained in winter, except that before the hairs drop out late in spring, they are not so intensely black. In all we have had an opportunity of examining, there are small tufts of white hairs irregularly disposed on the under surface, resembling those on the body of the mink. There are also a few scattered white hairs on the back and tail.

DIMENSIONS.

	Inches.	Lines.
Length of head and body - - - - - -	13	0
" tail (vertebræ) - - - - - -	9	1
" tail, including fur - - - -	13	0
Palm, to end of middle fore-claw - - - - -	1	7
Length of heel to the point of middle claw - -	2	7
" fur on the back - - - - - -	0	7
Breadth of tail with hair extended - - -	5	0

HABITS.

An opportunity was afforded us, many years since, of observing the habits of this species, in the northern part of the State of New-York. A seat under the shadow of a rock near a stream of water, was for several successive summers our favourite resort for retirement and reading. In the immediate vicinity were several large trees, in which were a number of holes, from which at almost every hour of the day were seen issuing this species of Black Squirrel. There seemed to be a dozen of them; they were all of the same glossy black colour, and although the Northern gray squirrel and its black variety were not rare in that neighbourhood, during a period of five or six years we never discovered any other than the present species in that locality; and after the lapse of twenty years, a specimen (from which our description was in part drawn up) was procured in that identical spot, and sent to us.

This species possesses all the sprightliness of the Northern gray squirrel, evidently preferring valleys and swamps to drier and more elevated situations. We observed that one of their favourite trees, to which they retreated on hearing the slightest noise, was a large white-pine (*Pinus strobus*) in the immediate vicinity. We were surprised at sometimes seeing a red squirrel, (*Sciurus Hudsonius*,) which had also given a preference to this tree, pursuing a Black Squirrel, threatening and scolding it vociferously, till the latter was obliged to make its retreat. When the Squirrels approached the stream, which ran within a few feet of our seat, they often stopped to drink, when, instead of lapping the water like the dog and cat, they protruded their mouths a considerable distance into the stream, and drank greedily; they would afterwards sit upright, supported by the tarsus, and with tail erect, busy themselves for a quartėr of an hour in wiping their faces with their paws, the latter being also occasionally dipped in the water. Their barking and other habits did not seem to differ from those of the northern gray squirrel.

GEOGRAPHICAL DISTRIBUTION.

Many of our specimens of the Black Squirrel were procured through the kindness of friends, in the counties of Rensselaer and Queens, New York. We have seen this species on the borders of Lake Champlain, at Ogdensburg, and on the eastern shores of Lake Erie; also near Niagara, on the Canada side. The individual described by Dr. Richardson, and which may be clearly referred to this species, was obtained by Captain Bayfield, at Fort William on Lake Superior. Black Squirrels exist through all our western forests, and to the northward of our great lakes; but whether they are of this species, or the black variety of the gray squirrel, we have not had the means of deciding. It is a well ascertained fact that the Black Squirrel disappears before the Northern gray squirrel. Whether the colour renders it a more conspicuous mark for the sportsman, or whether the two species are naturally hostile, we are unable to decide. It is stated by close observers that in some neighbourhoods where the Black Squirrel formerly abounded, the Northern gray squirrel now exclusively occupies its place.

GENERAL REMARKS.

We have admitted this as a true species, not so much in accordance with our own positive convictions, as in deference to the opinions of our naturalists, and from the consideration that if it be no more than a variety, it has by time and succession been rendered a permanent race. The only certain mode of deciding whether this is a true species or merely a variety, would be to ascertain whether male and female Black Squirrels and gray squirrels associate and breed together in a state of nature. When a male and a female, however different in size and colour, unite in a wild state and their progeny is prolific, we are warranted in pronouncing them of the same species. When on the contrary, there is no such result, we are compelled to come to an opposite conclusion.

We had great doubts for many years whether this species might not eventually prove another of the many varieties of the Northern gray squirrel, (*S. migratorius.*) Although these doubts have not been altogether removed by our recent investigations, they were considerably lessened on ascertaining the uniformity in size, shape, colour, and habits of all the individuals we have seen in a living state, as well as all the prepared specimens we have examined.

Much difficulty has existed among authors in deciding on the species to which the name of *S. niger* should be appropriated. The original description by LINNÆUS was contained in the single word "*niger*." If he had made no reference to any author, his description would have served quite well, as this was the only species of squirrel purely black, that was known at that day. He however made a reference to CATESBY, who figured the black variety of the Southern fox-squirrel, (*S. capistratus*,) and BRISSON, PENNANT, ERXLEBEN, and SCHREBER referred the species in the same manner to the description and figure of CATESBY. Our American writers on natural history, as well as Dr. RICHARDSON, have however adopted the name given by LINNÆUS, and applied it to this species. We consider it advisable to retain the name, omitting the reference to CATESBY.

It is difficult to decide, from the descriptions of Drs. HARLAN and GODMAN, whether they described from specimens of the black variety of the northern gray squirrel or from the present species.

Dr. RICHARDSON has, under the head of *Sciurus niger*, (see Fauna Boreali Americana, p. 191,) described a specimen from Lake Superior, which we conceive to be the black variety of the gray squirrel; but at the close of the same article (p. 192) he described another specimen from Fort William, which answers to the description of this species.

Migratory Squirrel

Drawn from Nature by J.J. Audubon F.R.S.F.L.S. Lith.d Printed & Col.d by J.T. Bowen, Philad.a

SCIURUS MIGRATORIUS.—Aud. and Bach.

Migratory Gray Squirrel.—Northern Gray Squirrel.

PLATE XXXV.—Male, Female, and Young.

S. S Carolinense robustior, S cinereo minor; cauda corpore multo longiore; variis coloribus.

CHARACTERS.

Larger than the Carolina Gray Squirrel; smaller than the Cat-squirrel; tail, much longer than the body; subject to many varieties of colour.

SYNONYMES.

Gray Squirrel, Pennant, Arct. Zool., vol. i., p. 185, Hist. Quad., No. 272.
Sciurus Cinereus, Harlan, Fauna, p. 173.
" Carolinensis, Godman, non Gmel.
" Leucotis, Gapper, Zool. Journ., London, vol. v., p. 206, (published about 1830.)
" " Bach., Proceedings of the Zoological Society, p. 91, London, 1838.
Common, or Little Gray Squirrel, Emmons, Report, 1842, p. 66.
Sciurus Leucotis, Dekay, Nat. Hist. N. Y., p. 57.
" Vulpinus, do. do. do. p. 59.

DESCRIPTION.

This Squirrel seems to have permanently twenty-two teeth. A large number of specimens procured at different seasons of the year, some of which from the manner in which their teeth were worn appeared to be old animals, presented the small front molars in the upper jaw. Even in an old male, obtained in December, with tufted ears, (the measurements of which will be given in this article,) the small molar existed. This permanency in teeth that have been usually regarded as deciduous, would seem to require an enlargement of the characters given to this genus; it will moreover be seen that several of our species are similar to this in their dental arrangement.

Incisors, strong and compressed, a little smaller than those of the Cat-squirrel, convex, and of a deep orange colour anteriorly. The upper ones have a sharp cutting edge, and are chisel-shaped; the lower are much longer and thinner. The anterior grinder, although round and small, is as long as the second; the remaining four grinders are considerably more excavated than those of the Cat-squirrel, presenting two transverse ridges of enamel. The lower grinders corresponding to those above have also elevated crowns.

The hair is a little softer than that of the Cat-squirrel, being coarsest on the forehead.

Nose, rather obtuse; forehead, arched; whiskers, as long as the head; ears, sharply rounded, concave on both sides, covered with hair; on the outside the hairs are longest. In winter the fur projects upward about three lines beyond the margin; in summer, however, the hairs covering the ears are very short, and do not extend beyond the margin.

COLOUR.

This species appears under many varieties; there are, however, two very permanent ones, which we shall attempt to describe.

1st, Gray variety.—The nose, cheeks, a space around the eyes extending to the insertion of the neck, the upper surface of the fore and hind feet, and a stripe along the sides, yellowish-brown; the ears on their posterior surface are in most specimens brownish-yellow; in about one in ten they are dull white, edged with brown. On the back, from the shoulders there is an obscure stripe of brown, broadest at its commencement, running down to a point at the insertion of the tail. In some specimens this stripe is wanting. On the neck, sides, and hips, the colour is light gray; the hairs separately are for one half their length dark cinereous, then light umber, then a narrow mark of black, and are tipped with white; a considerable number of black hairs are interspersed, giving it a yellowish-brown colour on the dorsal aspect, and a light gray tint on the sides; the hairs in the tail are light yellowish-brown from the roots, with three stripes of black, the outer one being widest, and broadly tipped with white; the whole under surface is white. The above is the most common variety.

There are specimens in which the yellowish markings on the sides and feet are altogether wanting. Dr. Godman, (vol. ii., p. 133,) supposed that the golden colour of the hind-feet is a very permanent mark. The specimens from Pennsylvania in our possession, and a few from the Upper Missouri, have generally this peculiarity, but many of those from New-York and New-England have gray feet, without the slightest mixture of yellow.

2d, Black variety.—This we have on several occasions seen taken with the gray variety from the same nest. Both varieties breed and rear their young together.

The black ones are of the same size and form as the gray ; they are dark brownish-black on the whole upper surface, a little lighter beneath. In summer their colour is less black than in winter. The hairs of the back and sides of the body, and of the tail, are obscurely annulated with yellow. There is here and there a white hair interspersed among the fur of the body, but no tuft of white as in *Sciurus niger*.

DIMENSIONS.

A Female in summer.

	Inches.	Lines.
Length of head and body - - - - - -	11	9
" tail (vertebræ) - - - - - -	10	0
" tail, to the tip - - - - - -	13	0
Height of ear - - - - - - - - -	0	7
Palm to the end of middle claw - - - -	1	10
Heel to the end of middle nail - - - - -	2	6
Length of fur on the back - - - - -	0	5
Breadth of tail with hairs extended - - - -	4	2

An old Male in winter pelage, obtained Dec. 16th.

Length of head and body - - - - - -	12	6
" tail (vertebræ) - - - - - -	11	0
" tail, to end of hair - - - - -	14	0
Height of ear - - - - - - - - -	0	7
" ear, to end of fur - - - - -	0	9
Heel to end of longest nail - - - - -	2	6
Length of fur on the back - - - - -	0	8

Weight 1 lb. 6 oz.

HABITS.

This appears to be the most active and sprightly species of Squirrel existing in our Atlantic States. It sallies forth with the sun, and is industriously engaged in search of food for four or five hours in the morning, scratching among leaves, running over fallen logs, ascending trees, or playfully skipping from bough to bough, often making almost incredible leaps from the higher branches of one tree to another. In the middle of the day it retires for a few hours to its nest, resuming its active labours and amusements in the afternoon, and continuing them without intermission till long after the setting of the sun. During the warm weather of spring and summer it prepares itself a nest on a tree, but not often at its summit. When constructing this summer-house it does not descend to the earth in search of materials, finding them ready at hand on the tree it intends to make its temporary residence. It first breaks off some dry sticks, if they can be procured; if, however, such materials are not within reach, it gnaws off green branches as large as a man's thumb and lays them in a fork of the stem, or of some large branch. It then proceeds to the extremities of the branches, and breaks off twigs and bunches of leaves, with which a compact nest is constructed, which on the inner side is sometimes lined with moss found on the bark of the tree. In the preparation of this nest both male and female are usually engaged for an hour in the morning during several successive days; and the noise they make in cutting the branches and dragging them with their leaves to the nest can be heard at a great distance. In winter they reside altogether in holes in trees, where their young in most instances are brought forth.

Although a family, to the number of five or six, probably the offspring of a single pair the preceding season, may occupy the same nest during winter, they all pair off in spring, when each couple occupies a separate nest, in order to engage in the duties of reproduction. The young, in number from four to six, are brought forth in May or June; they increase in size rapidly, and are sufficiently grown in a few weeks to leave the nest; at this time they may be seen clinging around the tree which contains their domicile; as soon as alarmed they run into the hole, but one of them usually returns to the entrance of it, and protruding his head out of the hollow, watches the movements of the intruder. In this stage of their growth they are easily captured by stopping up the entrance of the nest, and making an opening beneath; they can then be taken out by the hand protected by a glove. They soon become tolerably gentle, and are frequently kept in cages, with a wheel attached, which revolves as they bound forward, in which as if on a treadmill they exercise themselves for hours together.

Sometimes two are placed within a wheel, when they soon learn to accommodate themselves to it, and move together with great regularity.

Notwithstanding the fact that they become very gentle in confinement, no instance has come to our knowledge of their having produced young while in a state of domestication, although in a suitable cage such a result would in all probability be attained. This species is a troublesome pet; it is sometimes inclined to close its teeth on the fingers of the intruder on its cage, and does not always spare even its feeder. When permitted to have the freedom of the house, it soon excites the displeasure of the notable housewife by its habits of gnawing chairs, tables, and books.

During the rutting season the males (like deer and some other species) engage in frequent contests, and often bite and wound each other severely. The story of the conqueror emasculating the vanquished on these occasions, has been so often repeated, that it perhaps is somewhat presumptuous to set it down as a vulgar error. It may, however, be advanced, that the admission of such skill and refinement in inflicting revenge would be ascribing to the squirrel a higher degree of physiological and anatomical knowledge than is possessed by any other quadruped. From the observations we have been enabled to make, we are led to believe that the error originated from the fact that those parts in the male, which in the rutting season are greatly enlarged, are at other periods of the year diminished to a very small size; and that, in young males especially, they are drawn into the pelvis by the contraction of the muscles. A friend, who was a strenuous believer in this spiteful propensity ascribed to the squirrel, was induced to test the truth of the theory by examining a suitable number of squirrels of this species. He obtained in a few weeks upwards of thirty males; in none of these had any mutilation taken place. Two however out of this number were triumphantly brought forward as evidence of the correctness of the general belief. On examination it appeared that these were young animals of the previous autumn, with the organs perfect, but concealed in the manner above stated.

It is generally believed that this species lays up a great hoard of food as a winter supply; it may however be reasonably doubted whether it is very provident in this respect. The hollow trees in which these Squirrels shelter themselves in winter are frequently cut down, and but a very small supply of provisions has ever been found in their nests. On following their tracks in the snow, they cannot be traced to any hoards buried in the ground. We have sometimes observed them during a warm day in winter coming from great distances into the open fields, in search of a few dry hickory nuts which were still left suspended on the trees. If provisions had been laid up nearer home, they would hardly have undertaken these long journeys, or exposed themselves to so much danger in seeking a precarious supply. In fact, this species, in cold climates, seldom leaves its nest in winter, except on a warm sunny day; and in a state of inactivity and partial torpidity, it requires but little food.

Although this Squirrel is at particular seasons of the year known to search for the larvæ of different insects, which it greedily devours, it feeds principally on nuts, seeds, and grain, which are periodically sought for by all the species of this genus; among these it seems to prefer the shell-bark, (*Carya alba,*) and several species of hickory nuts, to any other kind of food. Even when the nuts are so green as to afford scarcely any nourishment, it may be seen gnawing off the thick pericarp or outer shell, which drops in small particles to the ground like rain, and then with its lower incisors it makes a small linear opening in the thinnest part of the shell immediately over the kernel. When this part has been extracted, it proceeds to another, till in an incredibly short space of time, the nut is cut longitudinally on its four sides, and the whole kernel picked out, leaving the dividing portions of the hard shell untouched.

At the season of the year when it feeds on unripe nuts, its paws and legs are tinged by the juices of the shells, which stain them an ochrey-red colour, that wears off, however, towards spring.

Were this species to confine its depredations to the fruit of the hickory, chesnut, beech, oak and maple, it would be less obnoxious to the farmer; but unfortunately for the peace of both, it is fond of the green Indian-corn and young wheat, to which the rightful owner imagines himself to have a prior claim. A war of extermination consequently ensues, and various inducements have been held out at different times to tempt the gunner to destroy it. In Pennsylvania an ancient law existed offering three pence a head for every squirrel destroyed, and in one year (1749) the sum of eight thousand pounds was paid out of the treasury in premiums for the destruction of these depredators. This was equal to 640,000 individuals killed. In several of the Northern and Western States the inhabitants, on an appointed day, are in the habit of turning out on what is called a squirrel hunt. They arrange themselves under opposite leaders, each party being stimulated by the ambition of killing the greatest number, and fastening on the other the expense of a plentiful supper. The hunters range the forest in every direction, and the accounts given us of the number of squirrels brought together at the evening rendezvous are almost incredible.

In addition to the usual enemies of this species in the Northern States, such as the weasel, fox, lynx, &c., the red-tailed hawk seems to regard it as his natural and lawful prey. It is amusing to see the skill and dexterity exercised by the hawk in the attack, and by the squirrel in attempting to escape. When the hawk is unaccompanied by his mate, he finds it no easy matter to secure the little animal; unless the latter be pounced upon whilst upon the ground, he is enabled by dodging and twisting round a branch to evade the attacks of the hawk for an hour or more, and frequently worries him into a reluctant retreat.

But the red-tails learn by experience that they are most certain of this prey when hunting in couples. The male is frequently accompanied by his mate, especially in the breeding season, and in this case the Squirrel is soon captured. The hawks course rapidly in opposite directions above and below the branch; the attention of the Squirrel is thus divided and distracted, and before he is aware of it the talons of one of the hawks are in his back, and with a shriek of triumph the rapacious birds bear him off, either to the aerie in which their young are deposited, to some low branch of a tree, or to a sheltered situation on the ground, where with a suspicious glance towards each other, occasionally hissing and grumbling for the choice parts, the hawks devour their prey.

This species of squirrel has occasionally excited the wonder of the populace by its wandering habits and its singular and long migrations. Like the lemming (*Lemmus Norvegicus*) of the Eastern continent, it is stimulated either by scarcity of food, or by some other inexplicable instinct, to leave its native haunts, and seek for adventures or for food in some (to it) unexplored portion of our land.

The newspapers from the West contain many interesting details of these migrations; they appear to have been more frequent in former years than at the present time. The farmers in the Western wilds regard them with sensations which may be compared to the anxious apprehensions of the Eastern nations at the flight of the devouring locust. At such periods, which usually occur in autumn, the Squirrels congregate in different districts of the far North-west; and in irregular troops bend their way instinctively in an eastern direction. Mountains, cleared fields, the narrow bays of some of our lakes, or our broad rivers, present no unconquerable impediments. Onward they come, devouring on their way every thing that is suited to their taste, laying waste the corn and wheat-fields of the farmer; and as their numbers are thinned by the gun, the dog, and the club, others fall in and fill up the ranks, till they occasion infinite mischief, and call forth more than empty threats of vengeance. It is often inquired, how these little creatures, that on common occasions have such an instinctive dread of water, are enabled to cross broad and rapid rivers, like the Ohio and Hudson for instance. It has been asserted by authors, and is believed by many, that they carry to the shore a suitable piece of bark, and seizing the opportunity of a favourable breeze, seat themselves upon this substitute for a boat, hoist their broad tails as a sail, and float safely to the opposite shore. This together with many other traits of intelligence ascribed to this species, we suspect to be apocryphal. That they do migrate at irregular, and occasionally at distant periods, is a fact sufficiently established; but in the only two instances in which we had opportunities of witnessing the migrations of these Squirrels, it appeared to us, that they were not only unskilful sailors but clumsy swimmers. One of these occasions, (as far as our recollection serves us), was in the autumn of 1808 or 1809; troops of Squirrels suddenly and unexpectedly made their appearance in the neighbourhood; among them were varieties not previously seen in those parts; some were broadly striped with yellow on the sides, and a few had a black stripe on each side, bordered with yellow or brown, resembling the stripes on the sides of the Hudson's Bay Squirrel, (*S. Hudsonius.*) They swam the Hudson in various places between Waterford and Saratoga; those which we observed crossing the river were swimming deep and awkwardly, their bodies and tails wholly submerged; several that had been drowned were carried downwards by the stream; and those which were so fortunate as to reach the opposite bank were so wet and fatigued, that the boys stationed there with clubs found no difficulty in securing them alive or in killing them. Their migrations on that occasion did not, as far as we could learn, extend farther eastward than the mountains of Vermont; many remained in the county of Rensselaer, and it was remarked that for several years afterwards squirrels were far more numerous there than before. It is doubtful whether any ever return to the West, as, finding forests and food suited to their taste and habits, they take up their permanent residence in their newly explored country, where they remain and propagate their species, until they are gradually thinned off by the increase of inhabitants, new clearings, and the dexterity of the sportsmen around them. The other instance occurred in 1819, when we were descending the Ohio river in a flat-boat, or ark, chiefly with the intention of seeking for birds then unknown to us. About one hundred miles below Cincinnati, as we were floating down the stream, we observed a large number of Squirrels swimming across the river, and we continued to see them at various places, until we had nearly reached Smithland, a town not more than about one hundred miles above the mouth of the Ohio.

At times they were strewed, as it were, over the surface of the water, and some of them being fatigued, sought a few moments' rest on our long "steering oar," which hung into the water in a slanting direction over the stern of our boat. The boys, along the shores and in boats, were killing the Squirrels with clubs in great numbers, although most of them got safe across. After they had reached the shore we saw some of them trimming their fur on the fences or on logs of drift-wood.

We kept some of these Squirrels alive; they were fed with hickory nuts, pecans, and ground or pea-nuts, (*Arachis hypogæa.*) Immediately after eating as much as sufficed for a meal, they hid away the remainder beneath the straw and cotton at the bottom of their cage in a little heap. A very tame and gentle one we had in a room at Shippingport, near Louisville, Kentucky, one night ate its way into a bureau, in which we had a quantity of arsenic in powder, and died next morning a victim to curiosity or appetite, probably the latter, for the bureau also contained some wheat.

GEOGRAPHICAL DISTRIBUTION.

This species exists as far to the north as Hudson's Bay. It was formerly very common in the New-England States, and in their least cultivated districts is still frequently met with. It is abundant in New-York and in the mountainous portions of Pennsylvania. We have observed it on the northern mountains of Virginia, and we obtained several specimens on the Upper Missouri. The black variety is more abundant in Upper Canada, in the western part of New-York, and in the States of Ohio and Indiana, than elsewhere. The Northern Gray Squirrel does not exist in any of its varieties in South Carolina, Georgia, Florida, or Alabama; and among specimens sent to us from Louisiana, stated to include all the squirrels existing in that State, we did not discover this species.

GENERAL REMARKS.

There exists a strong general resemblance among all our species of this genus, and it is therefore not surprising that there should have been great difficulty in finding characters to designate the various species. In the museums we examined in Europe, we observed that several species had been confounded, and we were every where told by the eminent naturalists with whom we conversed on the subject, that they could find no characters by which the different species could be distinguished. The little Carolina Gray Squirrel was first described by GMELIN. DESMAREST, who created a confusion among the various species of this genus, which is almost inextricable, confounded three species—the Northern Gray Squirrel, the Southern Carolina Squirrel, and the Cat-squirrel—under the name of *Sc. cinereus*, and gave them the diminutive size of ten inches six lines. His article was literally translated by HARLAN, including the measurements, (DESM., Mamm., p. 332; HARLAN'S Fauna, p. 173,) and he also apparently blended the three species—*S. cinereus*, *S. migratorius*, and *S. Carolinensis*. GODMAN called the Northern species *S. Carolinensis*, and LECONTE, who appears to have had a more correct view of the species generally than all previous authors, (see Appendix to McMURTRIE'S translation of CUVIER, vol. i., p. 433,) regarded the Carolina and the Northern Gray Squirrel as identical.

In 1833 and 1834 GAPPER, (Zoological Journal, vol. v., p. 201,) found in Upper Canada an individual, of what we suppose to be a variety of the Northern Gray Squirrel, with white ears, with the upper parts varied with mixture of white, black and ochre, and with a stripe of similar colour along the sides. Supposing it to be a species different from the common Gray Squirrel, he bestowed on it the characteristic name of *Sciurus Leucotis* (white eared). In our monograph of the genus SCIURUS, read before the Zoological Society, (Proceedings Zool. Soc., 1838, Op. Sup., cit., p. 91,) we adopted the name of GAPPER, without having seen his description, having been informed by competent naturalists that he had described this species.

Having, however, afterwards obtained a copy of the articles of GAPPER, and ascertained that he had described a variety that is very seldom met with, we were anxious to rid our nomenclature of a name which is very inappropriate to this species, and which is calculated constantly to mislead the student of nature.

GAPPER compared his specimen with the Northern Gray Squirrel, and finding that the latter species was gray, and not of an ochreous colour like the one he described, with ears not white but of the colour of the back, he regarded his variety as a different species. He designated the Northern Gray Squirrel as the Carolina Squirrel, the difference between the Northern and Southern Gray Squirrels not having been pointed out till it was done in our monograph four years afterwards.

As a general rule, we adhere to the views entertained by naturalists that it is best to retain a name once imposed, however inappropriate, unless likely to propagate important errors; in the present instance, however, we propose the name of *S. migratorius*, as applicable to the wide-ranging habits of this Squirrel, it being the only one in our country that appears to possess this peculiarity.

The name *leucotis* is appropriate only to the Southern Fox-squirrel, which has permanently, and in all its varieties, white ears.

We have been somewhat at a loss where to place the species given as the Fox-squirrel, *S. vulpinus* of DEKAY, (see Nat. Hist. New-York, p. 59,) and have marked our quotation with a doubt. His description does not apply very well to the Pennsylvania Fox-squirrel, (*S. cinereus*,) of which GMELIN'S *S. vulpinus* is only a synonyme. He states indeed, "We suspect that GODMAN'S Fox-squirrel as well as his Cat-squirrel, are varieties only of the Hooded-squirrel, and not to be referred to our Northern animal." We have, in our article on *S. cinereus*, noticed the errors contained in the above quotation, and only allude to it here as a possible clue to the species he had in view, viz., "not the species" given by GODMAN as *S. cinereus*, but another that agrees with the Northern Gray Squirrel "in every particular except the size." He further adds, that "its habits and geographical distribution are the same as in the preceding," meaning the Northern Gray Squirrel.

He evidently has reference to a larger species of the Gray Squirrel as existing in the same localities, with "the hair on the posterior surface of the ears projecting two lines beyond the margins," differing from the species he had just described as the Northern Gray Squirrel, which he characterized as having ears "covered with short hairs; no pencil of hairs at the tips." Although his figure resembles in several particulars that of the Cat-squirrel, (*S. cinereus*,) parts of his description and his account of the habits seem more appropriate to the tufted winter specimens of the present species. The appearance of the ears in specimens obtained in winter and summer pelage differs so widely that we ourselves were for many years misled by the tufts and large size of the old in winter. We recollect that in our school-boy days we were in the habit of obtaining many specimens of the Gray Squirrel during summer and autumn, which answered to the description of *S. migratorius*, having their ears clothed with short hairs which did not project beyond the margins on the posterior surface. During the following winter, however, we occasionally caught in a steel-trap a specimen much larger, very fat, and with ears tufted like that described as *S. vulpinus*; and we prepared the specimens under an impression that a new species had made its appearance in the neighbourhood. The following summer, however, we procured in that locality no other than the common Gray Squirrel, destitute of the fringes on the ears. We now resorted to a different mode of solving the problem. We obtained several young Northern Gray Squirrels, which we kept in cages; during the first winter their ears underwent no particular change. But in the month of December of the second year, when they had become very fat and had grown considerably larger, their ears on the posterior surface became fringed and exactly corresponded with the winter specimens we had previously obtained. As we could not feel a perfect confidence in our own notes made more than thirty years ago, we recently made inquiries from Dr. LEONARD, of Lansingburg, New-York, an accurate and intelligent naturalist, whose answer we subjoin:—"It is considered established by naturalists and observing sportsmen, that the Gray Squirrel, after the first year, has fringed ears in its winter pelage, and that of course there is but one species. Of ten prepared specimens, which I have recently examined, eight have bare ears, and two (one of them being of the black variety) have the ears fringed ; differing in no other respect, except the general fuller development of the hair, from the other specimens of their respective varieties."

We are moreover under an impression that the specimen of the Northern Gray Squirrel, from which DEKAY took his measurement, must have been a young animal. He gives head and body, eight inches : tail, eight inches five lines. Out of more than fifty specimens that we have measured in the flesh, there was not one that measured less than ten inches in body and eleven inches in tail.

The true *S. cinereus* or *S. vulpinus* has moreover not the same.geographical range as the Northern Gray Squirrel. It is not found in Canada, where the present species is common, nor in the most northerly parts of either New-York or the New-England States. We obtained several specimens from the New-York market, and as we have shown in our article on *S. cinereus*, it is occasionally found in the southern counties of the State ; but it is a very rare species north and east of Pennsylvania, and is principally confined to the Middle and some of the South-western States.

The Northern Gray Squirrel (*S. migratorius*) may be easily distinguished from the Carolina Gray Squirrel (*S. carolinensis*) by its larger size, broader tail, and lighter gray colours on the sides, and by its smaller persistent tooth.

S. cinereus or *S. vulpinus* differs from this species in being a little longer, having a much stouter body and legs, and a longer tail. It has, in proportion to its size, shorter ears, which are more rounded, and have the tufts or fringes in winter much shorter. The fur is also coarser, and it has in each upper jaw but four teeth, dropping its milk-tooth when very young, whilst the Northern Gray Squirrel (*S. migratorius*) has five on each side, which appear to be permanent.

Canada Porcupine

Drawn from Nature by J.J. Audubon, F.R.S.F.L.S.

Lith.d Printed & Col.d by J.T. Bowen Philad.a

GENUS HYSTRIX.—Linn.

DENTAL FORMULA.

$$Incisive\ \frac{2}{2};\quad Canine\ \frac{0-0}{0-0};\quad Molar\ \frac{4-4}{4-4} = 20.$$

Superior incisors, on the anterior portion, smooth, cuneiform at their extremity; inferior incisors, strong and compressed.

Molars, compound, with flat crowns, variously modified by plates of enamel, between which are depressed intervals.

Head, strong; snout, thick and tumid; ears, short and round; tongue, bristled with spiny scales; fore-feet, four-toed; hind-feet, five-toed; all the toes armed with powerful nails.

Spines on the body, sometimes intermixed with hair; tail, moderately long, in some species of the genus, prehensile.

Herbivorous, feeding principally on grain, fruits, roots, and the bark of trees—dig holes in the earth, or nestle in the hollows of trees.

The generic name is derived from the Greek word, ὑστριξ, (*hustrix*,) a porcupine—ὑς, (*hus*,) a hog, and θριξ, (*thrix*,) a bristle.

There are two species in North, and three in South America, one in Southern Europe, one in Africa, and one in India.

HYSTRIX DORSATA.—Linn.

Canada Porcupine.

PLATE XXXVI.—Male.

H. spinus brevibus, vellere sublatentibus; sine jubea; capite et collo setis longis vestitis; colore inter fulvum et nigrum variante.

CHARACTERS.

Spines, short, partially concealed by long hair; no mane; long bristles on the head and neck; colour, varying between light-brown and black.

SYNONYMES.

Hystrix Pilosus Americanus, Catesby, Cuv., App., p. 30, 1740.
The Porcupine from Hudson's Bay, Edwards' Birds, p. 52.
Hystrix Hudsonius, Brisson, Règne Animal, p. 128.
Hystrix Dorsata, Linn., Syst., Edwards, xii., p. 57.
" " Erxleben, p. 345.
" " Schreber, Säugethiere. p. 605.
L'Urson, Buffon, vol. xii., p. 426.
Canada Porcupine, Forst., Phil. Trans., vol. lxii., p. 374.
" " Penn., Quadrupeds, vol. ii., p. 126.
" " Arctic Zoology, vol. i., p. 109.
The Porcupine, Hearne's Journal, p. 381.
Erethizon Dorsatum, F. Cuv., in Mém. du Mus., ix., t. 20.
Porc-Epic Velu, Cuv., Règne Animal, i., p. 209.
Hystrix Dorsata, Sabine, Franklin's Journ., p. 664.
" " Harlan, Fauna, p. 109.
" " Godman, Nat. Hist., vol. ii., p. 160.
" Pilosus, Rich., Fauna Boreali Americana, p. 214.
" Hudsonius, Dekay, Nat. Hist. New-York, p. 77.

DESCRIPTION.

The body of this species is thick, very broad, cylindrical, and to a high degree clumsy. The back is much arched in a curve from the nose to the buttocks, when it declines in an angle to the tail.

The whole upper surface of the body from the nose to the extremity of the tail is covered by long and rather coarse hair, intermixed with a dense mass of spines or quills. These are of a cylindrical shape, very sharp at the extremity and pointed at the roots. The animal is capable of erecting them at pleasure, and they are detached by the slightest touch; they are barbed with numerous small reversed points or prickles, which, when once inserted in the flesh, will by the mere movement of the limbs work themselves deeper into the body. There seems to be in certain parts of the body of this species a regular gradation from hair to spines; on the nose for instance, the hair is rather soft, a little higher up it is succeeded by bristles intermixed with small spines. These spines continue to lengthen on the hinder parts of the head, to increase in size on the shoulders, and are longer and more rigid on the buttocks and thighs. In specimens of old animals, the whole upper surface of the body is covered by a mass of quills, with thin tufts of long hairs, six inches in length, on the forehead, shoulders, and along the sides.

Head, rather small for the size of the animal, and very short; nose, truncated, broad, flattish above, and terminating abruptly. The eyes are lateral and small; ears, small, rounded, covered by short fur, and concealed by the adjoining long hair; incisors, large and strong.

Legs, very short and rather stout; claws, tolerably long, compressed, moderately arched, and channelled beneath.

There are tufts of hair situated between the toes; palms, naked, and nearly oval, hard and tuberculous; on the fore-feet there are four short toes, the second, counting from the inside, longest, the third a little smaller, the first a size less, and the fourth smallest. On the hind-foot there are five toes, with claws corresponding to those on the fore-foot. The hairs are so thickly and broadly arranged along the sides of the soles that they give a great apparent breadth to the foot, enabling this clumsy animal to walk with greater ease in the snow. It is plantigrade, and like the bear, presses on the earth throughout the whole length of the soles. Tail, short and thick, covered above with spines, beneath with long rigid hairs; when walking or climbing, it is turned a little upwards. Four mammæ, all pectoral.

Whilst the whole upper surface of the body is covered with spines, the under surface is clothed with hair intermixed with fur of a softer kind. The hair on the throat and under the belly is rather soft; along the sides it is longer and coarser, and under the tail appears like strong bristles.

COLOUR.

Incisors, deep orange; whole upper surface, blackish-brown, interspersed with long hairs, many of them being eight inches in length; these hairs are for four-fifths of their length dark-brown, with the points from one to two inches white. There are also long white hairs interspersed under the fore-legs, on the chest, and along the sides of the tail.

The spines, or quills, which vary in length from one to four inches, are white from the roots to near their points, which are generally dark brown or black; frequently brown, and occasionally white. On some specimens the spines are so abundant and protrude so far beyond the hair that portions of the body, especially the hips, present a speckled appearance, owing to the preponderance of the long white quills tipped with black. The nails and the whole under surface are dark brown.

There is in this species a considerable difference both in size and colour of different specimens.

There are three specimens before us, that with slight variations answer to the above description and to the figure on our plate. Another, which we obtained at Fort Union on the Missouri, is of enormous size, measuring thirteen inches across the back; the long hairs on the shoulders, forehead, and sides of which, are light yellowish-brown, whilst another specimen from the same locality, which appears to be that of a young animal, is dull white, with brown nose, ears and rump. In every specimen, however, the hairs on the hips, upper surface of tail, and under surface of body, are dark blackish-brown. In all these cases, it is the long, overhanging, light-coloured hairs, that give the general whitish appearance.

The difference between these specimens is so striking, that whilst those from Lower Canada may be described as black, the others from the far West may be designated as light-gray. Except in size and colour, there are no especial marks of difference.

DIMENSIONS.

Length of head and body	29	inches.
Tail, (vertebræ)	7	do.
Tail, to end of fur	8½	do.
Breadth of nose	1⅛	do.
From heel to longest nail	3½	do.

We possess one specimen a little larger than the above, and several that are considerably smaller.

HABITS.

The Canada Porcupine, of all North American quadrupeds, possesses the strangest peculiarities in its organization and habits. In its movements it is the most sluggish of all our species. Although the skunk is slow of foot, he would prove no contemptible competitor with it in a trial of speed. Under such circumstances the inquiry arises, what protection has this animal against the attacks of the wolverene, the lynx, the wolf, and the cougar? and how long will it be before it becomes totally exterminated? But a wise Creator has endowed it with powers by which it can bid defiance to the whole ferine race, the grizzly bear not excepted. If the skunk presents to its enemies a formidable battery, that stifles and burns at the same time, the Porcupine is clothed in an impervious coat of mail bristling with bayonets.

We kept a living animal of this kind in a cage in Charleston for six months, and on many occasions witnessed the manner in which it arranged its formidable spines, in order to prove invulnerable to the attacks of its enemies.

It was occasionally let out of its cage to enjoy the benefit of a promenade in the garden. It had become very gentle, and evinced no spiteful propensities; when we called to it, holding in our hand a tempting sweet-potatoe or an apple, it would turn its head slowly towards us, and give us a mild and wistful look, and then with stately steps advance and take the fruit from our hand. It then assumed an upright position, and conveyed the potatoe or apple to its mouth with its paws. If it found the door of our study open, it would march in, and gently approach us, rubbing its sides against our legs, and looking up at us as if supplicating for additional delicacies. We frequently plagued it in order to try its temper, but it never evinced any spirit of resentment by raising its bristles at us; but no sooner did a dog make his appearance, than in a moment it was armed at all points in defence. It would bend its nose downward, erect its bristles, and by a threatening sideway movement of the tail, give evidence that it was ready for the attack.

A large, ferocious, and exceedingly troublesome mastiff, belonging to the neighbourhood, had been in the habit of digging a hole under the fence, and entering our garden. Early one morning we saw him making a dash at some object in the corner of the fence, which proved to be our Porcupine, which had during the night made its escape from the cage. The dog seemed regardless of all its threats, and probably supposing it to be an animal not more formidable than a cat, sprang upon it with open mouth. The Porcupine seemed to swell up in an instant to nearly double its size, and as the dog pounced upon it, it dealt him such a sidewise lateral blow with its tail, as to cause the mastiff to relinquish his hold instantly, and set up a loud howl in an agony of pain. His mouth, tongue, and nose, were full of porcupine quills. He could not close his jaws, but hurried open-mouthed out of the premises. It proved to him a lesson for life, as nothing could ever afterwards induce him to revisit a place where he had met with such an unneighbourly reception. Although the servants immediately extracted the spines from the mouth of the dog, we observed that his head was terribly swelled for several weeks afterwards, and it was two months before he finally recovered.

Cartwright, (Journal, vol. ii., p. 59,) gives a description of the destructive habits of the Porcupine, which in many particulars is so much in accordance with our own observations, that we will present it to our readers.

"The Porcupine readily climbs trees; for which purpose he is furnished with very long claws; and in the winter, when he mounts into a tree, I believe he does not come down until he has eaten the bark from the top to the bottom. He generally makes his course through the wood in a straight direction, seldom missing a tree, unless such as are old. He loves young ones best, and devours so much, (only eating the inner part of the rind,) that I have frequently known one Porcupine ruin nearly a hundred trees in a winter.

"A man who is acquainted with the nature of these animals will seldom miss finding them when the snow is on the ground. If he can but hit upon the rinding of that winter, by making a circuit around the barked trees he will soon come on his track, unless a very deep snow should have chanced to fall after his last ascent. Having discovered that, he will not be long ere he find the animal."

In reference to the manner in which the Porcupine defends itself with its quills, he makes the following observations: "It is a received opinion that a Porcupine can dart his quills at pleasure into a distant object, but I venture to affirm that this species cannot, (whatever any other may do,) for I have taken much pains to discover this fact. On the approach of danger he retreats into a hole, if possible, but where he cannot find one he seizes upon the best shelter that offers, sinks his nose between his fore-legs, and defends himself by a sharp stroke of his tail, or a sudden jerk of his back. As the quills are bearded at their points and not deeply rooted in the skin, they stick firmly into whatever they penetrate; great care should be taken to extract them immediately, otherwise by the muscular motion of the animal into which they are stuck, enforced by the beards of the quills, they soon work themselves quite through the part; but I never perceived the puncture to be attended with any worse symptoms than that of a chirurgical instrument."

We had on three occasions in the northern and western parts of New-York opportunities of witnessing the effects produced by the persevering efforts of this species in search after its simple food. In travelling through the forest from Niagara to Louisville a few years ago, we passed through two or three acres of ground where nearly all the young trees had on the previous winter been deprived of their bark, and were as perfectly killed as if a fire had passed through them. We were informed by our coachman, that in driving through this place during the winter he had on several occasions seen the Porcupine on one of these trees, and that he believed all the mischief had been done by a single animal. We perceived that it had stripped every slippery elm (*Ulmus fulva*) in the neighbourhood, left not a tree of the bass wood (*Tilia glabra*) alive, but had principally feasted on the hemlock, (*Abies Canadensis.*)

Mr. J. G. Bell, one of our companions in our recent journey to the West, met with some Porcupines that resorted to a ravine, in which about a hundred cotton-wood trees (*Populus angulatus*) were standing, that had been denuded of both the bark and leaves. They had remained in this locality until they had eaten not only the tender branches, but had devoured the bark of some of the largest trees, by which they killed nearly every one. They then were forced in their own defence to remove to new quarters. We were informed that in a similar ravine to the one just spoken of, no less than thirteen Porcupines were killed in a single season by a young hunter.

On a visit to the western portion of the county of Saratoga, New-York, in the winter of 1813, a farmer residing in the vicinity carried us in his sleigh to show us a Porcupine which he had frequently seen during the winter, assuring us that he could find it on the very tree where he had observed it the previous day. We were disappointed, finding that it had deserted the tree; we however traced it in the snow by a well beaten path, which it seemed to have used daily, to a beech tree not far distant, which we cut down, and at the distance of twenty feet from the root we found the object of our search in a hollow part. It growled at us, and was particularly spiteful towards a small dog that was with us. Our friend killed it by a blow on the nose, the only vulnerable part as he informed us. It seemed to have been confined to a space of about two acres of ground through the winter. It had fed principally on hemlock bark, and had destroyed upwards of a hundred trees. The observations made on this occasion incline us to doubt the correctness of the statement that the Canada Porcupine does not leave a tree until it has eaten off all the bark, and that it remains for a week or more on the same tree; we were on the contrary led to suppose that the individual we have just spoken of, retired nightly to its comfortable domicile and warm bed in the hollow beach, in which we discovered it.

The Porcupine we kept in Charleston did not appear very choice in regard to its food. It ate almost any kind of vegetable we presented to it. We gave it cabbages, turnips, potatoes, apples, and even bread, and it usually cut to pieces every thing we placed in the cage that it could not consume. We had a tolerably large sweet bay tree (*Laurus nobilis*) in the garden; the instant that we opened the door of the cage the Porcupine would make its way to this tree, and not only feed greedily on its bark, but on its leaves also. When it had once fixed itself on a tree it was exceedingly difficult to induce it to come down, and our efforts to force it from the tree were the only provocatives by which it could be made to growl at us. We occasionally heard it during the night, uttering a shrill note, that might be called a low querulous shriek.

As the spring advanced, we ascertained that the constitution of our poor Porcupine was not intended for a warm climate: when the hot weather came on, it suffered so much that we wished it back again in its Canadian wilds. It would lie panting in its cage the whole day, seemed restless and miserable, lost its appetite and refused food. We one evening placed it on its favourite bay tree ; it immediately commenced gnawing the bark, which we supposed a favourable symptom, but it fell off during the night, and was dead before morning.

Whilst on the Upper Missouri river in the year 1843, as our companion, Mr. J. G. Bell, was cautiously making his way through a close thicket of willows and brush-wood in search of a fine buck elk, that he with one of our men had seen enter into this cover when they were at least a mile distant, he could not avoid cracking now and then a dry stick or fallen branch. He could not see more than ten paces in any direction, from the denseness of the thicket, and, as he unfortunately trod upon a thicker branch than usual which broke with a crash, the elk brushed furiously out of the thicket, and was gone in a moment, making the twigs and branches rattle as he dashed them aside with (shall we say) "telegraphic" rapidity. Mr. Bell stood motionless for a minute, when as he was about to retreat into the open prairie, and join his companion after this unsuccessful termination of the elk hunt, his eyes were fixed by an uncouth mass on the ground almost at his feet ; it was a Porcupine ; it remained perfectly still, and when he approached did not attempt to retreat. Our friend was rather perplexed to know how to treat an enemy that would neither "fight nor fly," and seizing a large stick, he commenced operations by giving the Porcupine (which must have been by this time displeased at least, if not "fretful,") a severe blow with it on the nose. The animal immediately concealed the injured organ, and his whole head also, under his belly ; rolling himself up into a ball, with the exception of his tail, which he occasionally jerked about and flirted upwards over his back. He now remained still again, and Mr. Bell drew a good sized knife, with which he tried to kill him by striking at his side so as to avoid the points of the quills as much as he could. This fresh attack caused the Porcupine to make violent efforts to escape: he seized hold of the branches or roots within reach of his fore-feet, and pulled forwards with great force ; Mr. Bell then placed his gun before him, which stopped him ; then finding he could not lay hold of him nor capture him in any other way, he drew his ramrod, which had a large screw at the end for wiping out his gun, and commenced screwing it into the Porcupine's back. This induced the poor animal again to make violent efforts to escape, but by the aid of the screw and repeated thrusts with the knife, he soon killed the creature.

He was now anxious to rejoin his companion, but did not like to relinquish his game ; he therefore, not thinking it advisable to stop and skin it on the spot, managed to tie it by the fore-legs, and then dragged it on the ground after him until he arrived at the spot where the hunter was impatiently waiting for him. Here he skinned the Porcupine, and turned the skin entirely inside out, so that the quills were all within, and then no longer fearing to handle the skin, it was secured to the saddle of his horse, and the carcass thrown away.

A Porcupine that was confined for some time in the garret of a building in Broadway, New-York, in which PEALE's Museum was formerly kept, made its escape by gnawing a hole in a corner of the garret, and, (as was supposed,) got on to the roof, from whence it tumbled into the street, either by a direct fall from this elevation, or by pitching on to some roof in the rear of the main building, and thence into Murray-street. It was brought the next day to the museum for sale, as a great curiosity. The man who brought it, of course not knowing from whence it came, said that early in the morning, he (being a watchman) was attracted by a crowd in the Park, and on approaching discovered a strange animal which no one could catch; he got a basket, however, and captured the beast, which he very naturally carried off to the *watch-house*, thinking of course no place of greater security for any vagrant existed in the neighbourhood.

On an explanation before the keeper of the museum, instead of the police justices, and on payment of half a dollar, the Porcupine was again restored to his friends. He was now, however, watched more closely, and bits of sheet tin were frequently nailed in different parts of the room on which he had a predilection for trying his large teeth.

We have mentioned in our article on the Canada lynx, that one of those animals was taken in the woods in a dying state, owing to its mouth being filled with Porcupine quills. We have heard of many dogs, some wolves, and at least one panther, that were found dead, in consequence of inflammation produced by seizing on the Porcupine.

Its nest is found in hollow trees or in caves under rocks. It produces its young in April or May, generally two at a litter; we have however heard that three, and on one occasion four, had been found in a nest.

The Indians residing in the North, make considerable use of the quills of the Porcupine; moccasins, shot-pouches, baskets made of birch bark, &c., are ingeniously ornamented with them, for which purpose they are dyed of various bright colours.

The flesh of this species is sometimes eaten, and is said to have the taste of flabby pork.

The following information respecting the Porcupine was received by us from our kind friend WILLIAM CASE, Esq., of Cleveland, Ohio. "This animal was several years since (before my shooting days) very abundant in this region, the Connecticut Western Reserve; and no more than ten years ago one person killed seven or eight in the course of an afternoon's hunt for squirrels, within three or four miles of this city, while now probably one could not be found in a month. They are rapidly becoming extinct; the chief reason is probably the extreme hatred all hunters bear them on account of the injuries their quills inflict on their dogs. They do not hibernate, neither do I think they are particularly confined to their hollow trees during the coldest days in winter. Their movements from tree to tree in search of food (browse and bark) are rather slow and awkward: their track in the snow very much resembles that of a child (with the aid of imagination).

"They most delight in browsing and barking young and thrifty Elms, and are generally plenty in Elm or Bass-wood Swail."

GEOGRAPHICAL DISTRIBUTION.

This species, according to Richardson, has been met with as far north as the Mackenzie river, in latitude 67°. It is found across the continent from Labrador to the Rocky Mountains, and is tolerably abundant in the woody portions of the western part of Missouri. To us this has been rather a rare species in the Atlantic districts; we having seldom met with it in the Northern and Eastern States. It is found, however, in the northern and western parts of New-York, and is said to be increasing in some of the western counties of that State. Dr. Leonard, of Lansingburgh, recently obtained specimens from the mountains of Vermont. It exists sparingly in the mountains of the northern portion of Pennsylvania, and in a few localities in Ohio; we obtained it on the Upper Missouri. Lewis and Clarke have not enumerated it as one of the species inhabiting the west of the Rocky Mountains.

It does not exist in the southern parts of New-York or Pennsylvania. Dekay (Nat. Hist. of New-York, p. 79) states, that it is found in the northern parts of Virginia and Kentucky. We however sought for it without success in the mountains of Virginia, and could never hear of its existence in Kentucky.

N° 8.

Plate XXXVII.

Drawn on Stone by R. Trembly

Swamp Hare

Male

Drawn from Nature by J. J. Audubon, F.R.S. F.L.S.

Lith. Printed & Col.d by J. T. Bowen, Phil.

LEPUS AQUATICUS.—Bach.

Swamp-Hare.

PLATE XXXVII.—Male.

L. L Americani magnitudine; capite, auribus, caudaque longis; pedibus longis minus pilosis quam in L. sylvatico; supra fuscus; subtus albus.

CHARACTERS.

Size of the Northern hare; head, ears, and tail, long; feet, long, less covered with hair than those of the gray rabbit; general colour, dark grayish-brown above, white beneath.

SYNONYMES.

Lepus Aquaticus, Bach., Journal Acad. Nat. Sc., Philad., vol. vii., p. 2, p. 319, read March 21, 1837.
Lepus Douglassii, var. 1, Gray, Magazine Nat. Hist., London, November, 1837.

DESCRIPTION.

The body of this species is large, and formed both for strength and speed; the hairs do not hang as loosely on the surface as those of the Northern hare, but lie smooth and compact; the fur is coarser and more glossy than that of the gray rabbit.

Head, long, and moderately arched; skull, considerably larger than that of the Northern hare, (*L. Americanus*,) with a larger orbital cavity. The margins of the orbits project so as to produce a visible depression in the anterior part of the frontal bone; whiskers, half the length of the head; ears, long, shaped like those of the marsh-hare, clothed externally with a dense coat of very short hairs; internally, they are partially covered along the margins, but nearer the orifice are nearly naked.

The feet bear no resemblance to those of the Northern hare or those of the gray rabbit. Instead of being clothed, as in those species, with a compact mass of hair, they are formed like those of the marsh-hare; the toes, when spread, leaving distinct impressions on the earth. The fore-toes are long, and their claws large and considerably curved; on the hind-feet, the claws are very stout and broad, nearly double the size of those of the Northern hare.

The tail is rather long for the genus, upturned, and thickly clothed on both surfaces with long fur.

COLOUR.

Teeth, yellowish-white; the whole of the upper part of the body light brownish-yellow, blotched on the surface with black; in the winter, the whole of the back and the sides of the head become brownish-black, with here and there a mixture of reddish-brown visible on the surface; the fur beneath the long hairs is dark plumbeous, tipped with black. The long hairs, when examined singly, are dark-blue at the roots, then light buff, and are pointed with black. Behind the ears, rufous, with a stripe of a similar colour extending to the shoulders. A line around the eyes, light reddish-buff. Upper lip, chin, and belly, white, tinged with blue. Nails, in a winter specimen of a young male, dark-brown; in an old female procured in summer, yellowish; whiskers, black; inner surface of the ears, light grayish-white; outer surface, above, edged with black; under surface of the tail, pure white.

DIMENSIONS.

(The following measurements were taken by Dr. Lee, of Alabama, from a specimen in the flesh.)

Length from point of nose to insertion of tail	20	inches.
" of head	$4\frac{1}{2}$	do.
" of ears, posteriorly	$3\frac{7}{8}$	do.
Height to shoulder	11	do.
Length of the hind-foot	$4\frac{1}{2}$	do.
" " middle hind-claw	$5\frac{7}{8}$	do.
" of tail (vertebræ)	$2\frac{1}{8}$	do.
" of tail, including fur	3	do.

Weight of a female killed in the spring, (when suckling its young, and not in good condition,) 6lbs.

HABITS.

The habits of this animal are very singular, differing in one remarkable peculiarity from those of any other species of hare yet known, with the exception of the marsh-hare. Although the Swamp-Hare is occasionally seen on high grounds in the dense forest, it prefers low and marshy places, or the neighbourhood of streams and ponds of water, to which it is fond of resorting. It swims with great facility from one little islet to another, and is generally found seeking its food in wet places, or near the water, as it subsists on the roots of various kinds of aquatic plants, especially on a species of iris growing in the water.

Persons who have given us information on the subject of this hare, inform us, that when first started, and whilst running, its trampings are louder, and can be heard at a greater distance, than those of any other hare.

As it suddenly leaps or bounds from its hiding place ere it is seen, it is apt to startle the rambler who has intruded upon its solitary retreat, and he may be impressed with the belief that he has started a young deer. When chased by dogs, the Swamp-Hare runs with great swiftness, and is able to escape from them without difficulty; but it almost invariably directs its flight towards the nearest pond, as if led by instinct to seek an element in which all traces of its scent are soon lost to its eager pursuers. There is a specimen of the Swamp-Hare, which we added to the collection of the Academy of Natural Sciences, Philadelphia, considerably larger than the Northern hare; this individual, on being pursued by hounds, swam twice across the Alabama river, and was not captured till it had finally retreated to a hollow tree.

We have been informed that it is a very common habit of this species when pursued, to swim to the edge of some stream or pond, retreat beneath the overhanging roots of the trees that may be growing on its border, or seek for a secure shelter under the hollows made by the washing of the banks. The swiftness of foot possessed by this Hare, and the stratagems to which it is capable of resorting, might easily enable it to elude pursuit but for this habit of seeking for shelter as soon as it is chased, which is the cause of its being frequently captured.

When the waters in the swamps are low, it seeks the first hollow tree, where it is easily secured. In this manner, Major Lee informed us, that in his vicinity the boys and the domestics caught thirty or forty in three days.

The young of this hare are frequently found in nests formed of leaves and grasses, placed on hillocks in the swamps, or in the hollow of some fallen tree. We have been informed that it produces young at least twice in a season, and from four to six at a litter.

GEOGRAPHICAL DISTRIBUTION.

We have not heard of the existence of this animal to the east or north of the State of Alabama, but it is numerous in all the swamps of the western part of that State, is still more abundant in the State of Mississippi, and in the lower part of Louisiana, and is frequently brought by the Indians to the market of New Orleans. It was also obtained in Texas by Douglass and by J. W. Audubon. Gray states that it exists in California; we have however carefully inquired into the history of the specimen in the British Museum, which was received after the melancholy death of Douglass, and have reason to believe that the label was accidentally misplaced, and that it came from the eastern portion of Texas.

GENERAL REMARKS.

Although all our hares bear a strong resemblance to each other, particularly in their summer colours, yet all have different marks, by which they can, with a little attention, be distinguished. The present species, in its colour on the upper surface and in its aquatic habits, is closely assimilated to the marsh-hare; it differs, however, very widely in other respects.

The Swamp-Hare is a third larger than the marsh-hare; the largest specimen of the latter in more than fifty that we measured, was only fourteen inches long, whilst the largest Swamp-Hare was twenty-two inches, and we are informed that it is often much larger. The tail of the marsh-hare is exceedingly short, its vertebræ being not more than an inch long, whilst that of the present species is two inches and an eighth, being more than double the length. The ears differ in the same proportion. The under surface of the tail of the marsh-hare is ash-coloured mixed with brown, whilst that of the present species is pure white. Its feet are thinly covered with hair, and its toes (which are capable of being widely spread) are well adapted to enable it to swim, and to pass over marshy and muddy places.

The tracks of this species, and of the marsh-hare, in the mud, leave a distinct impression of the toes; whilst on the contrary the tracks of the gray rabbit, the Northern hare, and the Polar hare, exhibit no such traces, their feet being so thickly clothed with long hair that even the points of the nails are scarcely perceptible. The present species is larger than the gray rabbit, being very nearly the size of the Northern hare, which it probably exceeds in weight. Indeed, the Northern hare and this species, when divested of their hides, are very nearly equal in size; but the fur of the former being loose and long, whilst that of the present species lies compact and smooth, the Northern hare appears to be the larger of the two. This species differs from the gray rabbit in other particulars; whilst the points of the hair in the latter animal become whiter in winter, those of the Swamp-Hare become jet-black; whilst the gray rabbit strenuously avoids water, the present species plunges fearlessly into it, and finds it a congenial element.

SCIURUS FERRUGINIVENTRIS.—Aud. and Bach.

Red-Bellied Squirrel.

PLATE XXXVIII.—Male, Female, and Young.

Sc. Caroliniano paullulum minor; cauda corpore longiore; vellere, supra albo-cinereo. infra rufo, armis fuscis.

CHARACTERS.

A size smaller than the Carolina gray squirrel; tail, longer than the body; light gray above, reddish-brown on the shoulders, beneath, bright rufous.

SYNONYME.

Sciurus Ferruginiventris, Aud. and Bach., Jour. Acad. of Nat. Sc., Philadelphia, read October 5, 1841.

DESCRIPTION.

This species in form bears some resemblance to the Carolina gray squirrel, but differs widely from it in colour. The forehead is arched; nose, rather sharp, clothed with short fur; eyes, of moderate size; whiskers, as long as the head; ears, rather long, broad at base, ovate in shape.

The body is slender, seemingly formed for an agility equalling that of *Sciurus Hudsonius.* It is covered with a soft thick coat of fur, intermixed with longer hairs.

The feet are rather robust. Like all the squirrels, it has a blunt nail in place of a thumb, and the third toe, counting from the inner side, is longest; palms, nearly naked.

The tail is long, and capable of a distichous arrangement, but the hairs are not very thick or bushy.

COLOUR.

Teeth, yellow; nails, brown; point of nose and whiskers, black; ears, on the outer edges, tinged with brown, within gray; behind the ears on the neck a line of dull white. On the upper surface, the head, neck, back, and tail, are light gray, formed by hairs which are light plumbeous from

Red-Bellied Squirrel.

Drawn from Nature by J J Audubon, F.R.S. F.L.S.

Lith.d Printed & Col.d by J. T. Bowen, Philad.a

the roots to near the tips, where they have white and black annulations; most of the hairs are tipped with white. From the outer surface of the fore-legs there is a reddish-brown tinge, which extends over the shoulders and nearly meets on the back, gradually fading into the colours of the back and neck. The hairs on the tail are black at the roots, then yellowish, succeeded by a broad line of black tipped with white. The feet on the upper surface are grizzled with white and black. Sides of the face, chin, and throat, light-gray. All the rest of the under surface of the body, a line around the eyes, the neck, and the inner surface of the legs, are of a uniform bright rufous colour.

DIMENSIONS.

	Inches.	Lines.
Length of head and body - - - - -	8	9
" tail - - - - - - - -	10	0
Height of ear, posteriorly - - - - -	5	0
Length of tarsus - - - - - - - -	2	5

HABITS.

We are unfortunately without any information or account of the habits of this singularly marked and bright coloured Squirrel. We have represented three of them in our plate in different attitudes on a branch of mulberry.

GEOGRAPHICAL DISTRIBUTION.

Several specimens, differing a little in colour, which differences we have represented in our plate, were received from California; the precise locality was not given.

GENERAL REMARKS.

This species should perhaps be compared with the dusky squirrel (*S. nigrescens*) of Bennet, to which it bears some resemblance. From the description, however, which we made of the original specimen of *S. nigrescens*, deposited in the museum of the London Zoological Society, we have little hesitation in pronouncing this a distinct species.

To *Sciurus socialis* of Wagner, (Beiträge zur Kenntniss der warmblutigen Wirbelthiere Amerikas, p. 88, Dresden,) the present species also bears some distant resemblance, but in some of its markings differs widely from Wagner's animal.

SPERMOPHILUS TRIDECEM LINEATUS.—Mitchill.

Leopard-Spermophile.

PLATE XXXIX.—Male and Female.

Magnitudine Tamiæ Lysteri; supra striis octo longitudinalibus dilute fulvis cum striis novem fulvis alternatum distributis; harum quinque, stria media et duabus utrinque proximis guttis subalbidis *subquadratus* distinctis.

CHARACTERS.

Size of the chipping-squirrel (Tamias Lysteri); eight pale yellowish-brown stripes on the back, which alternate with nine broader yellowish-brown ones; the five uppermost being marked with a row of pale spots.

SYNONYMES.

Leopard Ground-Squirrel, Schoolcraft's Travels, p. 313, and Index, anno **1821**.
Sciurus Tridecem Lineatus, Mitchill, Med. Repository, 1821.
Arctomys Hoodii, Sabine, Linn. Trans., vol. xiii., p. 590, 1822.
" " Franklin's Journey, p. 663.
Striped and Spotted Ground-Squirrel, Say, Long's Expedition.
Spermophile, F. Cuvier, Hist. Nat. des Mamm.
Arctomys Tridecem Lineata, Harlan, Fauna, p. 164.
Hood's Marmot, Godman, vol. ii., p. 112.
Arctomys Hoodii, Fischer's Synopsis, p. 544.
Spermophilus Hoodii, Less., Mamm., p. 243, 654.
" " Desmarest, in Dict. des Sc. Nat., L. p. 139.
" " F. Cuv. et Geoff., Mamm., fasc. 46.
Arctomys Tridecem Lineata, Griffith, sp. 641.
Arctomys (Spermophilus) Hoodii, Rich., Fauna Boreali Americana, **p. 177, pl. 14.**

DESCRIPTION.

In form this species bears a considerable resemblance to the very common chipping squirrel of the Atlantic States; by its shorter ears, however, and by its longer nails, which are intended more for digging than climbing, it approaches the marmots. The head has a convex shape and is very much curved, especially from the forehead to the nose; the nose

N°8. Plate XXXIX

Drawn on Stone by R. Trembly

Leopard Spermophile

Drawn from Nature by J.J. Audubon, F.R.S. F.L.S.

Lith. Printed & Col.d by J.T. Bowen, Phil.

is obtuse, and with the exception of the nostrils and septum is completely covered with very short hairs. The mouth is far back; the cheek-pouches of moderate size. Whiskers, a little shorter than the head; eyes, large; ears, very short, consisting merely of a low short lobe behind and above the auditory opening; they are covered with very short hairs. The hair on the whole body is short, adpressed, and glossy.

Legs and feet, rather slender; nails, long, slightly arched, and channelled beneath toward their extremities. On the fore-feet the thumb has one joint, with an obtuse nail; the second toe is longest, (as in the spermophiles, and not the third, as in the squirrels;) the first and third are of equal length; the fourth shortest, and removed far back. The tail is linear; for an inch from the root the fur lies so close that it appears rounded, it then gradually widens, becomes flattened, and seems capable of a slight distichous arrangement. Mammæ, twelve, situated along the sides of the abdomen.

COLOUR.

A line around the eye and a spot beneath, inner and outer surfaces of the legs, and the whole under part of the body, of a pale yellowish colour; on the sides of the neck, the fore-legs near the shoulders, and on the hips, there are tinges of reddish-brown; the feet near the nails, and the under-jaw, are dull white. On the head, there are irregular and somewhat indistinct alternate stripes of brown and yellowish-white, being an extension of the stripes on the back, which, from the irregular blending of the colours, give it a spotted appearance.

On the back there are five longitudinal brown stripes, each having regular rows of square spots of yellowish-white; the dorsal stripe, which runs from the back part of the head, and extends for half an inch beyond the root of the tail, is a little the broadest. These dark-coloured stripes are separated from each other by straight and uniform lines of yellowish-white. There are also on each side, two less distinct brown stripes, that are not spotted. Thus the animal has five brown stripes that are spotted, and four that are plain and without spots, together with eigth yellowish-white stripes.

The hairs in the tail are yellowish-white at the roots, then broadly barred with black, and at the tips yellowish-white, giving it when distichally arranged a bar of black on each side of the vertebræ.

DIMENSIONS.

Head and body	- - - - - - -	$6\frac{1}{8}$	inches.
Tail (vertebræ)	- - - - - - -	$3\frac{1}{2}$	do.

Tail, to end of hair - - - - - - -	$4\frac{1}{2}$	inches.
From heel to end of nail - - - -	$1\frac{1}{3}$	do.
Longest claw on the fore-foot - - - -	$0\frac{3}{8}$	do.

Measurement of an old female.

Nose to root of tail - - - - - - -	$6\frac{3}{4}$	do.
Tail (vertebræ) - - - - - - -	$3\frac{1}{2}$	do.
Tail, to end of hair - - - - - - -	$4\frac{3}{8}$	do.
Fore-feet to end of claws - - - - - -	$0\frac{7}{8}$	do.
Heel to end of longest claw - - - -	$1\frac{3}{8}$	do.
Nose to opening of ear - - - - - -	$1\frac{1}{2}$	do.
Length of pouch, to angle of mouth - - -	$1\frac{7}{16}$	do.

Dr. RICHARDSON measured a male that was nine inches to the insertion of the tail. He remarks that the females are smaller than the males.

HABITS.

We believe it is generally supposed that "birds," with their varied and pleasing forms, gay and beautiful plumage, tuneful throats, and graceful movements through the air, present greater attractions to the student of nature than "quadrupeds," and awaken in him a stronger desire to acquire a knowledge of their natures and characters than he may entertain to study the habits of the mammalia.

In addition, however, to the fact that the latter are, like ourselves, viviparous, and approach our own organization, it should be remembered that all the productions of nature are the work of so infinite a wisdom, that they must, in every department of the physical world, excite our greatest interest and our admiration, even when examined superficially.

Among the quadrupeds, there are innumerable varieties of form and character; and although most animals are nocturnal, and therefore their habits cannot be studied with the same facility with which the manners and customs of the lively diurnal species of birds may be observed; yet when we follow them in their nightly wanderings, penetrate into their retreats, and observe the sagacity and extraordinary instincts with which they are endowed, we find in them matter to interest us greatly, and arouse our curiosity and astonishment.

Owls seem to us a dull and stupid race, principally because we only notice them during the day, which nature requires them to spend in sleep, the structure of their eyes compelling them to avoid the light, and seek concealment in hollow trees, in caves, and obscure retreats.

But we should recollect that the diurnal birds are, during night, the time for their repose, as dull and stupid as owls are during the day. We should therefore not judge the habits of quadrupeds by the same standard. In regard to their fur, and external markings, there are many that will strike even the most careless observer as eminently beautiful. The little animal which is here presented to you is one of this description. In the distribution of the tints that compose its gaudy dress, in the regularity of its lines and spots, and in the soft blendings of its various shades of colour, we have evidence that even species whose habitations are under ground, may present to the eye as rich and beautiful a vesture as is found in the garb of a majority of the lively songsters of the woods.

In the warm days of spring the traveller on our Western prairies is often diverted from the contemplation of larger animals, to watch the movements of this lively little species. He withdraws his attention for a moment from the bellowing buffalo herd that is scampering over the prairies, to fix his eyes on a lively little creature of exquisite beauty seated on a diminutive mound at the mouth of its burrow, which seems by its chirrupings and scoldings to warn away the intruder on its peaceful domains. On a nearer approach it darts into its hole ; but although concealed from view, and out of the reach of danger, its tongue, like that of other scolds of a more intelligent race, is not idle ; it still continues to vent its threats of resentment againt its unwelcome visitor by a shrill and harsh repetition of the word "seek—seek."

There is a great similarity in the habits of the various spermophili that compose the interesting group to which the present species belongs.

They live principally on the open prairies, make their burrows in the earth, and feed on roots and seeds of various kinds, which they carry in their pouches to their dark retreats under ground.

The holes of this species, according to Richardson, run nearly perpendicularly, and are so straight, that they will admit a stick to be inserted to the depth of four or five feet. He supposes that owing to the depth of their burrows, which the sun does not penetrate very early in spring, they do not make their appearance as early as some others, especially *S. Richardsonii.*

As soon as they feel the warmth of spring they come forth and go in quest of their mates ; at this period they seem fearless of danger, and are easily captured by the beasts and birds of prey that frequent the plains. The males are said to be very pugnacious at this season.

This is believed to be the most active and lively of all our known species of marmot-squirrels ; we recently observed one in New-York that played in a wheel in the manner of the squirrel. We saw in

Charleston a pair in a cage, that were brought from Missouri by an officer of the army. They were adults, had but recently been captured, and were rather wild. They seemed to keep up a constant angry querulous chattering; they were fed on various kinds of nuts and grains, but principally on corn-meal and pea or ground-nuts, (*Arachis hypogæa.*) They would come to the bars of the cage and take a nut from the hand, but would then make a hasty retreat to a little box in the corner of their domicile. On our placing a handful of filberts in front of the cage, they at first came out and carried off one by one to their store-house; but after we had retired so as not to be observed, they filled their pouches by the aid of their paws, and seemed to prefer this mode of transporting their provisions. As we were desirous of taking measurements and descriptions, we endeavoured to hold one in the hand by the aid of a glove, but it struggled so lustily and used its teeth so savagely that we were compelled to let it go.

This species frequently takes up its residence near the fields and gardens of the settlers; and in the neighbourhood of Fort Union and other places, was represented as particularly destructive to the gardens.

We found the Leopard-Spermophile quite abundant near Fort Union, on the Upper Missouri. Their burrows were made in a sandy gravelly soil; they were never deep or inclined downwards, but ran horizontally within about a foot of the surface of the earth. This difference in habit from those observed by RICHARDSON may be owing to the nature of the different soils. We dug some of their burrows and discovered that the holes ran in all directions, containing many furcations.

RICHARDSON states that "the males fight when they meet, and in their contests their tails are often mutilated." All the specimens, however, that we obtained, were perfect and in good order.

The Leopard-Spermophile has two more teats than are found in the majority of the species of this genus, and hence it may be expected to produce an additional number of young. RICHARDSON informs us that ten young were taken from a female killed at Carlton House. This was on the 17th May, and we from hence presume that they produce their young soon after this period.

GEOGRAPHICAL DISTRIBUTION.

We have not heard of the existence of this species farther to the north than latitude 55°. It was found by SAY at Engineer Encampment on the Missouri; we found it at Fort Union, latitude 40° 40'; and it is said to extend along the prairies on the Eastern side of the Rocky Mountains into Mexico.

GENERAL REMARKS.

The name *tridecem lineatus* (thirteen lined) is not particularly euphonious, nor very characteristic; yet as it has in conformity with long established usages existing among naturalists, been admitted into our standard works, we have concluded to adopt it.

The figures given by Sabine and F. Cuvier of this species are defective, each having been taken from a specimen in which the tail had been mutilated. That given by Richardson, Fauna boreali Americana, drawn by Landseer, is more characteristic.

MUS LEUCOPUS.—Rafinesque.

American White-Footed Mouse.

PLATE XL.—Male, Female, and Young.

Cauda elongata, villosa; auribus magnis; supra fulvo-fuscescens, subtus albus; pedibus albis.

CHARACTERS.

Tail, long and hairy; ears, large; yellowish brown above; feet and lower parts of the body, white.

SYNONYMES.

Mus Sylvaticus, Forster, Phil. Trans., vol. lxii., p. 380.
Field-Rat, Penn., Hist. Quad., vol. ii., p. 185.
" Arctic Zool., vol. i., p. 131.
Musculus Leucopus, Rafinesque, Amer. Month. Review, Oct. 1818, p. 444.
Mus Leucopus, Desmar. Mamm., esp. 493.
Mus Sylvaticus, Harlan, Fauna, p. 151.
Mus Agrarius, Godm., Nat. Hist., vol. ii., p. 88.
Mus Leucopus, Richardson, F. B. A., p. 142.
Arvicola Nuttallii, Harlan, variety.
Arvicola Emmonsii, Emm., Mass. Report, p. 61.
Mus Leucopus, Dekay, Nat. Hist. N. Y., pl. 1, p. 82.

DESCRIPTION.

Head, of moderate size; muzzle, sharp pointed; eyes, large; ears, large, membranous, rounded above, nearly naked. There are a few short hairs on the margins, on both surfaces, not sufficient to conceal the integument. Whiskers, longer than the head.

The form of this species is delicate and of fine proportions; the fur (which is not very long) is soft and fine, but not lustrous.

Feet, slender, and clothed with short adpressed hairs, covering the toes and nails; there are four toes on the fore-feet, with six tubercles on each palm; the thumb is rudimentary, and covered by a very small blunt nail. The nails are small, sharp, and hooked; the hind-feet are long, especially the tarsal bones; the toes are longer than on the fore-feet. The tail is round, slender, tapering, and thickly clothed with short

No 8 Plate XI

Drawn on Stone by Wm E. Hitchcock

White Footed Mouse

Drawn from Nature by J.J. Audubon, F.R.S. F.L.S.

Lith. Printed & Cold by J. T. Bowen, Philada.

hairs; no scales being visible like those on the common mouse, (*Mus musculus.*)

COLOUR.

Fur, from the roots to near the extremity, dark bluish-gray; on the upper parts, brownish yellow; being a little darker on the crown and back, and lighter on the sides; the colour of the cheeks and hips approaches reddish-brown. The above is the colour of this species through the winter and until it sheds its hair late in spring, when it assumes a bluish-gray tint, a little lighter than that of the common mouse. Whiskers, white and black; upper surface of the tail, the colour of the back. The lips, chin, throat, feet, legs, and the whole under surface of the body and tail, are pure white. On the sides, this colour extends high up along the flanks; there is a very distinct line of demarcation between the colours of the back and sides.

There are some varieties in this species; specimens which we examined, from Labrador, Hudson's Bay, and Oregon, were lighter in colour, and the white on the under surface extended farther toward the back, than on those from the Atlantic States; we also observed a striking difference in the length of their tails, some being longer than the body, whilst others were not much more than half the length. In size they also differ widely; we have seen some that are scarcely larger than the common mouse, whilst others are nearly double that size; they are considerably larger in Carolina than in the Eastern States.

DIMENSIONS.

Length of head and body - - - - - -	2¼	inches.
" " tail - - - - - - - - -	2½	do.
Another specimen.		
Length of head and body - - - - - -	3½	do.
" " tail - - - - - - - - -	3¼	do.

HABITS.

Next to the common mouse, this is the most abundant and widely diffused species of mouse in North America. We have received it (under various names) from every State in the Union, and from Labrador, Hudson's Bay, and the Columbia River. Being nocturnal in its habits, it is far more common than is generally supposed. In familiar localities, where we had never known of its existence, we found it almost the only species taken in traps at night.

The White-footed Mouse is an exceedingly active species. It runs, leaps, and climbs, with great facility. We have observed it taking up its abode in a deserted squirrel's nest, thirty feet from the earth; we have seen a family of five or six scamper from a hollow in an oak that had just been cut down; we have frequently found them in the loft of a corn-house or stable in Carolina; and at times have discovered their nests under stone-heaps or old logs, or in the ground.

In New-Jersey their favourite resorts are isolated cedars growing on the margins of damp places, where green briars (*Smilax rotundifolio* and *S. herbacea*) connect the branches with the ground, and along the stems of which they climb expertly.

When started from their nests in these trees, they descend along the vines in safety to the earth. When thus disturbed, however, if the nest is at some distance from the ground, they hesitate before they come down, and go out on a branch perhaps, to scrutinize the vicinity, and, if not farther molested, appear satisfied, and again retreat to their nests. They have been known to take possession of deserted bird's nests—such as those of the cat-bird, red-winged starling, song thrush, or red-eyed fly-catcher.

In the northern part of New-York we could always obtain specimens from under the sheaves of wheat that were usually stacked in the harvest fields for a few days before they were carried into the barn. We have also occasionally found their nests on bushes, from five to fifteen feet from the ground. They are in these cases constructed with nearly as much art and ingenuity as the nests of the Baltimore Oriole. There are several nests now lying before us, that were found near Fort Lee, New-Jersey. They are seven inches in length and four in breadth, the circumference measuring thirteen inches; they are of an oval shape and are outwardly composed of dried moss and a few slips of the inner bark of some wild grape-vine; other nests are more rounded, and are composed of dried leaves and moss. We have sometimes thought that two pair of these Mice might occupy the same nest, as we possess one, nine inches in length and eight inches in diameter, which has two entrances, six inches apart, so that in such a case the little tenants need not have interfered with each other. The entrance in all the nests is from below, and about the size of the animal.

When we first discovered this kind of nest we were at a loss to decide whether it belonged to a bird or a quadruped; on touching the bush, however, we saw the little tenant of this airy domicile escape. At our next visit she left the nest so clumsily and made her way along the ground so slowly, that we took her up in our hand, when we discovered that she had four young about a fourth grown, adhering so firmly to the teats that she dragged them along in the manner of the jumping mouse (*Meriones Americanus*), or of the Florida rat. We preserved this little family alive for eighteen months, during which time the female produced several broods of young. During the day they usually concealed themselves in their nests, but as soon as it was dark they became very active and playful, running up and down the wires of their cages, robbing each other's little store-houses of various grains that had been carried to them, and occasionally emitting the only sound we ever heard them utter—a low squeak resembling that of the common mouse. We have been informed by William Cooper, Esq., of Weehawken, New-Jersey, an intelligent and close observer, to whom science is indebted for many excellent papers on various branches of natural history, that this species, when running off with its young to a place of safety, presses its tail closely under its abdomen to assist in holding them on to the teats—a remarkable instance of the love of offspring.

The White-footed Mouse seems less carnivorous than most of its kindred species. We found it when in confinement always dragging to its nest any kind of meat we placed in the cage, but it was generally left there unconsumed. We have often caught it in traps set for larger animals and baited with meat. Its first object is to drag the meat to its little store-house of provisions; the bait, however, being tied with a string to the pan of the steel-trap is not so easily carried off; but without much loss of time the Mouse gnaws the string in two, and if not caught in the attempt, drags off the meat. Our friend, the late Dr. John Wright, of Troy, furnished us with information confirming the above; he says, "In trapping for a weasel last summer I tied bits of beef above each trap with twine. On my first visit to the traps I found the twine at one, cut, and the meat in the jaws of the trap. The next day the same thing was observed at one of the traps, but another held fast a specimen of the *Mus leucopus*. I am informed that the trapper is not unfrequently troubled in this manner."

We have known this Mouse to cut into pieces snares set for the ruffed grouse, placed in gaps left for the purpose in fences of brushwood.

In its wild state it is continually laying up little stores of grain and grass seeds. We have seen it carrying in its mouth acorns and chinquepins. In the Northern States, these little hoards are often composed wholly of wheat; in the South, of rice. This species, like all rats and mice, is fond of Indian-corn, from which it only extracts the choicest sweetest portions, eating the heart and leaving the rest untouched.

In thickly settled parts of the United States this Mouse avoids dwellings, and even outhouses, and either confines itself to the woods, or keeps near fences, stone-heaps, &c.; but in partially deserted houses, or in newly formed settlements, it seems to take the place of the common mouse. RICHARDSON states that in the fur countries it becomes an inmate of the dwelling houses. Dr. LEITNER, an eminent botanist, who, whilst acting as surgeon in the army, was unfortunately killed in the Florida war, informed us that whilst on a botanizing tour through Florida a few years ago, he was frequently kept awake during a portion of the night by the White-footed Mice which had taken possession of the huts of the Indians and the log cabins of the early white settlers. We are under an impression that in these localities the common cat, and the Norway rat, were both absent; as we have reason to believe that this species deserts premises whenever they are frequented by either of the above animals. We kept a pair of white Norway rats (Albino variety) separated by a partition from an interesting family of white-footed mice, but before we were aware of it, the rats gnawed through the partition and devoured all our little pets.

This is a timid and very gentle species; we have seldom known it to bite when taken into the hand, and have observed that in a state of confinement it suffered itself to be killed by the very carnivorous cotton-rat (*Sigmodon hispidum*) without making any resistance.

We are disposed to believe that this species produces at least two litters of young in a season, in the Northern States, and three, in the Southern. In the State of New-York we have seen the young every month from May to September; and in Carolina a female that was kept in confinement had young three times, first having three, at a second litter five, and having six at a third.

The White-footed Mouse has many enemies. Foxes, wild-cats, and owls, destroy it frequently; the house-cat strays into the fields and along fences in search of it. In Carolina some domesticated cats live in the fields and woods in a partially wild state, avoiding houses altogether; these subsist on birds and the smaller rodentia, and this species furnishes a considerable portion of their food; but we are disposed to regard the ermine (common weasel) as its most formidable and voracious persecutor. We believe that the White-footed Mouse does not always dig a burrow of its own, but that it takes possession of one dug by some other small species; in the Northern States, generally that of the chipping squirrel. Be this as it may, it is certain, that wherever the White-footed Mouse can enter, the ermine can follow, and he not only feeds upon it, but destroys whole families. An ermine at one time made its escape from us, carrying with it a small portion of a chain fastened around its neck: it was traced by a servant over the snow a mile into the woods, to a spot where it entered a very small hole. It was dug out, and the man brought us five or six Mice, of this species, that he found dead in the hole, having been killed, doubtless, by the ermine. From appearances, two only had been devoured; the remainder we observed had not been seized by the throat in the manner of the cat, but had the marks of the ermine's teeth in their skulls.

We do not regard this species as doing very extensive injury either to the garden or farm, in any of the Atlantic States of America. We suspect that its reputation in this respect, as well as that of the shrew-mole, has been made to suffer very unjustly, when in reality the author of the mischief is the little pine-mouse (*Arvicola pinetorum*, Le Conte), or perhaps Wilson's meadow-mouse, *Arvicola Pennsylvanica*, Ord, *A. hirsutus*, Emmons, and Dekay). The farmers and gardeners of the Northern and Eastern States, however, complain that this Mouse, which they generally call the "Deer-mouse," destroys many of their cabbage-plants and other young and tender vegetables, and gnaws the bark from young fruit trees; and if they have made no mistake in regard to the species, it must be much more destructive than we have heretofore considered it.

GEOGRAPHICAL DISTRIBUTION.

According to Richardson, this species is found as far north as Great Bear Lake. We saw in the London museums several specimens from Hudson's Bay; it extends across the continent to the Columbia River on the Pacific, from whence Mr. Townsend brought us several skins. We received specimens from Florida by Dr. Leitner; we found it west of the Mississippi at Fort Union, where it commits depredations in the garden attached to the Fort, and we have received specimens from Arkansas and Texas.

GENERAL REMARKS.

That a species so widely distributed, and subject to so many varieties in size, length of tail, and colour, should have been often described under different names, is not surprising. We have ourselves often been in a state of doubt on obtaining some striking variety. The name *Hypudæus gossipinus* of our friend, Major Le Conte, (see Appendix to McMurtrie's translation of Cuv. An. Kingd., vol. i., p. 434,) was intended for this species, as it is found in the Southern States. We were for several years disposed to regard it as distinct, and have, not without much hesitation, and after an examination of many hundred specimens, been induced to set it down as a variety only.

We have adopted the name given to it by RAFINESQUE, in deference to the opinions of RICHARDSON, who supposed that it applied to this species. RICHARDSON himself, however,—not RAFINESQUE,—gave a true description of it.

GODMAN, in describing *Mus agrarius*, we feel confident, had reference to this species. He had, however, never seen the European *Mus agrarius* of PALLAS, else he would not have made so great a mistake; we have on several occasions in Denmark and Germany compared them, and found that they scarcely bear any resemblance to each other. *Mus agrarius* has a short tail and short hairy ears. FORSTER, and HARLAN, refer this species to *Mus sylvaticus* of Europe. FORSTER's specimens came from Hudson's Bay at an early period, when it was customary to consider American species of Quadrupeds and Birds as mere varieties of those of Europe. HARLAN, instead of describing from an American specimen, literally translated DESMAREST's description of the European *Mus sylvaticus* and applied it to our species, (see Mam. p. 301,) in doing which, by neglecting to institute a comparison, he committed a great error.

We were favoured with the privilege of comparing specimens of *Mus sylvaticus* and *M. leucopus*, through the kindness of Prof. LICHTENSTEIN at the Berlin museum. Although there is a general resemblance, a moment's examination will enable the naturalist to discover sufficient marks of difference to induce him to separate the species. *Mus leucopus* has a little longer tail. Its ears are longer, but not so broad. The under surface of the tail of *Mus sylvaticus* is less white, and the white on the under surface of the body does not extend as high on the sides, nor is there any distinct line of separation between the colours of the back and under surface, which is a striking characteristic in the American species. But they may always be distinguished from each other at a glance by the following mark: in more than twenty specimens we examined of *Mus sylvaticus* we have always found a yellowish line edged with dark-brown on the breast. In many hundred specimens of *Mus leucopus* we have without a single exception found this yellow line entirely wanting, all of them being pure white on the breast, as well as on the whole under surface. We have no hesitation in pronouncing the species distinct.

Drawn on Stone by W.E. Hitchcock

Pennants Marten or Fisher.

Drawn from Nature by J.J. Audubon, F.R.S.F.L.S. Lith.d Printed & Col.d by J.T. Bowen, Philad.a

GENUS MUSTELA.—Cuv.

DENTAL FORMULA.

$$Incisive\ \frac{6}{6};\quad Canine\ \frac{1-1}{1-1};\quad Molar\ \frac{5-5}{6-6} = 38.$$

Head, small and oval; muzzle, rather large; ears, short and round; body, long, vermiform; tail, usually long and cylindrical; legs, short; five toes on each foot, armed with sharp, crooked, slightly retractile claws. No anal pouch, but a small gland which secretes a thickish offensive fluid. Fur, very fine.

This genus differs from the genus Putorius, having four carnivorous teeth on each side, in the upper jaw, instead of three, the number the true weasels exhibit, and, the last carnivorous tooth on the lower jaw, has a rounded lobe on the inner side, which renders this genus somewhat less carnivorous in its habits than Putorius; and consequently a slight diminution of the cruelty and ferocity displayed by animals of the latter genus, may be observed in those forming the present.

There are about twelve species of true Martens known, four of which inhabit North America.

The generic name Mustela, is derived from the Latin word *mustela*, a weasel.

MUSTELA CANADENSIS.—Schreber.

Pennant's Marten or Fisher.

Black Fox or Black Cat of the Northern Hunters.

PLATE XLI.—Male.

Capite et humeris cano fuscoque mixtis; naso, labiis, cruribus, et cauda, fusco-nigris.

CHARACTERS.

Head and shoulders, mixed with gray and brown; nose, lips, legs, and tail, dark brown.

SYNONYMES.

LE PEKAN, Buffon, vol. xiii., p. 304, A. D. 1749.
MUSTELA CANADENSIS, Schreber, Säugethiere, p. 492, 1775.
MUSTELA PENNANTI, Erxleben, Syst., p. 470, A. D. 1777.
FISHER, Penn., Arct. Zool., 4 vols., vol. i., p. 82, A. D. 1784.
MUSTELA CANADENSIS, Gmel., Linn., vol. i., p. 95, 1788.
WEJACK, Hearne's Journey.
FISHER, or BLACK FOX, Lewis and Clarke, vol. iii., p. 25.
FISHER, WEASEL, or PEKAN, Warden's United States.
MUSTELA PENNANTI Sabine, Frank. First Journey, p. 651.
MUSTELA CANADENSIS, Harlan, F., p. 65.
" " Godman, vol. i., p. 203.
MUSTELA GODMANI, Less., Mamm., p. 150.
MUSTELA CANADENSIS, Rich., F. B. A., p. 52.
PEKAN or FISHER, Dekay, Nat. Hist. N. Y., p. 31.

DESCRIPTION.

The head of this species bears a stronger resemblance to that of a dog than to the head of a cat. Its canine teeth, in the upper jaw, are so long, that with the slightest movement of the lip they are exposed. Head, broad and round, contracting rather suddenly toward the nose, which is acute. Eyes, rather small and oblique; ears, low, broad, semicircular, and far apart, covered on both surfaces with short soft fur; whiskers, half the length of the head; body, long, and formed for agility and strength.

The pelage consists of a short fine down next the skin, intermixed with longer and coarser hairs about an inch and a half in length; these hairs are longer on the posterior parts of the animal than on the shoulders.

The feet are robust. Fore-feet, shorter than the hind-feet, thickly clothed with rather fine and short hairs; nails, long, strong, curved, and sharp; soles, hairy; the toes on all the feet are connected at the base by a short hairy web; the callosities consequently make only a slight impression when the animal is walking or running on the snow.

Tail, long, bushy, and gradually diminishing to a point toward the extremity.

This species has so strong a smell of musk, (like the pine marten,) that we have found the skin somewhat unpleasant to our olfactories, several years after it had been prepared as a specimen.

COLOUR.

Fur on the back, from the roots to near the extremity, chesnut-brown, tipped with reddish-brown and light gray. On the head, shoulders, and fore part of the back, there are so many long whitish hairs interspersed, that they produce a somewhat hoary appearance. Whiskers, nose, chin, ears, legs, feet, and tail, dark-brown; margins of the ears, light-brown; hips and posterior part of the back, darker than the shoulders; eyes, yellowish-brown; nails, light horn-colour.

In some specimens, we have seen a white spot on the throat, and a line of the same colour on the belly; others, (as was the case with the one from which our drawing was made,) have no white markings on the body. We have seen a specimen, nearly white, with a brown head. Another, obtained in Buncombe county, North Carolina, was slightly hoary on the whole upper surface.

DIMENSIONS.

From point of nose to root of tail	23	inches.
Tail, (vertebræ)	12	do.
" to end of hair	14½	do.
Breadth of head	3½	do.
Height of ear	1	do.
Breadth of ear	2	do.
From point of nose to eye	2	do.
" heel to point of longest nail	4¼	do.

Weight, 8½ lbs.

HABITS.

Although this species is represented as having been rather common in every part of the Northern and Middle States, in the early periods of our history, and is still met with in diminished numbers, in the thinly settled portions of our country; very little of its history or habits has been written, and much is still unknown. We have occasionally met with it, but it has been to us far from a common species. Even in the mountainous portions of the Northern and Eastern States, the Fisher, thirty years ago, was as difficult to procure as the Bay lynx. It has since become still more rare, and in places where it was then known, scarcely any vestige of the knowledge of its former existence can now be traced.

Dr. DEKAY (Nat. Hist. N. Y., p. 32, 1843,) states that, "in Hamilton county, (N. Y.,) it is still numerous and troublesome." On an excursion we made in the State of New-York, 1827, we heard of it occasionally near the head waters of Lake Champlain, along the St. Lawrence, and to the west as far as Lake Erie, but it was every where represented as a species that was fast disappearing.

Whilst residing in the northern part of our native State, (New-York,) thirty-five years ago, the hunters were in the habit of bringing us two or three specimens of this Marten in the course of a winter. They obtained them by following their tracks in the snow, when the animals had been out in quest of food on the previous night, thus tracing them to the hollow trees in which they were concealed, which they chopped down. They informed us that as a tree was falling, the Fisher would dart from the hollow, which was often fifty feet from the ground, and leap into the snow, when the dogs usually seized and killed him, although not without a hard struggle, as the Fisher was infinitely more dangerous to their hounds than either the gray or the red fox. They usually called this species the Black Fox.

A servant, on one occasion, came to us before daylight, asking us to shoot a raccoon for him, which, after having been chased by his dogs the previous night, had taken to so large a tree that he neither felt disposed to climb it nor to cut it down. On our arrival at the place, it was already light, and the dogs were barking furiously at the foot of the tree. We soon perceived that instead of being a raccoon, the animal was a far more rare and interesting species, a Fisher. As we were anxious to study its habits we did not immediately shoot, but teased it by shaking some grape vines that had crept up nearly to the top of the tree. The animal not only became thoroughly frightened, but seemed furious; he leaped from branch to branch, showing his teeth and growling at the same time; now and then he ran half way down the trunk of the tree, elevating his back in the manner of an angry cat, and we every moment expected to see him leap off and fall among the dogs. He was brought down after several discharges from the gun. He seemed extremely tenacious of life, and was game to the last, holding on to the nose of a dog with a dying grasp. This animal proved to be a male; the body measured twenty-five inches, and the tail, including the fur, fifteen. The servant who had traced him, informed us that he appeared to have far less speed than a fox, that he ran for ten minutes through a swamp in a straight direction, and then took to a tree.

The only opportunity that was ever afforded us of judging of the speed of the Fisher occurred near the Virginia Gray-Sulphur Springs, in 1839. We had ascended Peter's Mountain in search of rare plants for our herbarium; out of health and fatigued, we had for some time been seated on a rock to rest, when we observed a gray squirrel pass within ten feet of us, seemingly in a great fright, and with all the speed it could command, with a Fisher in full pursuit. They were both too much occupied with their own affairs to take any notice of us. The Fisher seemed to make more rapid progress than the squirrel, and we feel confident that if the latter had not mounted a tree it would have been overtaken before it could have advanced many feet farther; it ran rapidly up the sides of a cucumber tree, (*Magnolia acuminati,*) still pursued by its hungry foe. The squirrel leaped lightly among the smaller branches, on which its heavier pursuer seemed unwilling to trust himself. At length the affrighted animal pitched from one of the topmost boughs and landed on its feet unhurt among the rocks beneath. We expected every moment to see the Fisher give us a specimen also of his talent at lofty tumbling, but he seemed to think that the "better part of valour was discretion," and began to run down the stem of the tree. At this point we interfered. Had he imitated the squirrel in its flying leap, he might have been entitled to the prey, provided he could overtake it; but as he chose to exercise some stratagy and jockeying in the race, when the chances were so much in his favour, we resolved to end the chase by running to the foot of the tree which the Fisher was descending. He paused on the opposite side as if trying to ascertain whether he had been observed; we were without a gun, but rattled away with a knife on our botanizing box, which seemed to frighten the Marten in his turn, most effectually;—the more noise we created the greater appeared to be his terror; after ascending to the top of the tree he sprang to another, which he rapidly descended, till within twenty feet of the earth, when he jumped to the ground, and with long leaps ran rapidly down the side of the mountain, and was out of sight in a few moments.

This scene occurred in the morning of, a warm day in the month of July, a proof that this species is not altogether nocturnal in its habits. We are, however, inclined to believe that the above was only an exception to the general character of the animal.

Species that are decidedly nocturnal in their habits, frequently may be seen moving about by day during the period when they are engaged in providing for their young. Thus the raccoon, the opossum, and all our hares, are constantly met with in spring, and early summer, in the morning and afternoon, whilst in the autumn and winter they only move about by night.

In the many fox-hunts, in which our neighbours were from time to time engaged, not far from our residence at the north, during the period when we obtained the information concerning their primitive mode of enjoying that amusement, which we have laid before our readers, in pages 49 and 50, (where we also spoke of Pennant's Marten as not being very scarce at that time in Rensselaer county, N. Y.,) we never heard of their having encountered a single Fisher in the day-time; but when they traversed the same grounds at night, in search of raccoons, it was not unusual for them to discover and capture this species. We were informed by the trappers that they caught the Fisher in their traps only by night.

The specimen, from which the figure in our plate was drawn, was taken alive in some part of the Alleghany Mountains, in the State of Pennsylvania, and we soon afterwards received a letter from our esteemed friend, Spencer F. Baird, Esq., of Carlisle, in that State, informing us of its having been captured, which enabled us, through that gentleman, to purchase it. We received it at New-York, in good condition, in a case tinned inside, with iron bars in front, to prevent the animal from making its escape, as it was so strong and so well supplied with sharp teeth that it could easily have eaten its way out of a common wooden box. In Mr. Baird's note he says: "All the account I was able to procure respecting this species was the following:—It was found in company with an older one, in Peters' Mountain, six miles above Harrisburgh, about five weeks ago. (His letter is dated Carlisle, March 16th, 1844.) After a most desperate resistance the old one was killed, after having beaten off the dogs, to whose assistance the hunters were obliged to come. This individual ran up a tree, and being stoned by the hunters, jumped off from a height of about forty feet! when being a little stunned by the leap, the men ran up quickly, threw their coats over it, and thus secured it. The old one was said to have been about the size of a pointer dog. The young one is very savage, and emits a rather strong musky odour."

We kept this individual alive for some days, feeding it on raw meat, pieces of chicken, and now and then a bird. It was voracious, and very spiteful, growling, snarling and spitting when approached, but it did not appear to suffer much uneasiness from being held in captivity, as, like many other predacious quadrupeds, it grew fat, being better supplied with food than when it had been obliged to cater for itself in the woods.

The older one, which, as Mr. BAIRD mentions, was killed by the dogs and the hunters, was a female, and no doubt was the mother of the one that was captured, and probably died in the hope of saving her young.

On several occasions we have seen the tracks of the Fisher in the snow; they resemble those of the pine marten, but are double their size. To judge by them, the animal advances by short leaps in the manner of a mink.

Pennant's Marten appears to prefer low swampy grounds: we traced one which had followed a trout stream for some distance, and ascertained that it had not gone into the water. Marks were quite visible in different places where it had scratched up the snow by the side of logs and piles of timber, to seek for mice or other small quadrupeds, and we have no doubt it preys upon the Northern hare, gray rabbit, and ruffed grouse, as we observed a great many tracks of those species in the vicinity. It further appears, that this animal makes an occasional meal on species which are much more closely allied to it than those just mentioned. In a letter we received from Mr. FOTHERGILL, in which he furnished us with notes on the habits of some of the animals existing near Lake Ontario, he informs us that "a Fisher was shot by a hunter named MARSH, near Port Hope, who said it was up in a tree, in close pursuit of a pine marten, which he also brought with it." Mr. FOTHERGILL stuffed them both at the time.

LEWIS and CLARK state, that in Oregon the Fisher captures not only the squirrel, but the raccoon, and that in pursuing them it leaps from tree to tree.

RICHARDSON remarks that, "the Fisher is said to prey much on frogs in the summer season; but I have been informed that its favourite food is the Canada porcupine, which it kills by biting in the belly." He says also, "it will feed on the hoards of frozen fish laid up by the residents."

We can scarcely conceive in what manner it is able to overturn the porcupine, so as to bite it on the belly, as it is large and heavy, and is armed with bristles at all points.

It is stated by Dr. DEKAY, on the authority of a person who resided many years near Lake Oneida, New-York, that the name (Fisher) "was derived from its singular fondness for the fish used to bait traps."

An individual of this species, which had been caught in a steel-trap, was brought to Charleston and exhibited in a menagerie. It had been taken only a few months, and was sullen and spiteful; when fed, it gulped down a moderate quantity of meat in great haste, swallowing it nearly whole, and then retired in a growling humour to a dark corner of its cage.

The Fisher is represented as following the line of traps set by the trappers, and in the manner of the wolverene, robbing them of their bait. The season for hunting this species is stated, by Dr. DEKAY, to commence in the western part of New-York, about the 10th of October, and to last till the middle of May; and he says the ordinary price paid for each skin is a dollar and a half.

This species brings forth once a year, depositing its young in the trunk of a large tree, usually some thirty or forty feet from the ground. Dr. RICHARDSON observes that it produces from two to four young at a litter; DEKAY confines the number to two. We once saw three extracted from the body of a female on the 20th of April, in the northern part of New-York.

GEOGRAPHICAL DISTRIBUTION.

This species inhabits a wide extent of country. To the north it exists, according to RICHARDSON, as far as Great Slave Lake, latitude 63°. It is found at Labrador, and extends across the continent to the Pacific. It is stated by all our authors that it does not exist further south than Pennsylvania. This is an error, as we saw it on the mountains of Virginia. We had an opportunity of examining a specimen obtained by Dr. GIBBES, of Columbia, South Carolina, from the neighbourhood of Ashville, Buncombe county, North Carolina. We have seen several skins procured in East Tennessee, and we have heard of at least one individual that was captured near Flat-Rock, in that State, latitude 35°.

We have also seen many skins from the Upper Missouri; and the Fisher is enumerated, by LEWIS and CLARKE, as one of the species existing on the Pacific Ocean, in the vicinity of the Columbia River.

GENERAL REMARKS.

Notwithstanding the fact that on the large plate of this animal in our folio edition we gave to LINNÆUS the credit of having first applied a scientific name to this species, we must now transfer it to SCHREBER, by whom, LINNÆUS having been unacquainted with it, it was described in 1775. It was described two years afterwards by ERXLEBEN, and in 1788, by GMELIN, &c. It is probable that, by some mistake, the habits of the mink have been ascribed to the Fisher; hence its English name seems to be inappropriate; but as it appears to be entitled to it, by right of long possession, we do not feel disposed to change it. We are, however, not quite sure of its having no claim to the name by its mode of living. Its partially webbed feet seem indicative of aquatic habits; it is fond of low swampy places, follows streams, and eats fish when in captivity. We feel pretty confident that it does not dive after the finny tribes, but it is not improbable that it surprises them in shallow water; and we are well informed that, like the raccoon, it searches under the banks of water-courses for frogs, &c.

By the Canadian hunters and trappers, it is universally called the Pekan. In New-England and the Northern counties of New-York it is sometimes named the Black Fox, but more frequently is known as the Fisher. According to DEKAY, it is called the Black Cat by the inhabitants of the western portion of New-York.

GENUS MEPHITIS.—Cuv.

DENTAL FORMULA.

$$Incisive \frac{6}{6}; \ Canine \frac{1-1}{1-1}; \ Molar \frac{4-4}{5-5} = 34.$$

Canine teeth, very strong, conical; two small anterior cheek-teeth, or false molars, above, and three below, on each side. The superior tuberculous teeth, very large, as broad as they are long; inferior molars having two tubercles on the inner side.

Head, short; nose, somewhat projecting; snout, in most of the species blunt.

Feet, with five toes; toes of the fore-feet, armed with long, curved nails, indicating the habit of burrowing in the earth; heel very little raised in walking.

Hairs on the body, usually long, and on the tail, very long.

The anal glands secrete a liquor which is excessively fetid. The various species of this genus burrow in the ground, or dwell in fissures of rocks, living on poultry, bird's eggs, small quadrupeds, and insects. They move slowly, and seldom attempt to run from man, unless they chance to be near their burrows. They are to a considerable extent gregarious; large families being occasionally found in the same hole.

In the recent work of Dr. Lichtenstein, (Ueber die Gattung Mephitis, Berlin, 1838,) seventeen species of this genus are enumerated, one of which is found at the Cape of Good Hope, two in the United States of America, and the remainder in Mexico and South America.

The generic name Mephitis, is derived from the Latin word *Mephitis*, a strong odour.

Common American Skunk

Drawn from Nature by J.J. Audubon F.R.S.F.L.S. Lith. Printed & Colᵈ by J.T. Bowen, Philadᵃ

MEPHITIS CHINGA.—Tiedimann

Common American Skunk.

PLATE XLII.—Female.

Magnitudine F. cati ; supra nigricans, stries albis longitudinalibus insigneta ; cauda longa villosissima.

CHARACTERS.

Size of a cat ; general colour, blackish-brown, with white longitudinal stripes on the back ; many varieties in its white markings ; tail, long and bushy.

SYNONYMES.

Ouinesque, Sagard Theodat, Canada, p. 748.
Enfant du Diable, Charlevoix, Nouv. France, iii., p. 133.
Skunk-Weasel, Pennant's Arctic Zool., vol. i., p. 85.
Skunk, Hearne's Journey, p. 377.
Mephitis Chinga, Tiedimann, Zool. i., p. 361, (Anh. 37,) 1808.
Pole-Cat Skunk, Kalm's Travels, vol. ii., p. 378.
Vivera Mephitis, Gmel. (L.) Syst. Nat., p. 88.
Mustela Americana, Desm. Mam., p. 186, A. D. 1820.
Mephitis Americana, Sab., Frank. Journal, p. 653.
" " Harlan, Fauna, p. 70.
The Skunk, Godm., Nat. Hist., vol. i., p. 213.
Mephitis Americana, Var. Hudsonica, Rich., F. B. A., p. 55.
" Chinga, Lichtenstein, Darstellung neuer oder wenig bekannter Säugethiere, Berlin, 1827–34, xlv. Tafel, 1st figure.
" Chinque, Licht., Ueber die Gattung Mephitis, p. 32, Berlin, 1838.
" Americana, Dekay, Nat. Hist. N. Y., pt. 1, p. 29.

DESCRIPTION.

This species in all its varieties has a broad fleshy body, resembling that of the wolverene ; it stands low on its legs, and is much wider at the hips than at the shoulders. Fur, rather long and coarse, with much longer, smooth and glossy hairs, interspersed.

The head is small compared with the size of the body ; forehead, somewhat rounded ; nose, obtuse, covered with short hair to the snout, which is naked; eyes, small; ears, short, broad and rounded, clothed with hair on both surfaces; whiskers, few and weak, extending a little beyond the eyes; feet, rather broad, and covered with hair concealing the nails, which on the fore-feet are robust, curved, compressed, and acute; palms, naked. The trunk of the tail is nearly half as long as the body. Hair on the tail, very long and bushy, containing from within an inch of the root to the extremity, no mixture of the finer fur. The glands are situated on either side of the rectum: the ducts are about an inch in length, and are of a somewhat pyriform shape. The inner membrane is corrugated; the principal portion of the glands is a muscular tendinous substance. The sac is capable of containing about three drachms. When the tail is erected for the purpose of ejecting the nauseous fluid, the open orifices of the ducts are perceptible on a black disk surrounding the anus. The exit from the duct at the anus when distended will admit a crow-quill.

COLOUR.

This species varies so much in colour that there is some difficulty in finding two specimens alike; we have given a representation on our plate of the colour which is most common in the Middle States, and which Dr. Harlan described as *Mephitis Americana*, our specimen only differing from his in having a longitudinal stripe on the forehead.

The under fur on all those portions of the body which are dark coloured, is dark brown; in those parts which are light coloured, it is white from the roots. These under colours, however, are concealed by a thick coat of longer, coarser hairs, which are smooth and glossy.

There is a narrow white stripe commencing on the nose and running to a point on the top of the head; a patch of white, of about two inches in length, and of the same breadth, commences on the occiput and covers the upper parts of the neck; on each side of the vertebræ of the tail there is a broad longitudinal stripe for three-fourths of its length; the tail is finally broadly tipped with white, interspersed with a few black hairs. The colour on every other part of the body is blackish-brown.

Another specimen from the same locality has a white stripe on the forehead; a large white spot on the occiput, extending downwards, diverging on the back, and continuing down the sides to within two inches of the extremity of the tail, leaving the back, the end of the tail, and the whole of the under surface, blackish-brown.

The young on the plate are from the same nest; one has white stripes on the back, with a black tail; the other has no stripes on the back, but the end of its tail is white.

In general we have found the varieties in a particular locality marked with tolerable uniformity. To this rule, however, there are many exceptions.

In the winter of 1814 we caused a burrow to be opened in Rensselaer county, N. Y., which we knew contained a large family of this species. We found eleven: they were all full grown, but on examining their teeth and claws, we concluded that the family was composed of a pair of old ones, with their large brood of young of the previous season. The male had a white stripe on the forehead; and from the occiput down the whole of the back had another white stripe four inches in breadth; its tail was also white. The female had no white stripe on the forehead, but had a longitudinal stripe on each side of the back, and a very narrow one on the dorsal line; the tail was wholly black. The young differed very widely in colour; we could not find two exactly alike; some were in part of the colour of the male, others were more like the female, whilst the largest proportion were intermediate in their markings, and some seemed to resemble neither parent. We recollect one that had not a white hair, except the tip of the tail and a minute dorsal line.

On the other hand, we had in February (the same winter) another family of Skunks, captured with a steel-trap placed at the mouth of their burrow; they were taken in the course of ten days, and we have reason to believe none escaped. In this family there was a very strong resemblance. The animals which we considered the old pair, had two longitudinal stripes on the back, with a spot on the forehead; in the young, the only difference was, that in some of the specimens the white line united on the back above the root of the tail, whilst in others it extended down along the sides of the tail till it nearly reached the extremity; and in some of the specimens the tail was tipped with white, in others, black. We had an opportunity near Easton, Pennsylvania, of seeing an old female Skunk with six young. We had no knowledge of the colour of the male. The female, however, had two broad stripes, with a very narrow black dorsal line; the young differed considerably in their markings, some having black, and others white, tails.

In the sand-hills near Columbia, South Carolina, we met along the sides of the highway four half-grown animals of this species; they all had a narrow white line on each side of the back, and a small white spot on the forehead; the tails of two of them were tipped with white; the others had the whole of their tails black.

From all the observations we have been able to make in regard to the colours of the different varieties of this species, we have arrived at the conclusion, that when a pair are alike in colour the young will bear a strong resemblance in their markings to the old. When on the contrary the parents differ, the young assume a variety of intermediate colours.

DIMENSIONS.

From point of nose to root of tail - - -	17	inches.
Tail (vertebræ) - - - - - - -	8¾	do.
Tail, to end of hair - - - - - -	12¾	do.
Distance between eyes - - - - - -	1⅛	do.
From point of nose to corner of mouth - -	1⅜	do.

Weight, 6½ pounds.

HABITS.

There is no quadruped on the continent of North America the approach of which is more generally detested than that of the Skunk: from which we may learn that, although from the great and the strong we have to apprehend danger, the feeble and apparently insignificant may have it in their power to annoy us almost beyond endurance.

In the human species we sometimes perceive that a particular faculty has received an extraordinary development, the result of constant devotion to one subject; whilst in other respects the mind of the individual is of a very ordinary character. The same remark will hold good applied to any particular organ or member of the body, which, by constant use, (like the organs of touch in the blind man,) becomes so improved as to serve as a substitute for others: but in the lower orders of animals this prominence in a particular organ is the result of its peculiar conformation, or of instinct. Thus the power of the rhinoceros is exerted chiefly by his nasal horn, the wild boar relies for defence on his tusks, the safety of the kangaroo depends on his hind-feet, which not only enable him to make extraordinary leaps, but with which he deals vigorous blows, the bull attacks his foes with his horns, the rattlesnake's deadly venom is conveyed through its fangs, and the bee has the means of destroying some of its enemies by its sting, whilst in every other power for attack or self-defence these various creatures are comparatively feeble.

The Skunk, although armed with claws and teeth strong and sharp enough to capture his prey, is slow on foot, apparently timid, and would be unable to escape from many of his enemies, if he were not possessed of a power by which he often causes the most ferocious to make a rapid retreat, run their noses into the earth, and roll or tumble on the ground as if in convulsions; and, not unfrequently, even the bravest of our boasting race is by this little animal compelled suddenly to break off his train of thought, *hold his nose*, and run, as if a lion were at his heels!

Among the first specimens of natural history we attempted to procure was the Skunk, and the sage advice to "look before you leap," was impressed on our mind, through several of our senses, by this species.

It happened in our early school-boy days, that once, when the sun had just set, as we were slowly wending our way home from the house of a neighbour, we observed in the path before us a pretty little animal, playful as a kitten, moving quietly along: soon it stopped, as if waiting for us to come near, throwing up its long bushy tail, turning round and looking at us like some old acquaintance: we pause and gaze; what is it? It is not a young puppy or a cat; it is more gentle than either; it seems desirous to keep company with us, and like a pet poodle, appears most happy when only a few paces in advance, preceding us, as if to show the path: what a pretty creature to carry home in our arms! it seems too gentle to bite; let us catch it. We run towards it; it makes no effort to escape, but waits for us; it raises its tail as if to invite us to take hold of its brush. We seize it instanter, and grasp it with the energy of a miser clutching a box of diamonds; a short struggle ensues, when—faugh! we are suffocated; our eyes, nose, and face, are suddenly bespattered with the most horrible fetid fluid. Imagine to yourself, reader, our surprise, our disgust, the sickening feelings that almost overcome us. We drop our prize and take to our heels, too stubborn to cry, but too much alarmed and discomfited just now, to take another look at the cause of our misfortune, and effectually undeceived as to the real character of this seemingly mild and playful little fellow.

We have never felt that aversion to the musky odour imparted by many species of the ferine tribe of animals, that others evince; but we are obliged to admit that a close proximity to a recently killed Skunk, has ever proved too powerful for our olfactories. We recollect an instance when sickness of the stomach and vomiting were occasioned, in several persons residing in Saratoga county, N. Y., in consequence of one of this species having been killed under the floor of their residence during the night. We have seen efforts made to rid clothes which have been sprinkled by a Skunk, of the offensive odour: resort was had to burying them in the earth, washing, and using perfumes; but after being buried a month they came forth almost as offensive as when they had first been placed in the ground; and as for the application of odoriferous preparations, it seemed as if all the spices of Araby could neither weaken nor change the character of this overpowering and nauseating fluid. Washing and exposure to the atmosphere certainly weaken the scent, but the wearer of clothes that have been thus infected, should he accidentally stand near the fire in a close room, may chance to be mortified by being reminded that he is not altogether free from the consequences of an "unpleasant" hunting excursion. We have, however, found chloride of lime a most effectual disinfectant when applied to our recent specimens. That there is something very acrid in the fluid ejected by the Skunk, cannot be doubted, when we consider its effects. Dr. RICHARDSON states that he knew several Indians who lost their eyesight in consequence of inflammation produced by its having been thrown into them by the animal. The instant a dog has received a discharge of this kind on his nose and eyes, he appears half distracted, plunging his nose into the earth, rubbing the sides of his face on the leaves and grass, and rolling in every direction. We have known several dogs, from the eyes of which the swelling and inflammation caused by it did not disappear for a week; still we have seen others, which, when on a raccoon hunt, did not hesitate, in despite of the consequences, to kill every Skunk they started, and although severely punished at the time, they showed no reluctance to repeat the attack the same evening, if a fresh subject presented itself.

This offensive fluid is contained in two small sacs situated on each side of the root of the tail, and is ejected through small ducts near the anus. We have on several occasions witnessed the manner in which this secretion is discharged. When the Skunk is irritated, or finds it necessary to defend himself, he elevates his tail over his back, and by a strong muscular exertion ejects it in two thread-like streams in the direction in which the enemy is observed. He appears to take an almost unerring aim, and almost invariably salutes a dog in his face and eyes. Dr. RICHARDSON states that he ejects this noisome fluid for upwards of four feet; in this he has considerably underrated the powers of this natural syringe of the Skunk, as we measured the distance on one occasion, when it extended upwards of fourteen feet. The notion of the old authors that this fluid is the secretion of the kidneys, thrown to a distance by the aid of his long tail, must be set down among the vulgar errors, for in that case whole neighbourhoods would be compelled to breath a tainted gale, as Skunks are quite common in many parts of the country.

The Skunk, in fact, is a very cleanly animal, and never suffers a drop of this fluid to touch his fur; we have frequently been at the mouth of his burrow, and although a dozen Skunks might be snugly sheltered within, we could not detect the slightest unpleasant smell. He is as careful to avoid soiling himself with this fluid, as the rattlesnake is, not to suffer his body to come in contact with his poisonous fangs.

Should the Skunk make a discharge from this all-conquering battery during the day, the fluid is so thin and transparent that it is scarcely perceptible; but at night it has a yellowish luminous appearance; we have noticed it on several occasions, and can find no more apt comparison than an attenuated stream of phosphoric light. That the spot where a Skunk has been killed will be tainted for a considerable time, is well known. At a place where one had been killed in autumn, we remarked that the scent was still tolerably strong after the snows had thawed away the following spring. Generally, however, the spot thus scented by the Skunk is not particularly offensive after the expiration of a week or ten days. The smell is more perceptible at night and in damp weather, than during the day or in a drought.

The properties of the peculiarly offensive liquor contained in the sacs of the Skunk, have not, so far as we are advised, been fully ascertained. It has, however, been sometimes applied to medical purposes. Professor Ives, of New-Haven, administered to an asthmatic patient a drop of this fluid three times a day. The invalid was greatly benefited: all his secretions, however, were soon affected to such a degree, that he became highly offensive both to himself and to those near him. He then discontinued the medicine, but after having been apparently well for some time, the disease returned. He again called on the doctor for advice,—the old and tried recipe was once more recommended, but the patient declined taking it, declaring that the remedy was worse than the disease!

We were once requested by a venerable clergyman, an esteemed friend, who had for many years been a martyr to violent paroxysms of asthma, to procure for him the glands of a Skunk; which, according to the prescription of his medical adviser, were kept tightly corked in a smelling bottle, which was applied to his nose when the symptoms of his disease appeared.

For some time he believed that he had found a specific for his distressing complaint; we were however subsequently informed, that having uncorked the bottle on one occasion while in the pulpit during service, his congregation finding the smell too powerful for their olfactories, made a hasty retreat, leaving him nearly alone in the church.

We are under an impression, that the difficulty of preparing specimens of this animal may be to a considerable extent obviated, by a proper care in capturing it. If it has been worried and killed by a dog, skinning a recent specimen is almost insupportable; but if killed by a sudden blow, or shot in a vital part, so as to produce instant death, the Skunk emits no unpleasant odour, and the preparation of a specimen is even less unpleasant than stuffing a mink. We have seen several that were crushed in deadfalls, that were in nowise offensive. We had one of their burrows opened to within a foot of the extremity, where the animals were huddled together. Placing ourselves a few yards off, we suffered them successively to come out. As they slowly emerged and were walking off, they were killed with coarse shot aimed at the shoulders. In the course of half an hour, seven, (the number contained in the burrow,) were obtained; one only was offensive, and we were enabled without inconvenience to prepare six of them for specimens.

The Skunk does not support a good character among the farmers. He will sometimes find his way into the poultry-house, and make some havoc with the setting hens; he seems to have a peculiar penchant for eggs, and is not very particular whether they have been newly laid, or contain pretty large rudiments of the young chicken; yet he is so slow and clumsy in his movements, and creates such a commotion in the poultry-house, that he usually sets the watch-dog in motion, and is generally detected in the midst of his depredations; when, retiring to some corner, he is either seized by the dog, or is made to feel the contents of the farmer's fowling piece. In fact the poultry have far more formidable enemies than the Skunk. The ermine and brown weasel are in this respect rivals with which his awkward powers cannot compare; and the mink is a more successful prowler.

The Skunk is so slow in his actions, that it is difficult to discover in what manner he obtains food to enable him always to appear in good condition. In the northern part of New-York the gray rabbit frequently retires to the burrow of the fox, Maryland marmot, or Skunk. Many of them remain in these retreats during the day. We have seen the tracks of the Skunk in the snow, on the trail of the gray rabbit, leading to these holes, and have observed tufts of hair and patches of skin scattered in the vicinity, betokening that the timid animal had been destroyed. We on one occasion marked a nest of the ruffed grouse, (*T. umbellus,*) with the intention of placing the eggs under a common hen a few days before they should hatch, but upon going after them we found they had been eaten, and the feathers of the grouse were lying about the nest. Believing the depredator to have been an ermine, we placed a box-trap near the spot baited with a bird; and on the succeeding night caught a Skunk, which we doubt not was the robber. This species also feeds on mice, frogs, and lizards; and during summer no inconsiderable portion of its food consists of insects, as its excrements usually exhibit the legs and backs of a considerable number of beetles.

On dissecting a specimen which we obtained from the middle districts of Carolina, we ascertained that the animal had been a more successful collector of entomological specimens than ourselves, as he had evidently devoured on the night previous a greater number (about a dozen) of a very rare and large beetle, (*Scarabæus tityus*,) than we had been able to find in a search of ten years.

The Skunk being very prolific, would, if allowed to multiply around the farm-yard, prove a great and growing annoyance. Fortunately there are nocturnal animals that are prowling about as well as he. The dog, although he does not eat this species, scarcely ever fails to destroy a Skunk whenever he can lay hold of him. A wolf that had been sent from the interior of Carolina to Charleston, to be prepared as a specimen, we observed was strongly tainted with the smell of this animal, and we concluded from hence, that as a hungry wolf is not likely to be very choice in selecting his food, he will, if nothing better offers, make a meal on it. Whilst riding along the border of a field one evening, we observed a large bird of some species darting to the ground, and immediately heard a struggle, and were saluted by the odour from the "Enfant du diable," as old CHARLEVOIX has designated the Skunk. We visited the spot on the following day, and found a very large animal of this species partly devoured. We placed a fox-trap in the vicinity, and on the following morning found our trap had captured a large horned owl, which had evidently caused the death of the Skunk, as in point of offensive effluvia there was no choice between them This species is generally very easily taken in traps. It will not avoid any kind of snare—is willing to take the bait, whether it be flesh, fish, or fowl, and proves a great annoyance to the hunters whose traps are set for the fisher and marten. The burrows of the Skunk are far less difficult to dig out than those of the fox. They are generally found on a flat surface, whilst the dens of the fox are more frequently dug on the side of a hill. They have seldom more than one entrance, whilst those of the fox have two, and often three. The gallery of the burrow dug by the Skunk runs much nearer the surface than that excavated by the fox. After extending seven or eight feet in a straight line, about two feet beneath the surface, there is a large excavation containing an immense nest of leaves. Here during winter may be found lying, from five to fifteen individuals of this species. There are sometimes one or two galleries diverging from this bed, running five or six feet further; in which, if the burrow has been disturbed, the whole family may generally be found, ready to employ the only means of defence with which Nature has provided them.

This animal generally retires to his burrow about December, in the Northern States, and his tracks are not again visible until near the tenth of February. He lays up no winter store; and like the bear, raccoon, and Maryland marmot, is very fat on retiring to his winter quarters, and does not seem to be much reduced in flesh at his first appearance toward spring, but is observed to fall off soon afterwards. He is not a sound sleeper on these occasions; on opening his burrow we found him, although dull and inactive, certainly not asleep, as his black eyes were peering at us from the hole, into which we had made an opening, seeming to warn us not to place too much reliance on the hope of finding this striped "weasel asleep."

In the upper districts of Carolina and Georgia, where the Skunk is occasionally found, he, like the raccoon in the Southern States, does not retire to winter quarters, but continues actively prowling about during the night through the winter months.

A large Skunk, which had been in the vicinity of our place, near New-York, for two or three days, was one morning observed by our gardener in an old barrel with only one head in, which stood upright near our stable. The animal had probably jumped into it from an adjoining pile of logs to devour an egg, as our hens were in the habit of laying about the yard. On being discovered, the Skunk remained quietly at the bottom of the barrel, apparently unable to get out, either by climbing or by leaping from the bottom. We killed him by throwing a large stone into the open barrel;—he did not make the least effort to eject the nauseous fluid with which he was provided. Had he not been discovered, he would no doubt have died of starvation, as he had no means of escaping. At times, especially during the summer season, the Skunk smells so strongly of the fetid fluid contained in his glands, that when one or two hundred yards distant it is easily known that he is in the neighbourhood.

We doubt not the flesh of the Skunk is well tasted and savoury. We observed it cooked and eaten by the Indians. The meat was white and fat, and they pronounced it better than the opossum,—infinitely superior to the raccoon, (which they called rank meat,) and fully equal to roast pig. We now regret that our squeamishness prevented us from trying it.

We have seen the young early in May; there were from five to nine in a litter.

The fur is rather coarse. It is seldom used by the hatters, and never we think by the furriers; and from the disagreeable task of preparing the skin, it is not considered an article of commerce.

GEOGRAPHICAL DISTRIBUTION.

This species has a tolerably wide range, being found as far to the north as lat. 56° or 57°. We have met with it both in Upper and Lower Canada, where it however appeared less common than in the Atlantic States. It is exceedingly abundant in every part of the Northern States. In New-England, New-York, and Pennsylvania, it is more frequently met with than in Maryland, Virginia, and the more Southern States. It is not uncommon on both sides of the Virginia Mountains, and is well known in Kentucky, Indiana, and Illinois. It is not unfrequently met with in the higher portions of South Carolina, Georgia, and Alabama. In the alluvial lands of these three States, however, it is exceedingly rare. We possess in the Charleston Museum two specimens procured in Christ Church Parish, by Professor Edmund Ravenel, that were regarded as a great curiosity by the inhabitants. It becomes more common a hundred miles from the seaboard, and is not unfrequently met with in the sand-hills near Columbia. To the south we have traced it to the northern parts of Florida, and have seen it in Louisiana. To the west it has been seen as far as the banks of the Mississippi. Lewis and Clark, and others, frequently saw Skunks west of the Rocky Mountains, near their winter encampments, but we have as yet had no means of ascertaining that they were of this species.

GENERAL REMARKS.

Although we do not regard the distribution of colours in the American Skunk, as of much importance in deciding on the species, and hence, have rejected as mere varieties, all those that can only be distinguished from each other by their markings, we nevertheless differ very widely from Baron Cuvier (Ossemens Fossiles, iv.) and others, who treat all the American Mephites as mere varieties. We have examined and compared many specimens in the museums of Europe and America, and possess others from Texas and other portions of the United States, and we feel confident that both in North and South America several very distinct species exist. We will endeavour, as we proceed in the present work, to investigate their characters, and describe those species that are found within the range to which we have restricted our inquiries. We have in the museums of London examined and compared the species described by Bennet, (Proceeds. Zool. Soc., 1833, p. 39,) as *M. nasuta*, which appears to have been previously described by Dr. Lichtenstein, of Berlin, under the name of *M. mesoleuca*, (Darst. der Säugeth. tab. 44, fig. 2,) as also several species characterized by Gray, (Magazine of Nat. Hist., 1837, p.

581,) that are very distinct from the present. In the immense collection existing in the museum at Berlin, one of the best regulated museums in Europe, and which is particularly rich in the natural productions of Mexico, Texas, California, and South America, several species are exhibited that cannot be referred to our Skunk. We are under obligations to Dr. LICHTENSTEIN for a valuable work, (Darstellung neuer oder wenig bekannter Säugethiere, Berlin, 1827–1834,) which contains figures and descriptions of a number of new species of Skunks. Also a monograph, (Ueber die Gattung Mephitis, Berlin, 1838,) in sixty-five pages quarto, with plates, which contains much learned research, and has greatly extended our previous knowledge of the genus. He describes seventeen species, all, with one African exception, belonging to North and South America. North of Texas, however, he recognises only two species, the present, and *Mephitis interrupta*, of RAFINESQUE; the latter, however, still requires a more careful comparison. All our American authors have applied the name *Mephitis Americana*, of DESMAREST, to our present species. It is now ascertained, however, that TIEDIMANN described it twelve years earlier under the name of *M. chinga*, which, according to the rigid rules to which naturalists feel bound to adhere, must be retained, and we therefore have adopted it.

Drawn on Stone by W. E. Hitchcock

Hare Squirrel.

Drawn from Nature by J. J. Audubon, F.R.S.F.L.S.

Lith.d Printed & Col.d by J. T. Bowen, Philad.a

SCIURUS LEPORINUS.—Aud. and Bach

Hare Squirrel.

PLATE XLIII.

S. magnitudine S. cinereum inter et S. migratorium intermedius; cauda corpore longiore, crassa maximeque disticha; vellere supra ex cinereo fusco; subtus albo.

CHARACTERS.

Intermediate in size between the Northern gray squirrel and the cat squirrel. Tail, longer than the body, large and distichous; colour, grayish-brown above, white beneath.

SYNONYME.

Sciurus Leporinus, Aud. and Bach., Proceedings of the Acad. of Nat. Sci., Philadelphia, 1841, p. 101.

DESCRIPTION.

Head, of moderate size; nose, blunt, covered with short hairs; forehead, arched; eyes, large; whiskers, numerous, extending to the ears; ears, broad at base, rounded at the edges, and forming an obtuse angle at the extremity, clothed with sharp hairs on both surfaces. Body, stout, covered by a coat of thick but rather short hair, coarser than that of the Northern gray squirrel; limbs, large, and rather long; tail, distichous, but not very bushy.

COLOUR.

Teeth, orange; whiskers, black; nose, dark brown; ears, light brown; behind the ears, a tuft of soft cotton-like, whitish fur. The hairs of the back are cinereous at the roots, then light brown, and are tipped with brown and black, giving it so much the colour of the English hare that we determined to borrow from it our specific name. On the sides, the colour is a shade lighter than on the back; the tail, which, from the broad white tips of the hair, has a white appearance, is brown at the roots, and three times annulated with black. The upper lips, chin, neck, and whole under surface, including the inner surface of the legs, white; the hair being of this colour from the roots. Feet, dull yellowish-white. On the outer surface of the hind-leg, above the heel, a small portion of the fur is brown; there is also a spot of the same colour on the upper surface of the hind-foot.

DIMENSIONS.

	Inches.	Lines.
Length of head and body - - - - - -	11	9
" " tail - - - - - - - -	12	6
Height of ear, - - - - - - - -	0	8
Heel to end of middle claw - - - - -	2	9
Breadth of tail with hairs extended - - -	5	6

HABITS.

This species, which is one of our most beautifully furred Squirrels, is especially remarkable for its splendid tail, with its broad white border. We know nothing of its habits, as it was brought from California, without any other information than that of its locality.

We have represented two of these Squirrels in our plate, on a branch of hickory, with a bunch of nearly ripe nuts attached.

GEOGRAPHICAL DISTRIBUTION.

The range of this Squirrel through California, is, as well as its habits, totally unknown to us. It will not be very long, however, we think, before a great deal of information respecting that portion of our continent, so rich in rare and new species, may be expected, and we should not be surprised to find it extending toward the south-western portions of Texas, where several species of Squirrels that we have not obtained, are said to exist.

GENERAL REMARKS.

This species, in its general appearance, so much resembles some varieties of *Sciurus migratorius* and *S. cinereus*, that had it not been for its distant western locality, we should at first have been tempted to set it down, without further examination, as one or other of those species. There can, however, be no doubt, from its differing in size and in so many details of colour from all other species, that it must be regarded as distinct. It should be further observed, that *S. migratorius* has never been found south of Missouri, and that *S. cinereus* is not found west of the Mississippi. Indeed, the geographical range of the latter terminates several hundred miles to the eastward of that river, and it would be contrary to all our past experience, that a species existing in one part of our continent should be found in another, separated by an extent of several thousand miles of intermediate country, in no portion of which is it known to exist.

GENUS PSEUDOSTOMA.—Say.

DENTAL FORMULA.

$$\textit{Incisive}\ \frac{2}{2};\quad \textit{Canine}\ \frac{0-0}{0-0};\quad \textit{Molar}\ \frac{4-4}{4-4} = 20.$$

Incisors, naked, truncated; molars, destitute of radicles; crowns, simple, oval; anterior ones, double.

Head, large and depressed; nose, short; mouth, small.

The cheek-pouches are large, and open exterior to the mouth.

The eyes are small and far apart. The external ear is very short; auditory openings, large. Body, sub-cylindrical; tail, rather short, round, tapering slightly, clothed with short hairs.

Legs, short, with five toes to each foot.

Burrowing in sandy soils, feeding on grasses, roots, nuts, &c., which they convey to their burrows in their capacious cheek-pouches, are habits common to this genus.

There are about six well determined species of Pouched Rats, all existing in North America.

The generic name is derived from ψευδο, (*pseudo,*) false, and στομα, (*stoma,*) a mouth, in allusion to the false mouths or cheek-pouches of the genus.

PSEUDOSTOMA BURSARIUS.—Shaw.

Canada Pouched Rat.

PLATE XLIV.—Males, Female and Young.

P. supra, rufo-fuscus; subtus, cinereo-fuscus; pedibus, albis.

CHARACTERS.

Reddish-brown above, ashy-brown beneath; feet, white.

N° 9.

Plate XLIV

Canada Pouched Rat.

Drawn from Nature by J. J. Audubon F.R.S. F.L.S.

Lith. Printed & Col^{d} by J. T. Bowen, Philada.

SYNONYMES.

MUS BURSARIUS, Shaw, Descript. of the M. Bursarius in Linn. Transact., vol. v., p. 227 to 228.
MUS BURSARIUS, Shaw's Gen. Zool., vol. ii., p. 100, pl. 138, (figures with cheek-pouches unnaturally inverted.)
MUS BURSARIUS, Mitchill, Silliman's Journal, vol. iv., p. 183.
MUS SACCATUS, Mitchill, N. Y., Medical Repository, Jan. 1821.
SACCOPHORUS BURSARIUS, Kuhl, Beit., p. 66.
CRICETUS BURSARIUS, Desm. in Nouv. Dict., 14, p. 177.
" " F. Cuv. in Dict. des Sc. Nat., t. xx., p. 257.
PSEUDOSTOMA BURSARIUS, Say, in Long's Expedi., vol. i., p. 406.
" " Godm., vol. ii., p. 90, fig. 2.
" " Harlan, p. 133.
GEOMYS ? BURSARIUS, Rich., F. B. A., p. 203.

DESCRIPTION.

Head, large ; nose, broad and obtuse, covered with hair, with the exception of the margins of the nostrils, which are naked ; the nostrils are small oblong openings a line apart, and are on their superior margins considerably vaulted.

The incisors protrude beyond the lips ; they are very large and truncated ; in the superior jaw they are each marked by a deep longitudinal groove near the middle, and by a smaller one at the inner margin ; in the young, they exhibit only a single groove. The molars penetrate to the base of their respective alveoles without any division into roots, their crowns are simply discoidal, transversely oblong-oval, margined by the enamel ; the posterior tooth is rather more rounded than the others, and that of the upper-jaw has a small prominent angle on its posterior face ; the anterior tooth is double, in consequence of a profound duplicature in its side, so that its crown presents two oval disks, of which, the anterior one is smallest, and in the lower-jaw somewhat angulated. (SAY.)

Eyes, small ; ears, very short, and scarcely visible ; whiskers, not numerous, shorter than the head.

The cheek-pouches are very large, extending from the sides of the mouth to the shoulders, and are internally lined with short soft hairs ; the body is broad and stout, sub-cylindrical, and has a clumsy appearance, not unlike that of the shrew-mole. It is thickly clothed on both the upper and lower surfaces with soft hair, that on the back being in some parts half an inch long, whilst on the under surface it is much shorter, and more compact.

The feet have five toes each ; the fore-feet are robust, with large, elongated, compressed, and hooked nails ; the middle nail is much the longest, the fourth is next in length, the second shorter, the fifth still shorter, and the first very short; there is a large callous protuberance on the hinder-part of the palms. On the hind-feet the toes are short, and the nails are very short, concave beneath, and rounded at tip; the middle nail is longest, the second almost as long, the fourth a little shorter, the first still shorter, the fifth very short. This Rat is plantigrade, and presses on the earth from the heel to the toes.

The tail is for one-third of its length from the root clothed with hair, but toward the extremity is naked.

COLOUR.

Incisors, yellow; nostrils, light pink; eyes, black. The fur is plumbeous from the roots to near the extremity, where it is broadly tipped with reddish-brown; on the under surface it is a little paler, owing to the ends of the hairs being but slightly tipped with brown.

The head and the dorsal line are a shade darker than the surrounding parts.

Moustaches, white and black; nails, and all the feet, white.

The colours here described are those which this species exhibits during winter and the early part of summer. Immediately after shedding its hair it takes the colour of the young, light-plumbeous, which gradually deepens at the approach of winter.

DIMENSIONS.

From nose to root of tail	9¾	inches.
" " to ear	2	do.
" " to end of pouch	4¼	do.
Tail	2¼	do.
Depth of pouch	3	do.
Fore-foot with longest claw	1⅝	do.
Distance between the eyes	⅞	do.

Weight of largest specimen, 14 oz.

HABITS.

During a visit which we made to the Upper Missouri in the spring and summer of 1843, we had many opportunities of studying the habits of this species. In the neighbourhood of St. Louis, at the hospitable residence of the late Pierre Chouteau, Esq., we procured several of them alive. In that section of country they are called "Muloes."

They are considered by the gardeners in that vicinity as great plagues, devouring every tap-root vegetable, or grass, within their reach, and perforating the earth in every direction, not only at night, but oftentimes during the day.

Having observed some freshly thrown up mounds in Mr. CHOUTEAU's garden, several servants were called, and set to work to dig out the animals, if practicable, alive; and we soon dug up several galleries worked by the Muloes, in different directions. One of the main galleries was about a foot beneath the surface of the ground, except where it passed under the walks, in which places it was sunk rather lower. We turned up this entire gallery, which led across a large garden-bed and two walks, into another bed, where we discovered that several fine plants had been killed by these animals eating off their roots just beneath the surface of the ground. The burrow ended near these plants under a large rose-bush. We then dug out another principal burrow, but its terminus was amongst the roots of a large peach-tree, some of the bark of which had been eaten off by these animals. We could not capture any of them at this time, owing to the ramifications of their galleries having escaped our notice whilst following the main burrows. On carefully examining the ground, we discovered that several galleries existed that appeared to run entirely out of the garden into the open fields and woods beyond, so that we were obliged to give up the chase. This species throws up the earth in little mounds about twelve or fifteen inches in height, at irregular distances, sometimes near each other, and occasionally ten, twenty, even thirty paces asunder, generally opening near a surface well covered with grass or vegetables of different kinds.

The Pouched Rat remains under ground during cold weather in an inactive state, most probably dormant, as it is not seen to disturb the surface of the earth until the return of spring, when the grass is well grown.

The earth when thrown up is broken or pulverized, and as soon as the animal has completed his galleries and chambers, he closes the aperture on the side towards the sun, or on the top, although more usually on the side, leaving a sort of ring or opening about the size of his body.

Possessed of an exquisite sense of hearing, and an acute nose, at the approach of any one travelling on the ground the "Muloes" stop their labours instantaneously, being easily alarmed; but if you retire some twenty or thirty paces to *leeward* of the hole, and wait there for a quarter of an hour or so, you will see the "Gopher" (another name given to these animals by the inhabitants of the State of Missouri), raising the earth with his back and shoulders, and forcing it out before and around him, leaving an aperture open during the process. He now runs a few steps from the hole and cuts the grass, with which he fills his cheek-pouches, and then retires into his burrow to eat it undisturbed.

You may see the Pseudostoma now and then sitting on its rump and basking in the rays of the sun, on which occasion it may easily be shot if you are prompt, but if missed it disappears at once, is seen no more, and will even dig a burrow to a considerable distance, in order to get out of the ground at some other place where it may not be observed.

This species may be caught in steel-traps, or common box-traps, with which we procured two of them. When caught in a steel-trap, they frequently lacerate the leg by which they are held, which is generally the hind one, by their struggles to get free. They are now and then turned up by the plough, and we have known one caught in this manner. They sometimes destroy the roots of young fruit-trees to the number of one or two hundred in the course of a few days and nights; and they will cut those of full grown trees of the most valuable kinds, such as the apple, pear, peach and plum. This species is found to vary in size very greatly on comparing different individuals, and they also vary in their colour according to age, although we found no difference caused by sex.

The commonly received opinion is, that these rats fill their pouches with the earth from their burrows, and empty them at the entrance. This is, however, quite an erroneous idea. Of about a dozen, which were shot dead in the very act of rising out of their mounds and burrows, none had any earth in their sacs; but the fore-feet, teeth, nose, and the anterior and upper portion of their heads, were found covered with adherent earth. On the contrary, most of them had their pouches filled with either blades of grass or roots of different trees; and we think of these pouches, that their being hairy within, rather corroborates the idea that they are only used to convey food to their burrows. This species appears to raise up the earth very much in the manner of the common shrew-mole.

When running, the tails of these animals drag on the ground, and they hobble along at times with their long front claws bent underneath their feet as it were, backwards, and never by leaps. They can travel almost as fast backwards as forwards. When turned on their backs they have great difficulty in regaining their natural position, and kick about in the air for a minute or two with their legs and claws extended, before they can turn over. They can bite severely, as their incisors by their size and sharpness plainly indicate; and they do not hesitate to attack their enemies or assailants with open mouth, squealing when in a rage like the common Norway or wharf rat, (*Mus decumanus.*) When they fight among themselves they make great use of their snouts, somewhat in the manner of hogs. They cannot travel faster when above ground than a man walks; they feed frequently whilst seated on their rump, using their fore-feet and long claws somewhat in the manner of squirrels. When sleeping they place their heads beneath the shoulder on the breast, and look like a round ball or a lump of earth. They clean their hair, whiskers, and body, in the same manner as rats, squirrels, &c.

We kept four of these animals alive for several weeks, and they never during that time drank any thing, although we offered them both water and milk. We fed them on cabbages, potatoes, carrots, &c., of which they ate a larger quantity than we supposed them capable of consuming. They tried constantly to make their escape, by gnawing at the floor of the apartment. They slept on any of the clothing about the room which would keep them warm; and these mischievous pets cut the lining of our hunting coat, so that we were obliged to have it repaired and patched. We had left a handkerchief containing sundry articles, tied as we thought securely, but they discovered it, and on opening it one of them caught hold of our thumb, with (luckily) only one of his incisors, and hung on until we shook it off violently. While confined thus in our room, these animals gnawed the leather straps of our trunks, and although we rose frequently from our bed at night to stop their career of destruction, they began to gnaw again as soon as we were once more snugly ensconced beneath the counterpane. Two of them entered one of our boots, and probably not liking the idea of returning by the same way, ate a hole at the toes, by which they made their exit. We have given in our plate four figures of this singular species.

The nest of the Canada Pouched Rat is usually rounded, and is about eight inches in diameter. It is well lined with soft substances as well as with the hair of the female. It is not placed at the end of a burrow, nor in a short gallery, but generally is one that is in the centre of sundry others diverging to various points, at which the animal can escape if pursued, and most of which lead to the vicinity of grounds where their favourite food is abundant.

The female brings forth from five to seven young at a litter, about the end of March or early in April. They are at a very early period able to run about, dig burrows, and provide for themselves.

GEOGRAPHICAL DISTRIBUTION.

The *Pseudostoma bursarius* has a wide geographical range. We found it in all those places we visited, east of the Rocky Mountains and west of the Mississippi, where the soil and food suited its habits. It has been observed as far to the north as lat. 52°. It abounds in Michigan and Illinois. Farther to the south it extends along the western prairies, and it was observed near the shores of the Platte, Arkansas, Canadian, and Red Rivers, to lat. 34°, and probably ranges still further to the south.

There are Pouched Rats in Texas and Mexico, but we are at present unable to determine whether they are of this species.

GENERAL REMARKS.

The first naturalist who gave a specific name to this Pouched Rat was Dr. SHAW, in the Linnæan Transactions, accompanied by a figure representing it as having only three toes. The drawing had been made by Major DAVIES. Subsequently (in 1801) he again described and figured it in his General Zoology, vol. ii., p. 100, pl. 138. The pouches in both cases are inverted, and hanging down like long sacks on each side. These would be very inconvenient, as the animal could not place its nose on the earth or fill its sacks, with such an unnatural appendage dangling at its mouth. The error seems to have originated from the whim or ignorance of an Indian. It is recorded, that in 1798 one of this species was presented by a Canadian Indian to the Lady of Governor PRESCOTT. Its pouches had been inverted, filled, and greatly distended with earth ; and from this trival circumstance an error originated which has been perpetuated even to the present day.

RAFINESQUE, who was either careless or unscrupulous in forming new genera and species, and whose writings are so erroneous that we have seldom referred to him, contributed to create still farther confusion among the species of this genus. He arranged them under two genera : GEOMYS, with cheek-pouches opening into the mouth, and DIPLOSTOMA, with cheek-pouches opening exterior to the mouth. This last genus he characterizes by its having no tail, and only four toes on each foot. (Am. Monthly Magazine, 1817.)

We consider it unfortunate that our friend Dr. RICHARDSON should have adopted both these genera, and given several species under each. We have examined nearly all the original specimens from which his descriptions were taken, and feel confident that they all belong to the genus PSEUDOSTOMA, of SAY.

In regard to the present species, Dr. RICHARDSON was undecided under what genus it should be placed. The opportunities afforded us for making a careful examination, leave no room for any doubt on that subject.

That there are several species of pouched rats on both sides of the

Rocky Mountains from Mexico to Canada, cannot be doubted, but the difficulty of distinguishing the species is greater than is usually supposed.

They possess similar habits, specimens belonging to the same species are found of different sizes and of different colours; all the species have short ears and tails. They live under the earth; and many persons who have for years resided in their immediate vicinity, although they daily observe traces of their existence, have never seen the animals.

American naturalists have sometimes been reminded by their European brethren, of the duty devolving on them of investigating the habits and describing the species of animals existing in their country. The charge of our having hitherto depended too much on Europeans to effect this laudable object, is true to a considerable extent. It should, however, be borne in mind, that this vast country belongs to many nations; that large portions of it are either unpeopled deserts or are roamed over by fierce savage tribes; that the Northern regions visited by RICHARDSON are exclusively under the control of Great Britain and that the vast chain of the Rocky Mountains presents more formidable barriers than the oceans which separate Europe from the Western shores of America.

It is not, therefore, surprising, that in order to become acquainted with some rare species, American naturalists are obliged to seek access to European museums, instead of the imperfect private collections of their own country.

In the United States, east of the Rocky Mountains, we are not aware of the existence of more than two species of Pouched Rat,—the present species and another existing in Georgia and Florida. It is, however, not improbable that *Pseudostoma Mexicanus* may yet be found in Texas.

GENUS ARVICOLA.—Lacépède.

DENTAL FORMULA.

$$Incisive\ \frac{2}{2};\ Canine\ \frac{0-0}{0-0};\ Molar\ \frac{3-3}{3-3} = 16.$$

Incisors, in the upper jaw, large and cuneiform; in the inferior jaw, sharp.

Molars, compound, flat on their crowns, the enamel forming angular ridges on the surface.

Fore-feet, having the rudiments of a thumb, and four toes, furnished with weak nails.

Hind-feet, with five toes, hairy on their borders, armed with claws.

Ears, clothed with hair; tail, cylindrical and hairy, shorter than the body. From eight to twelve pectoral and ventral mammæ.

The old family of Mus has undergone many subdivisions. It formerly included many of our present genera. The Arvicolæ, by the structure of their teeth, and the hairy covering of their ears and tail, the latter being besides short, may advantageously be separated from the rest.

They burrow in the earth, and feed on grain, bulbous roots and grasses; some are omnivorous, they do not climb, are not dormant in winter, but seek their food during cold weather, eating roots, grasses, and the bark of trees.

There have been about forty species of Arvicola described; some of these, however, are now arranged under other genera. Some of the species are found in each quarter of the world: about seven species inhabit North America.

The generic name is derived from two Latin words, *arvus*, a field, and *colo*, I inhabit.

No 9. Plate XLV

Wilson's Meadow Mouse.

Drawn from Nature by J.J. Audubon, F.R.S. F.L.S.

Lith. Printed & Col'd by J.T. Bowen, Philada

ARVICOLA PENNSYLVANICA.—Ord

WILSON'S MEADOW-MOUSE.

PLATE XLV.—Two figures.

A. supra, cervinus; subtus, subalbicans; auriculis abreviatis rotundatisque.

CHARACTERS.

Brownish fawn-colour above; beneath, grayish-white; eyes, small; ears, short and round.

SYNONYMES.

SHORT-TAILED MOUSE, Forster, Phil. Trans., vol. lxii., p. 380, No. 18.
MEADOW MOUSE, Pennant's Arctic Zoology, vol. i., p. 133.
THE CAMPAGNOL OR MEADOW MOUSE of PENNSYLVANIA, Warden's Description of the U. S., vol. v., p. 625.
ARVICOLA PENNSYLVANICA, Ord, Guthrie's Geography.
" " " in Wilson's Ornithology, vol. vi., pl. 50, fig. 3.
" PENNSYLVANICA, Harlan, F. A., p. 144.
ARVICOLA ALBO-RUFESCENS, Emmons, Mass. Reports, p. 60, variety.
ARVICOLA HIRSUTUS, Emmons, Mass. Report.
" " Dekay, Nat. Hist. N. Y., p. 86.

DESCRIPTION.

Body, robust, cylindrical, broadest across the shoulders; diminishing towards the loins; fur, on the whole body, long and fine, but not lustrous; on the upper surface (in winter specimens) half an inch long, but not more than half that length beneath.

Head, large and conical; forehead, arched; nose, rather blunt; incisors, projecting; eyes, small, situated equidistant from the auditory opening and the point of the nose; the longest whiskers, about the length of the head; nostrils, lateral; nose, bilobate, clothed with short hairs; lips, fringed with longer hairs; mouth, beneath, not terminal; ears, large, rounded, membranous, concealed by the fur, naked within, except along the margins, where they have a few long soft hairs; auricular opening, large. The neck is so short that the head and shoulders seem united, like those of the shrew-mole.

Fore-feet slender, having four toes and a thumb, which is furnished with a sharp nail; nails, small, compressed, slightly hooked and sharp. The toes have five tubercles; the second toe from the thumb is longest, the third a little shorter, the first still shorter, and the outer one shortest.

The hind-feet are a little longer than the fore-feet; the third and fourth toes from the inner side are nearly of equal length, the second toe is a little shorter, the fifth still shorter, and the first is shortest. The soles of the hind-feet have five distinct tubercles; all the feet are clothed with short, adpressed hairs. The tail is short, scaly, cylindrical, slightly clothed with rigid hair extending beyond the vertebræ.

COLOUR.

Teeth, dark orange; fur, from the roots to near the tips, on every part of the body, dark plumbeous. The colour differs a shade or two between winter and summer. It may be characterized as brownish-gray above, a little darker on the back. The lips, chin, throat, and abdomen, are light bluish-gray. Feet, dark-brown; tail, brown above, and a shade lighter beneath; eyes, black; whiskers, white and black.

DIMENSIONS.

Length of head and body - - - - - -	5	inches.
" " tail - - - - - - - - -	1¾	do.

Another specimen.

Length of head and body - - - - - -	5½	do.
" " tail (vertebræ) - - - - - -	1½	do.
" " " including fur - - - - -	1¾	do.

HABITS.

We have had opportunities in New-York, Pennsylvania, and the New-England States, of learning some of the habits of this species. It is, in fact, the common Meadow-Mouse of the Northern and Eastern States.

Wherever there is a meadow in any of these States, you may find small tortuous paths cut through the grass, appearing as if they had been partially dug into the earth, leading to the roots of a stump, or the borders of some bank or ditch. These are the work of this little animal. Should you dig around the roots, or upturn the stump, you may find a family of from five to ten of this species, and will see them scampering off in all directions; and although they do not run fast, they have so many hiding places, that unless you are prompt in your attack, they are likely to escape you. Their galleries do not run under ground like those of the shrew mole, or the mischievous pine-mouse, (of LECONTE,) but extend along the surface sometimes for fifty yards.

The food of this species consists principally of roots and grasses. During summer it obtains an abundant supply of herds-grass, (*Phleum pratense*,) red-top, (*Agrostis vulgaris*,) and other plants found in the meadows; and when the fields are covered with snow it still pursues its summer paths, and is able to feed on the roots of these grasses, of which there is always a supply so abundant that it is generally in good condition. It is also fond of bulbs, and feeds on the meadow-garlic, (*Allium Canadense*,) and red lily, (*Lilium Philadelphicum*.)

We doubt whether this active little arvicola ever does much injury to the meadows; and in the wheat-fields it is not often a depredator, as it is seldom seen on high ground. Still, we have to relate some of its habits that are not calculated to win the affections of the farmer. In very severe winters, when the ground is frozen, and there is no covering of snow to protect the roots of its favourite grasses, it resorts for a subsistence to the stems of various shrubs and fruit trees, from which it peals off the bark, and thus destroys them. We possessed a small but choice nursery of fruit trees, which we had grafted ourselves, that was completely destroyed during a severe winter by this Meadow-Mouse, the bark having been gnawed from the wood for several inches from the ground upwards. Very recently our friend, the late Dr. WRIGHT, of Troy, sent us the following observations on this species :—

"Two or three winters ago several thousand young fruit trees were destroyed in two adjoining nurseries near our city; the bark was gnawed from them by some small animal, for the space of several inches, the lowest part of the denuded surface being about ten inches from the ground. I examined the premises the following spring. The ground had been frozen very hard all winter, owing to the small quantity of snow that had fallen. I supposed that some little animal that subsists on the roots of grasses, had been cut off from its ordinary food by the stony hardness of the ground, and had attacked the trees from the top of the snow. I looked around for the destroyer, and found a number of the present species, and no other. I strongly suspect that this animal caused the mischief, as it is very abundant and annoys the farmer not a little.

"A few years ago a farmer gave me permission to upset some stacks of corn on a piece of low land: I found an abundance of this species in shallow holes under them, and discovered some distance up between the stalks, the remains of cobs and kernels, showing that they had been doing no friendly work for the farmer."

We suspect, however, that the mischief occasioned to the nursery by this species is infinitely greater than that arising from any depredations it commits on wheat or corn-fields.

The nests of this arvicola are always near the surface; sometimes two or three are found under the same stump. We have frequently during summer observed them on the surface in the meadows, where they were concealed by the overshadowing grasses. They are composed of about a double handful of leaves of soft grasses, and are of an oval shape, with an entrance on the side.

Wilson's Meadow-Mouse swims and dives well. During a freshet which covered some neighbouring meadows, we observed several of them on floating bunches of grass, sticks, and marsh weeds, sitting in an upright posture as if enjoying the sunshine, and we saw them leaving these temporary resting places and swimming to the neighbouring high grounds with great facility; a stick thrown at them on such occasions will cause them to dive like a musk-rat.

This species does not, in any part of the United States, visit dwellings or outhouses, although Richardson states that it possesses this habit in Canada. We have scarcely ever met with it on high grounds, and it seems to avoid thick woods.

It produces young three or four times during the summer, from two to five at a birth. As is the case with the Florida rat and the white-footed mouse, the young of this species adhere to the teats, and are in this way occasionally dragged along by the mother. We would, however, here remark, that this habit, which is seen in the young of several animals, is by no means constant. It is only when the female is suddenly surprised and driven from her nest whilst suckling her young, that they are carried off in this manner. The young of this species that we had in confinement, after satisfying themselves, relinquished their hold, and permitted the mother to run about without this incumbrance.

This species is easily caught in wire-traps baited with a piece of apple, or even meat; we have occasionally found two in a trap at the same time. When they have become accustomed to the confinement of a cage they are somewhat familiar, feed on grass and seeds of different kinds, and often come to the bars of the cage to receive their food.

They frequently sit erect in the manner of marmots or squirrels, and while in this position clean their faces with their paws, continuing thus engaged for a quarter of an hour at a time. They drank a good deal of water, and were nocturnal in their habits. During the day-time they constantly nestled under some loose cotton, where they lay, unless disturbed, until dusk, when they ran about their place of confinement with great liveliness and activity, clinging to the wires and running up and down in various directions upon them, as if intent on making their escape.

GEOGRAPHICAL DISTRIBUTION.

We have found this species in all the New-England States, where it is very common. It is abundant in all the meadows of the State of New-York. It is the most common species in the neighbourhood of Philadelphia. We have found it in Maryland and Delaware. It exists in the valleys of the Virginia Mountains; and we obtained a number of specimens from our friend, Edmund Ruffin, Esq., who procured them on the Pamunkey River, in Hanover county, in that State, where it is quite abundant. We have traced it as far south as the northern boundary of North Carolina; and to the north have met with it in Upper and Lower Canada. Forster obtained it from Hudson's Bay, and Richardson speaks of it as very abundant from Canada to Great Bear Lake, latitude 65°.

To the west it exists along the banks of the Ohio, but we were unable to find it in any part of the region lying between the Mississippi and the Rocky Mountains.

GENERAL REMARKS.

We are fully aware of the difficulty of finding characters by which the various species of this genus may be distinguished. We cannot speak positively of Wilson's diminutive figure of the Meadow-Mouse, (American Ornithology, vol. vi., plate 50, fig. 3; description given, p. 59, in the article on the barn-owl,) but the accurate description of it by Ord, which is creditable to him as a naturalist, cannot possibly apply to any other species than this. It is the most common arvicola near Philadelphia, and no part of the description will apply to either of the only two other species of this genus existing in that vicinity.

We had an opportunity, at the museum of Zurich, to compare specimens of this species with the campagnol or meadow-mouse of Europe, *Mus agrestis* of Linnæus, and *Arvicola vulgaris* of Desmarest, to which Godman, (Nat. Hist., vol. ii., p. 88,) referred it. There is a strong general resemblance, but the species are distinct. The European animal has longer and narrower ears, protruding beyond the fur; its tail is shorter, and the body is more ferruginous on the upper surface than in our species.

In the last work published on American quadrupeds, the writer endeavours to show that this species, (which he has named *A. hirsutus*,) differs from *A. Pennsylvanica.* The following remarks are made at p. 87:—"Upon the suggestion that it might possibly be the *Pennsylvanicus* of Ord and Harlan, it was shown to both those gentlemen, who pronounced it to be totally distinct." To this we would observe, without the slightest design of undervaluing the scientific attainments of the respectable naturalists here referred to, that it was taxing their memories rather too much, to expect them, after the lapse of fifteen or twenty years, during which time their minds had been directed to other pursuits, to be as well qualified to decide on a species as they were when they first described it, (with all the specimens before them,) and when the whole subject was fresh in their minds. In regard to Dr. Harlan, he candidly wrote in answer to our inquiries respecting this and several other species, that having been long engaged in other investigations, and never having preserved specimens, he could not rely on his present judgment with any degree of accuracy. His description, moreover, being contained in two and a half lines, cannot be depended on, and is equally applicable to a considerable number of species. In regard to referring subjects, requiring such minute investigation, to the memory, when the period at which the specimens were examined has long passed, we have in mind the reply of Johnson, the great philologist, to an inquiry for information in regard to the derivation of a word, and of Newton, when asked for a solution of some knotty point in the higher branches of science: the former referred the inquirer to his "Dictionary,"—the latter, to his "Principia." The description of Mr. Ord is full and accurate, and by this we are quite willing to abide. We, moreover, are perfectly satisfied, that when that gentleman has an opportunity of comparing specimens of the several species found in the vicinity of Philadelphia with his own description, he will refer the species described and figured as *A. hirsutus* to his *A. Pennsylvanica.*

The arvicola *Albo-rufescens* of Emmons is evidently a variety of this species. We obtained a specimen from a nest in the northern part of New-York, which answered in every particular to his description. From the same nest two others were taken, with white rings round their necks, and three marked like the common *Arvicola Pennsylvanica,* differing in no respect from *Arvicola hirsutus.*

No. 10 Plate XLVI

American Beaver.

Drawn from Nature by J. J. Audubon, F.R.S.F.L.S.

Lith. Printed & Col^{d} by J. T. Bowen, Philada.

GENUS CASTOR.—Linn.

DENTAL FORMULA.

$$\textit{Incisive}\,\frac{2}{2};\quad \textit{Canine}\,\frac{0-0}{0-0};\quad \textit{Molar}\,\frac{4-4}{4-4}=20.$$

Incisors very strong. In the upper jaw their anterior surface is flat and their posterior surface angular. The molars differ slightly from each other in size, and have one internal and three external grooves. In the lower jaw the incisors present the same appearance as those of the upper; but are smaller. In the molars there are three grooves on the inner side, with one on the external.

Eyes, small; ears, short and round; five toes on each foot. On the fore-feet the toes are short and close; on the hind-feet long and palmated. Tail, large, flat and scaly. Mammæ, four, pectoral: a pouch near the root of the tail, in which an unctuous matter is secreted.

There is but one well established species known to belong to this genus.

The generic name is derived from the Latin word *Castor*, a beaver.

CASTOR FIBER.—Linn.

(VAR. AMERICANUS.)

American Beaver.

PLATE XLVI.

C. Arct. monace major, supra badius, infra dilutior; cauda plana, ovata, squamosa.

CHARACTERS.

Larger than the ground-hog, (Arctomys monax;) of a reddish-brown colour, with a short downy grayish fur beneath; tail, flat, scaly, and oval.

SYNONYMES.

Castor Fiber, Linn., 12th ed., p. 78.
Castor, Sagard Theodat, Canada, p. 767.
Beaver, Castor, Pennant, Arc. Zool., vol. i., p. 98.
Castor Ordinaire, Desm., Mamm.
Castor Americanus, F. Cuvier.
Castor Fiber, Lewis and Clarke's Expedition, vol. i.
The Beaver, Hearne's Journal, vol. viii., p. 245.
Beaver, Cartwright's Journal, vol. i., p. 62.
" Catesby, App., p. 29.
Castor Fiber, Harlan, Fauna, p. 122.
" " Godman, vol. ii., p. 21.
" " Americanus, Richardson, F. B. A., p. 105.
" " Emmons, Mass. Reports, p. 51.
" " Dekay, pl. 1, p. 72.

DESCRIPTION.

The shape of the body bears a considerable resemblance to that of the musk-rat; it is, however, much larger, and the head is proportionally thicker and broader. It is thick and clumsy, gradually enlarging from the head to the hips, and then is somewhat abruptly rounded off to the root of the tail.

Nose, obtuse and divided; eyes, small; ears, short, rounded, well clothed with fur, and partially concealed by the longer surrounding hairs; moustaches, not numerous, but very rigid like hogs' bristles, reaching to the ears; neck, rather short. The fur is of two kinds. The upper and longer hair is coarse, smooth, and glossy; the under coat is dense, soft and silky. Fore-feet, short and rather slender; toes, well separated and very flexible. The fore-feet are used like hands to convey food to the mouth. The fore-claws are strong, compressed, and channelled beneath. The middle toe is the longest, those on each side a little shorter, and the outer and inner ones shortest.

The hind-feet bear some resemblance to those of the goose. They are webbed beyond the roots of the nails, and have hard and callous soles. In most of the specimens we have seen, there is a double nail on the second inner toe. The palms and soles are naked. When walking, the whole heel touches the ground. The Beaver is accustomed to rest itself on its hind-feet and tail; and when in this sitting position contracts its fore-claws in the manner of the left hand figure represented in the plate. The upper surface of all the feet, with the exception of the nails, which are naked, is thickly covered with short adpressed hairs.

The tail is very broad and flat, tongue-shaped, and covered with angular scales. The root of the tail is for an inch covered with fine fur. The glandular sacs containing the castoreum, a musky unctuous substance, are situated near the anus.

COLOUR.

Incisors, on their outer surface, orange; moustaches, black; eyes, light-brown. The soft under down is light grayish-brown. The upper fur on the back is of a shining chesnut colour; on the under surface, and around the mouth and throat, a shade lighter. Nails, brown; webs between the toes, and tail, grayish-brown. We have seen an occasional variety. Some are black; and we examined several skins that were nearly white.

DIMENSIONS.

Male, represented in the plate.—Rather a small specimen.

From nose to root of tail,	23	inches.
Tail,	10	do.
From heel to end of middle claw,	$5\frac{1}{2}$	do.
Greatest breadth of tail,	$3\frac{1}{4}$	do.
Thickness of tail,	$\frac{7}{8}$	do.

Weight, $11\frac{1}{4}$ lbs.

HABITS.

The sagacity and instinct of the Beaver have from time immemorial been the subject of admiration and wonder. The early writers on both continents have represented it as a rational, intelligent, and moral being, requiring but the faculty of speech to raise it almost to an equality, in some respects, with our own species. There is in the composition of every man, whatever may be his pride in his philosophy, a proneness in a greater or less degree to superstition, or at least credulity. The world is at best but slow to be enlightened, and the trammels thrown around us by the tales of the nursery are not easily shaken off. Such travellers into the northern parts of Sweden, Russia, Norway, and Lapland, as Olaus Magnus, Jean Marius, Rzaczynsky, Leems, &c., whose extravagant and imaginary notions were recorded by the credulous Gesner, who wrote marvellous accounts of the habits of the Beavers in Northern Europe, seem to have worked on the imaginations and confused the intellects of the early explorers of our Northern regions—La Hontan, Charlevoix, Theodat, Ellis, Beltrami, and Cartwright. These last, excited the enthusiasm of Buffon, whose romantic stories have so fastened themselves on the mind of childhood, and have been so generally made a part of our education, that we now are almost led to regret that three-fourths of the old accounts of this extraordinary animal are fabulous; and that with the exception of its very peculiar mode of constructing its domicile, the Beaver is, in point of intelligence and cunning, greatly exceeded by the fox, and is but a few grades higher in the scale of sagacity than the common musk-rat.

The following account was noted down by us as related by a trapper named Prevost, who had been in the service of the American Fur Company for upwards of twenty years, in the region adjoining the spurs of the Rocky Mountains, and who was the "Patroon" that conveyed us down the Missouri river in the summer and autumn of 1843. As it confirms the statements of Hearne, Richardson, and other close observers of the habits of the Beaver, we trust that although it may present little that is novel, it will from its truth be acceptable and interesting to our readers. Mr. Prevost states in substance as follows.

Beavers prefer small clear-water rivers, and creeks, and likewise resort to large springs. They, however, at times, frequent great rivers and lakes. The trappers believe that they can have notice of the approach of winter weather, and of its probable severity, by observing the preparations made by the Beavers to meet its rigours; as these animals always cut their wood in good season, and if this be done early, winter is at hand.

The Beaver dams, where the animal is at all abundant, are built across the streams to their very head waters. Usually these dams are formed of mud, mosses, small stones, and branches of trees cut about three feet in length and from seven to twelve inches round. The bark of the trees in all cases being taken off for winter provender, before the sticks are carried away to make up the dam. The largest tree cut by the Beaver, seen by Prevost, measured eighteen inches in diameter; but so large a trunk is very rarely cut down by this animal. In the instance just mentioned, the branches only were used, the trunk not having been appropriated to the repairs of the dam or aught else by the Beavers.

In constructing the dams, the sticks, mud and moss are matted and interlaced together in the firmest and most compact manner; so much so that even men cannot destroy them without a great deal of labour. The mud and moss at the bottom are rooted up with the animal's snout, somewhat in the manner hogs work in the earth, and clay and grasses are stuffed and plastered in between the sticks, roots, and branches, in so workmanlike a way as to render the structure quite water-tight. The dams are sometimes seven or eight feet high, and are from ten to twelve feet wide at the bottom, but are built up with the sides inclining towards each other, so as to form a narrow surface on the top. They are occasionally as much as three hundred yards in length, and often extend beyond the bed of the stream in a circular form, so as to overflow all the timber near the margin, which the Beavers cut down for food during winter, heap together in large quantities, and so fasten to the shore under the surface of the water, that even a strong current cannot tear it away; although they generally place it in such a position that the current does not pass over it. These piles or heaps of wood are placed in front of the lodges, and when the animal wishes to feed he proceeds to them, takes a piece of wood, and drags it to one of the small holes near the principal entrance running above the water, although beneath the surface of the ground. Here the bark is devoured at leisure, and the wood is afterwards thrust out, or used in repairing the dam. These small galleries are more or less abundant according to the number of animals in the lodges. The larger lodges are, in the interior, about seven feet in diameter, and between two and three feet high, resembling a great oven. They are placed near the edge of the water, although actually built on or in the ground. In front, the Beavers scratch away the mud to secure a depth of water that will enable them to sink their wood deep enough to prevent its being impacted in the ice when the dam is frozen over, and also to allow them always free egress from their lodges, so that they may go to the dam and repair it if necessary. The top of the lodge is formed by placing branches of trees matted with mud, grasses, moss, &c., together, until the whole fabric measures on the outside from twelve to twenty feet in diameter, and is six or eight feet high, the size depending on the number of inhabitants. The outward coating is entirely of mud or earth, and smoothed off as if plastered with a trowel. As Beavers, however, never work in the day-time, no person we believe has yet seen how they perform their task, or give this hard-finish to their houses. This species does not use its fore-feet in swimming, but for carrying burthens: this can be observed by watching the young ones, which suffer their fore-feet to drag by the side of the body, using only the hind-feet to propel themselves through the water. Before diving, the Beaver gives a smart slap with its tail on the water, making a noise that may be heard a considerable distance, but in swimming, the tail is not seen to work, the animal being entirely submerged except the nose and part of the head; it swims fast and well, but with nothing like the speed of the otter, (*Lutra Canadensis.*)

The Beavers cut a broad ditch all around their lodge, so deep that it cannot freeze to the bottom, and into this ditch they make the holes already spoken of, through which they go in and out and bring their food. The beds of these singular animals are separated slightly from each other, and are placed around the wall, or circumference of the interior of the lodge; they are formed merely of a few grasses, or the tender bark of trees; the space in the centre of the lodge being left unoccupied. The Beavers usually go to the dam every evening to see if repairs are needed, and to deposite their ordure in the water near the dam, or at least at some distance from their lodge.

They rarely travel by land, unless their dams have been carried away by the ice, and even then they take the beds of the rivers or streams for their roadway. In cutting down trees they are not always so fortunate as to have them fall into the water, or even towards it, as the trunks of trees cut down by these animals are observed lying in various positions; although as most trees on the margin of a stream or river lean somewhat towards the water, or have their largest branches extended over it, many of those cut down by the Beavers naturally fall in that direction.

It is a curious fact, says our trapper, that among the Beavers there are some that are lazy and will not work at all, either to assist in building lodges or dams, or to cut down wood for their winter stock. The industrious ones beat these idle fellows, and drive them away; sometimes cutting off a part of their tail, and otherwise injuring them. These "Paresseux" are more easily caught in traps than the others, and the trapper rarely misses one of them. They only dig a hole from the water running obliquely towards the surface of the ground twenty-five or thirty feet, from which they emerge when hungry, to obtain food, returning to the same hole with the wood they procure, to eat the bark.

They never form dams, and are sometimes to the number of five or seven together; all are males. It is not at all improbable, that these unfortunate fellows have, as is the case with the males of many species of animals, been engaged in fighting with others of their sex, and after having been conquered and driven away from the lodge, have become idlers from a kind of necessity. The working Beavers, on the contrary, associate, males, females, and young together.

Beavers are caught and found in good order at all seasons of the year in the Rocky Mountains; for in those regions the atmosphere is never warm enough to injure the fur; in the low-lands, however, the trappers rarely begin to capture them before the first of September, and they relinquish the pursuit about the last of May. This is understood to be along the Missouri, and the (so called) Spanish country.

Cartwright, (vol. i., p. 62,) found a Beaver that weighed forty-five pounds; and we were assured that they have been caught weighing sixty-one pounds before being cleaned. The only portions of their flesh that are considered fine eating, are the sides of the belly, the rump, the tail, and the liver. The tail, so much spoken of by travellers and by various authors, as being very delicious eating, we did not think equalled their descriptions. It has nearly the taste of beef marrow, but is rather oily, and cannot be partaken of unless in a very moderate quantity, except by one whose stomach is strong enough to digest the most greasy substances.

Beavers become very fat at the approach of autumn ; but during winter they fall off in flesh, so that they are generally quite poor by spring, when they feed upon the bark of roots, and the roots of various aquatic plants, some of which are at that season white, tender, and juicy. During winter, when the ice is thick and strong, the trappers hunt the Beaver in the following manner : A hole is cut in the ice as near as possible to the aperture leading to the dwelling of the animal, the situation of which is first ascertained; a green stick is placed firmly in front of it, and a smaller stick on each side, about a foot from the stick of green wood ; the bottom is then patted or beaten smooth and even, and a strong stake is set into the ground to hold the chain of the trap, which is placed within a few inches of the stick of green wood, well baited, and the Beaver, attracted either by the fresh bark or the bait, is almost always caught. Although when captured in this manner, the animal struggles, diving and swimming about in its efforts to escape, it never cuts off a foot in order to obtain its liberty ; probably because it is drowned before it has had time to think of this method of saving itself from the hunter. When trapping under other circumstances, the trap is placed within five or six inches of the shore, and about the same distance below the surface of the water, secured and baited as usual. If caught, the Beavers now and then cut off the foot by which they are held, in order to make their escape.

A singular habit of the Beaver was mentioned to us by the trapper, PREVOST, of which we do not recollect having before heard. He said that when two Beaver lodges are in the vicinity of each other, the animals proceed from one of them at night to a certain spot, deposit their castoreum, and then return to their lodge. The Beavers in the other lodge, scenting this, repair to the same spot, cover it over with earth, and then make a similar deposit on the top. This operation is repeated by each party alternately, until quite a mound is raised, sometimes to the height of four or five feet.

The strong musky substance contained in the glands of the Beaver, is called castoreum ; by trappers, bark-stone ; with this the traps are baited.

A small stick, four or five inches long, is chewed at one end, and that part dipped in the castoreum, which is generally kept in a small horn. The stick is then placed with the anointed end above water, and the other end downwards. The Beaver can smell the castoreum at least one hundred yards, makes towards it at once, and is generally caught.

Where Beavers have not been disturbed or hunted, and are abundant, they rise nearly half out of water at the first smell of the castoreum, and become so excited that they are heard to cry aloud, and breathe hard to catch the odour as it floats on the air. A good trapper used to catch about eighty Beavers in the autumn, sixty or seventy in the spring, and upwards of three hundred in the summer, in the mountains; taking occasionally as many as five hundred in one year. Sixty or seventy Beaver skins are required to make a pack weighing one hundred pounds; which when sent to a good market, is worth, even now, from three to four hundred dollars.

The Indians occasionally destroy Beaver-dams in order to capture these animals, and have good dogs to aid them in this purpose. The Mountain Indians, however, are not trappers.

Sometimes the Indians of the Prairies break open Beaver lodges in the summer-time, as, during winter, they are usually frozen hard. The Beaver is becoming very scarce in the Rocky Mountains, so much so, that if a trapper now secures one hundred in the winter and spring hunt, he is considered fortunate.

Formerly, when the fur was high in price, and the animals abundant, the trading companies were wont to send as many as thirty or forty men, each with from six to twelve traps and two good horses: when arrived at a favourable spot to begin their work, these men erected a camp, and each one sought alone for his game, the skins of which he brought to camp, where a certain number of men always remained to stretch and dry them.

The trappers subsist principally upon the animals they kill, having a rifle and a pair of pistols with them. After a successful hunt, on meeting each other at the camp, they have a "frolic" as they term it.

Some old and wary Beavers are so cunning, that on finding the bait they cover it over, as if it were on the ground, with sticks, &c., deposit their own castoreum on the top, and manage to remove the trap. This is often the case when the Beaver has been hunted previously. In places where they have remained undisturbed, but few escape the experienced trapper. The trappers are not very unfrequently killed by the Indians, and their occupation is one involving toil and hazard. They rarely gain a competence for their old age, to say nothing of a fortune, and in fact all the articles they are of necessity obliged to purchase in the "Indian country," cost them large sums, as their price is greatly increased by the necessary charges for transportation to the remote regions of the West.

When at Fort Union, we saw a trapper who had just returned from an unfortunate expedition to the mountains; his two horses had been stolen, and he lost his gun and rifle in coming down the river in a slender canoe, and was obliged to make for the shore, dig a hole wherein to deposit the few furs he had left, and travel several hundred miles on foot with only berries and roots for his food. He was quite naked when he reached the Fort.

The Beaver which we brought from Boston to New-York was fed principally on potatoes and apples, which he contrived to peel as if assisted with a knife, although his lower incisors were his only substitute for that useful implement. While at this occupation the animal was seated on his rump, in the manner of a ground-hog, marmot, or squirrel, and looked like a very large wood-chuck, using his fore-feet, as squirrels and marmots are wont to do.

This Beaver was supplied every day with a large basin filled with water, and every morning his ordure was found to have been deposited therein. He generally slept on a good bed of straw in his cage, but one night having been taken out and placed at the back of the yard in a place where we thought he would be secure, we found next morning to our surprise that he had gnawed a large hole through a stout pine door which separated him from that part of the yard nearest the house, and had wandered about until he fell into the space excavated and walled up outside the kitchen window. Here he was quite entrapped, and having no other chance of escape from this pit, into which he had unluckily fallen, he gnawed away at the window-sill and the sash, on which his teeth took such effect that on an examination of the premises we found that a carpenter and several dollars' worth of work were needed, to repair damages. When turned loose in the yard in the day-time he would at times slap his tail twice or thrice on the brick pavement, after which he elevated this member from the ground, and walked about in an extremely awkward manner. He fell ill soon after we had received him, and when killed, was examined by Dr. JAMES TRUDEAU, who found that he would shortly have died of an organic disease.

It is stated by some authors that the Beaver feeds on fish. We doubt whether he possesses this habit, as we on several occasions placed fish before those we saw in captivity, and although they were not very choice in their food, and devoured any kind of vegetable, and even bread, they in every case suffered fish to remain untouched in their cages.

The food of this species, in a state of nature, consists of the bark of several kinds of trees and shrubs, and of bulbous and other roots. It is particularly fond of the bark of the birch, (*Betula*,) the cotton-wood, (*Populus*,) and of several species of willow, (*Salix*;) it feeds also with avidity on the roots of some aquatic plants, especially on those of the *Nuphair luteum*. In summer, when it sometimes wanders to a distance from the water, it eats berries, leaves, and various kinds of herbage.

The young are born in the months of April and May; those produced in the latter month are the most valuable, as they grow rapidly and become strong and large, not being checked in their growth, which is often the case with those that are born earlier in the season. Some females have been taken in July, with young, but such an event is of rare occurrence. The eyes of the young Beaver are open at birth. The dam at times brings forth as many as seven at a litter, but from two to five is the more usual number. The young remain with the mother for at least a year, and not unfrequently two years, and when they are in a place of security, where an abundance of food is to be procured, ten or twelve Beavers dwell together.

About a month after their birth, the young first follow the mother, and accompany her in the water; they continue to suckle some time longer, although if caught at that tender age, they can be raised without any difficulty, by feeding them with tender branches of willows and other trees. Many Beavers from one to two months old are caught in traps set for old ones. The gravid female keeps aloof from the male until after the young have begun to follow her about. She resides in a separate lodge till the month of August, when the whole family once more dwell together.

GEOGRAPHICAL DISTRIBUTION.

According to Richardson the Beaver exists on the banks of the Mackensie, which is the largest river that discharges itself into the Polar Sea: he speaks of its occurring as high as 67½ or 68° north latitude, and states that its range from east to west extends from one side of the continent to the other. It is found in Labrador, Newfoundland, and Canada, and also in some parts of Maine and Massachusetts. There can be no doubt that the Beaver formerly existed in every portion of the United States. Catesby noticed it as found in Carolina, and the local names of Beaver Creek, Beaver Dam, &c., now existing, are evidences that the animal was once known to occupy the places designated by these compounds of its name. We have, indeed, examined several localities, some of which are not seventy miles from Charleston, where we were assured the remains of old Beaver dams existed thirty-five years ago. BARTRAM, in his visit to Florida in 1778, (Travels, p. 281,) speaks of it as at that time existing in Georgia and East Florida. It has since then become a scarce species in all the Atlantic States, and in some of them has been entirely extirpated. It, however, may still be found in several of the less cultivated portions of many of our States. Dr. DEKAY was informed that in 1815 a party of St. Regis Indians obtained three hundred Beavers in a few weeks, in St. Lawrence county, N. Y. In 1827 we were shown several Beaver-houses in the north-western part of New-York, where, although we did not see the animals, we observed signs of their recent labours. DEKAY supposes, (N. Y. Fauna, p. 78,) that the Beaver does not at present exist south of certain localities in the state of New-York. This is an error. Only two years ago we received a foot of one, the animal having been caught not twenty miles from Ashville in North Carolina. We saw in 1839 several Beaver-lodges a few miles west of Peter's Mountain in Virginia, on the head waters of the Tennessee River, and observed a Beaver swimming across the stream. There is a locality within twenty miles of Milledgeville, Georgia, where Beavers are still found. Our friend, Major LOGAN, residing in Dallas county, Alabama, informed us that they exist on his plantation, and that within the last few years a storekeeper in the immediate vicinity purchased twenty or thirty skins annually, from persons residing in his neighbourhood.

We were invited to visit this portion of Alabama to study the habits of the Beaver, and to obtain specimens. Some years ago we shot one near Henderson, Kentucky, in Canoe Creek ; it was regarded as a curiosity, and probably none have been seen in that section of the country since. We have heard that the Beaver was formerly found near New-Orleans, but we never saw one in Louisiana. This species exists on the Arkansas River, in the streams running from the Rocky Mountains, and along their whole range on both sides ; we have traced it as far as the northern boundaries of Mexico, and it is no doubt found much farther south along the mountain range. Thus it appears that the Beaver once existed on the whole continent of North America, north of the Tropic of Cancer, and may still occur, although in greatly diminished numbers, in many localities in the wild and uncultivated portions of our country ; we are nevertheless under the impression, that in the Southern States the Beaver was seldom found in those ranges of country where the musk-rat does not exist ; hence we think it could never have been abundant in the alluvial lands of Carolina and Georgia, as the localities where its dams formerly existed are on pure running streams, and not on the sluggish rivers near the sea-coast.

GENERAL REMARKS.

It is doubted by some authors whether the American Beaver is identical with the Beaver which exists in the north of Europe; F. CUVIER, KUHL, and others, described it under the names of *C. Americanus*, *C. Canadensis*, &c. From the amphibious habits of this animal, and its northern range on both continents, strong arguments in favour of the identity of the American and European species might be maintained, even without adopting the theory of the former connexion of the two adjacent continents. We carefully compared many specimens (American and European) in the museums of Europe, and did not perceive any difference between them, except that the American specimens were a very little larger than the European. We saw a living Beaver in Denmark that had been obtained in the north of Sweden; in its general appearance and actions it did not differ from those we have seen in confinement in America. It has been argued, however, that the European animal differs in its habits from the American, and that along the banks of the Weser, the Rhone, and the Danube, the Beavers are not gregarious, and that they burrow in the banks like the musk-rat. But change of habit may be the result of altered circumstances, and is not in itself sufficient to constitute a species. Our wild pigeon (*Columba migratoria*) formerly bred in communities in the Northern States; we once saw one of their breeding places near Lake Champlain, where there were more than a hundred nests on a single tree. They still breed in that portion of the country, but the persecutions of man have compelled them to adopt a different habit, and two nests are now seldom found on a tree.

The banks of the European rivers, (on which the Beaver still remains although scarcely more than a straggler can be found along them now,) have been cultivated to the water's edge, and necessity, not choice, has driven the remnant of the Beaver tribe to the change of habit we have referred to. But if the accounts of travellers in the north of Europe are to be relied on, the habits of the Beaver are in the uncultivated portions of that country precisely similar to those exhibited by the animal in Canada. We consider the account of these animals given us by HEARNE, (p. 234,) as very accurate. He speaks of their peculiarly constructed huts, their living in communities, and their general habits. In the account of Swedish Lapland, by Professor LEEMS, published in Danish and Latin, Copenhagen, 1767, we have the following notice of the European species; (we quote from the English translation in PINKER-

TON's Voyages, vol. i., p. 419.) "The Beaver is instinctively led to build his house near the banks of lakes and rivers. He saws with his teeth birch trees, with which the building is constructed; with his teeth he drags the wood along to the place destined for building his habitation; in this manner one piece of timber is carried after another, where they choose. At the lake or river where their house is to be built, they lay birch stocks or trunks, covered with their bark, in the bottom itself, and forming a foundation, they complete the rest of the building, with so much art and ingenuity as to excite the admiration of the beholders. The house itself is of a round and arched figure, equalling in its circumference the ordinary hut of a Laplander. In this house the floor is for a bed, covered with branches of trees, not in the very bottom, but a little above, near the edge of a river or lake; so that between the foundation and flooring on which the dwelling is supported, there is formed as it were a cell, filled with water, in which the stalks of the birch tree are put up; on the bark of this, the Beaver family who inhabit this mansion feed. If there are more families under one roof, besides the laid flooring, another resembling the former is built a little above, which you may not improperly name a second story in the building. The roof of the dwelling consists of branches very closely compacted, and projects out far over the water. You have now, reader, a house consisting and laid out in a cellar, a flooring, a hypocaust, a ceiling, and a roof, raised by a brute animal, altogether destitute of reason, and also of the builder's art, with no less ingenuity than commodiousness."

It should be observed that LEEMS, who was a missionary in that country, gave this statement as related to him by the Laplanders who reside in the vicinity of the Beavers, and not from his own personal observations. This account, though mixed up with some extravagancies and the usual vulgar errors, (which we have omitted,) certainly proves that the habits of the Beaver in the northern part of Europe are precisely similar to those of that animal on the northern continent of America.

GENUS MELES.—Brisson.

DENTAL FORMULA.

$$Incisive\ \frac{6}{6};\quad Canine\ \frac{1-1}{1-1};\quad Molar\ \frac{4-4}{5-5} = 34.$$

OR,

$$Incisive\ \frac{6}{6};\quad Canine\ \frac{1-1}{1-1};\quad Molar\ \frac{4-4}{4-4} = 32.$$

The canine teeth in this genus are rather large and strong. In addition to the four persistent molars on each side in the upper jaw, there is an additional small molar which is deciduous, dropping out when the animal is quite young.

Nose, somewhat elongated, obtuse at the point; tongue, smooth; ears, short and round; eyes, small; body, thick-set; legs, short. Mammæ, six, two on the lower part of the chest and four on the abdomen. There are transverse glandular follicles between the anus and the root of the tail, which discharge a fetid matter.

The feet are five-toed, and are armed with strong nails. The fore-feet are longer than the hind-feet.

Three species of this genus have been described; one inhabits Europe, one India, and one America.

The generic name is derived from the Latin word *Meles*, a badger.

MELES LABRADORIA.—Sabine.

American Badger.

PLATE XLVII.—Male.

Supra fusco-ferruginea; infra, subalbida; capite, fascia longitudinale alba; cruribus et pedibus nigris.

American Badger

CHARACTERS.

Colour above, hoary-yellowish-brown ; a broad white longitudinal line dividing the head above into two equal parts ; dull white, beneath ; legs and feet, black.

SYNONYMES.

CARCAJOU, Buffon, tom. vi., p. 117, pl. 23.
COMMON BADGER, Pennant's Arctic Zool., vol. i., p. 71.
BADGER, Var. B. AMERICAN, Penn. Hist. Quad., vol. ii., p. 15.
URSUS TAXUS, Schreber, Säugeth., p. 520.
" LABRADORIUS, Gmel., vol. i., p. 102.
PRAROW, Gass. Journal, p. 34.
BLAIREAU, Lewis and Clarke, vol. i., pp. 50, 137, 213.
TAXUS LABRADORICUS, Long's Expedition, vol. i., p. 261.
MELES LABRADORIA, Sabine, Franklin's First Journey, p. 649.
AMERICAN BADGER, Harlan, F., p. 57.
" " Godm., vol. i., p. 179.
BLAIREAU D'AMERIQUE, F. Cuvier, Hist. Nat. des Mamm.
MELES LABRADORIA, Richardson, F. B. A., pl. 2.
" " Waterhouse, Trans. Zool. Soc., London, vol. ii., p. 1, p. 343.

DESCRIPTION.

There is a very striking difference between the teeth of this species and those of the European Badger, (*Meles vulgaris ;*) besides which, the present species has one tooth less than the latter on each side in the lower jaw. We have ascertained, by referring to three skulls in our possession, that the dentition of the American Badger corresponds so minutely with the scientific and accurate account given of it by WATERHOUSE, in the Transactions of the Zool. Society of London, vol. ii., part 5, p. 343, that we are willing to adopt his conclusions.

He says: "The subgeneric name, TAXIDEA, may be applied to the American Badger, and such species as may hereafter be discovered with incisors $\frac{6}{6}$; canines $\frac{1-1}{1-1}$; false molars $\frac{2-2}{2-2}$, the posterior false molar of the lower jaw, with an anterior large tubercle, and a posterior smaller one; molars $\frac{2-2}{2-2}$; the carnassière and the grinding molars of the upper jaw each of a triangular form, or nearly so, and about equal in size. The modification observable in the form of the molars of the upper jaw of TAXIDEA, furnishes us with an interesting link between MEPHITIS and MELES, whilst the former of these genera links the Badger with MUSTELA and its subgenera."

The body of this species is thick, heavy, flat, very broad and fleshy.

46

and its whole structure indicates that it is formed more for strength than speed.

Head, of moderate size, and conical; the skull, between the ears, broad, giving it somewhat the appearance of a pug-faced dog. Tip of the nose, hairy above; ears, short, and of an oval shape, clothed on both surfaces with short hairs; whiskers, few, not reaching beyond the eyes. The fur on the back is (in winter) three inches long, covering the body very densely; on the under surface it is short, and so thin that it does not conceal the colour of the skin. There is, immediately below the tail, a large aperture leading into a kind of sac. Although there seems to be no true glandular apparatus, this cavity is covered on its sides by an unctuous matter; there is a second and smaller underneath, in the midst of which the anus opens, and on each side of the anus is a pore from which an unctuous matter escapes, which is of a yellow colour and offensive smell. Legs, short; feet, robust, palmated to the outer joint; nails, long and strong, slightly arched, and channelled underneath toward their extremities; palms, naked. The heel is well clothed with hair; the tail is short, and is covered with long bushy hairs.

COLOUR.

Hair on the back, at the roots dark-gray, then light-yellow for two-thirds of its length, then black, and broadly tipped with white; giving it in winter a hoary-gray appearance; but in summer it makes a near approach to yellowish-brown. The eyes are bright piercing black. Whiskers, upper lips, nose, forehead, around the eyes, and to the back of the head, dark yellowish-brown. There is a white stripe running from the nose over the forehead and along the middle of the neck to the shoulder. Upper surface of ear, dark brown; inner surface and outer edge of ear, white; legs, blackish-brown; nails, pale horn-colour; sides of face, white, which gradually darkens and unites with the brown colour above; chin and throat, dull white; the remainder of the under surface is yellowish-white; tail, yellowish-brown.

We have noticed some varieties in this species. In one of the specimens before us the longitudinal white line does not reach below the eyes, leaving the nose and forehead dark yellowish-brown. In two of them the under surface of the body is yellowish-white, with a broad and irregular longitudinal line of white in the centre; whilst another and smaller specimen has the whole of the under surface pure white, shaded on the sides by a line of light yellow.

DIMENSIONS.

A male in winter pelage.

From point of nose to root of tail,	21 inches.
Tail, (vertebræ,)	4 do.
" to end of hair,	$6\frac{1}{2}$ do.
Nose to root of ear,	$3\frac{5}{8}$ do.
Between the ears,	4 do.
Height of "	$1\frac{1}{4}$ do.
Breadth of "	$1\frac{3}{8}$ do.
Length of head,	$4\frac{5}{8}$ do.
Breadth of body,	$10\frac{1}{8}$ do.
Length of fore-leg to end of claw,	$7\frac{3}{4}$ do.

Weight, $16\frac{1}{2}$ lbs.

A living specimen, (examined in a menagerie at Charleston, S. Carolina.)

Length of head and body,	30 inches.
" tail, (vertebræ,)	5 do.
" " to end of hair,	$7\frac{1}{2}$ do.
Breadth of body,	12 do.
Heel to end of nail,	4 do.

Weight, 23 lbs.

A stuffed specimen in our collection.

Length of head and body,	31 inches.
" tail, (vertebræ,)	$5\frac{1}{2}$ do.
" " to end of hair,	$7\frac{1}{2}$ do.
" heel to end of nail,	$4\frac{1}{2}$ do.

HABITS.

During our stay at Fort Union, on the Upper Missouri River, in the summer of 1843, we purchased a living Badger from a squaw, who had brought it from some distance to the Fort for sale; it having been caught by another squaw at a place nearly two hundred and fifty miles away, among the Crow Indians. It was first placed in our common room, but was found to be so very mischievous, pulling about and tearing to pieces every article within its reach, trying to dig up the stones of the hearth, &c., that we had it removed into an adjoining apartment. It was regularly fed morning and evening on raw meat, either the flesh of animals procured by our hunters, or small birds shot during our researches through the adjacent country. It drank a good deal of water, and was rather cleanly in its habits. In the course of a few days it managed to dig a hole under the hearth and fire-place nearly large and deep enough to conceal its body, and we were obliged to drag it out by main force whenever we wished to examine it. It was provoked at the near approach of any one, and growled continuously at all intruders. It was not, however, very vicious, and would suffer one or two of our companions to handle and play with it at times.

At that period this Badger was about five months old, and was nearly as large as a full grown wood-chuck or ground-hog, (*Arctomys monax.*) Its fur was of the usual colour of summer pelage, and it was quite a pretty looking animal. We concluded to bring it to New-York alive, if possible, and succeeded in doing so after much trouble, it having nearly made its escape more than once. On one occasion when our boat was made fast to the shore for the night, and we were about to make our "camp," the Badger gnawed his way out of the box in which he was confined, and began to range over the batteau; we rose as speedily as possible, and striking a light, commenced a chase after it with the aid of one of the hands, and caught it by casting a buffalo robe over it. The cage next day was wired, and bits of tin put in places where the wooden bars had been gnawed through, so that the animal could not again easily get out of its prison. After having become accustomed to the box, the Badger became quite playful and took exercise by rolling himself rapidly from one end to the other, and then back again with a reversed movement, continuing this amusement sometimes for an hour or two.

On arriving at our residence near New-York, we had a large box, tinned on the inside, let into the ground about two feet and a half and filled to the same depth with earth. The Badger was put into it, and in a few minutes made a hole, in which he seemed quite at home, and where he passed most of his time during the winter, although he always came out to take his food and water, and did not appear at all sluggish or inclined to hibernate even when the weather was so cold as to make it necessary to pour hot water into the pan that was placed within his cage, to enable him to drink, as cold water would have frozen immediately, and in fact the pan generally had a stratum of ice on the bottom which the hot water dissolved when poured in at feeding-time.

Our Badger was fed regularly, and soon grew very fat; its coat changed completely, became woolly and of a buff- brown colour, and the fur by the month of February had become indeed the most effectual protection against cold that can well be imagined.

We saw none of these animals in our hunting expeditions while on our journey up the Missouri River, and observed only a few burrowing places which we supposed were the remains of their holes, but which were at that time abandoned. We were informed that these animals had burrows six or seven feet deep running beneath the ground at that depth to the distance of more than thirty feet. The Indians speak of their flesh as being good; that of the one of which we have been speaking, when the animal was killed, looked very white and fat, but we omitted to taste it.

Before taking leave of this individual we may remark, that the change of coat during winter from a *hairy* or furry texture to a woolly covering, is to be observed in the Rocky-mountain sheep, (*Ovis montana*,) and in other animals exposed in that season to intense cold. Thus the skin of *Ovis montana*, when obtained pending the change from winter to summer pelage, will have the outside hairs grown out beyond the wool that has retained the necessary warmth in the animal during the cold weather. The *wool* begins to drop out in early spring, leaving in its place a coat of hair resembling that of the elk or common deer, thus giving as a peculiarity of certain species a *change* of pelage, quite different in character from the ordinary thickening of the coat or hair, common to all furred animals in winter, and observed by every one,—for instance, in the horse, the cow, &c., which shed their winter coats in the spring.

We had an opportunity in Charleston of observing almost daily for a fortnight, the habits of a Badger in a menagerie; he was rather gentle, and would suffer himself to be played with and fondled by his keeper, but did not appear as well pleased with strangers; he occasionally growled at us, and would not suffer us to examine him without the presence and aid of his keeper.

In running, his fore-feet crossed each other, and his body nearly touched the ground. The heel did not press on the earth like that of the bear, but was only slightly elevated above it. He resembled the Maryland marmot in running, and progressed with about the same speed. We have never seen any animal that could exceed him in digging. He would fall to work with his strong feet and long nails, and in a minute bury himself in the earth, and would very soon advance to the end of a chain ten feet in length. In digging, the hind, as well as the fore-feet, were at work, the latter for the purpose of excavating, and the former, (like paddles,) for expelling the earth out of the hole, and nothing seemed to delight him more than burrowing in the ground; he seemed never to become weary of this kind of amusement; when he had advanced to the length of his chain he would return and commence a fresh gallery near the mouth of the first hole; thus he would be occupied for hours, and it was necessary to drag him away by main force. He lived on good terms with the raccoon, gray fox, prairie wolf, and a dozen other species of animals. He was said to be active and playful at night, but he seemed rather dull during the day, usually lying rolled up like a ball, with his head under his body for hours at a time.

This Badger did not refuse bread, but preferred meat, making two meals during the day, and eating about half a pound at each.

We occasionally saw him assuming rather an interesting attitude, raising the fore-part of his body from the earth, drawing his feet along his sides, sitting up in the manner of the marmot, and turning his head in all directions to make observations.

The Badger delights in taking up his residence in sandy prairies, where he can indulge his extravagant propensity for digging. As he lives upon the animals he captures, he usually seeks out the burrows of the various species of marmots, spermophiles, ground-squirrels, &c., with which the prairies abound; into these he penetrates, enlarging them to admit his own larger body, and soon overtaking and devouring the terrified inmates. In this manner the prairies become so filled with innumerable Badger-holes, that when the ground is covered with snow they prove a great annoyance to horsemen.

RICHARDSON informs us that early in the spring when they first begin to stir abroad they may be easily caught by pouring water into the holes, the ground at that time being so frozen that the water cannot escape through the sand, but soon fills the hole and its tenant is obliged to come out.

The Badger, like the Maryland marmot, is a rather slow and timid animal, retreating to its burrow as soon as it finds itself pursued. When once in its snug retreat, no dexterity in digging can unearth it. RICHARDSON states that "the strength of its fore-feet and claws is so great, that one which had insinuated only its head and shoulders into a hole, resisted the utmost efforts of two stout young men, who endeavoured to drag it out by the hind-legs and tail, until one of them fired the contents of his fowling-piece into its body."

This species is believed to be more carnivorous than that of Europe, (*Meles taxus.*) RICHARDSON states that a female which he had killed had a small marmot nearly entire, together with some field-mice, in its stomach, and that it had at the same time been eating some vegetables. As in its dentition it approaches the skunk, which is very decidedly carnivorous in habit, we should suppose that its principal food in its wild state is meat.

From November to April the American Badger remains in its burrow, scarcely ever showing itself above ground; here it passes its time in a state of semi-torpidity. It cannot, however, be a very sound sleeper in winter, as not only the individual which we examined in Charleston, but even that which we kept in New-York, continued tolerably active through the winter. During the time of their long seclusion they do not lose much flesh, as they are represented to be very fat on coming abroad in spring. As this, however, is the pairing season, they, like other animals of similar habits, soon become lean.

The American Badger is said to produce from three to five young at a litter.

Several European writers, and among the more recent, GRIFFITH, in his Animal Kingdom, have represented the Badger as leading a most gloomy and solitary life; but we are not to suppose from the subterranean habits of this species that it is necessarily a dull and unhappy creature. Its fat sides are certainly no evidence of suffering or misery, and its form is well adapted to the life it is destined to lead. It is, like nearly all our quadrupeds, nocturnal in its habits, hence it appears dull during the day, and cannot endure a bright light. To a being constituted like man, it would be a melancholy lot to live by digging under ground, shunning the light of day, and only coming forth under the shadow of night; but for this life the Badger was formed, and he could not be happy in any other. We believe that a wise Providence has created no species which, from the nature of its organization, must necessarily be miserable; and we should, under all circumstances, rather distrust our short-sighted views than doubt the wisdom and infinite benevolence of the Creator.

GEOGRAPHICAL DISTRIBUTION.

The American Badger has a very extensive range. It has been traced as far north as the banks of Peace River, and the sources of the River of the Mountains, in latitude 58°. It abounds in the neighbourhood of Carlton-House, and on the waters that flow into Lake Winnepeg.

LEWIS and CLARK, and TOWNSEND, found it on the open plains of the Columbia, and also on the prairies east of the Rocky Mountains.

We have not been able to trace it within a less distance from the Atlantic than the neighbourhood of Fort Union. To the south we have seen specimens which were said to have come from the eastern side of the Rocky Mountains, in latitude 36°. There is a specimen in the collection of the London Zoological Society, the skull of which was described and figured by WATERHOUSE, that was stated to have been received from Mexico. It is probable that the Flacoyole of FERNANDEZ, which was described as existing in Mexico, is the same species. There is also another specimen in the museum of the Zoological Society of London, that was brought by DOUGLASS, which is believed to have come from California.

It is very doubtful whether it exists on the eastern side of the American continent.

We are not aware that it has ever been found either in upper or Lower Canada, and we could obtain no knowledge of it in our researches at Labrador.

GENERAL REMARKS.

The difference between the European and American species of Badger is so great that it is unnecessary to institute a very particular comparison. Our species may be distinguished from that of Europe by its muzzle being hairy above, whilst it is naked in the other; the fore-limbs are stouter, and the claws stronger; its head is also more conical in form. The European species has more conspicuous ears; it has three broad white marks, one on the top of the head, and one on each side, and between them are two broad black lines, which include the eyes and ears; and the whole of the throat and under-jaw are black; whilst the throat and lower-jaw of the American species are white; there is also a broad white patch separating the black colours between the sides of the forehead and ear. There are several other marks of difference which it is unnecessary to particularize, as the species are now universally admitted to be distinct.

Sabine supposed the American Badger to be a little the smallest. There is a considerable difference among different individuals of both species, but we have on an average found the two species nearly equal in size. Mr. SABINE's American specimen was a small one, measuring two feet two inches in body. BUFFON's specimen was two feet four inches. One of ours was two feet seven. On the other hand, SHAW gives the length of head and body of the European species as about two feet. FISCHER in his synopsis gives it as two and one-third, and CUVIER as two and a half. We have not found any European specimen measuring more than two feet six inches.

It was for a long time supposed, and was so stated by BUFFON, that there was no true species of Badger in America; that author, however, afterwards received a specimen that was said to have come from Labrador, which was named by GMELIN after the country where it was supposed to be common. The name "*Labradoria*" will be very inappropriate should our conjectures prove correct, that it is unknown in that country. BUFFON's specimen had lost one of its toes; hence he described it as four-toed. GMELIN, who gave it a scientific name, made "*Palmis tetradactylis*" one of its specific characters.

SCHREBER first considered the American as a distinct species from the European Badger; CUVIER seems to have arrived at a different conclusion; SHAW gave tolerably good figures of both species on the same plate, pointing out their specific differences; and SABINE entered into a minute comparison. RICHARDSON (F. B. A.) added considerably to our knowledge of the history and habits of the American Badger; and our esteemed friend, G. R. WATERHOUSE, Esq., has given descriptions and excellent figures of the skull and teeth, in which the distinctive marks in the dentition of the two species are so clearly pointed out, that nothing farther remains to be added in that department.

We have compared specimens of the *Blaireux* of LEWIS and CLARK found on the plains of Missouri, with those obtained by TOWNSEND near the Columbia, and also with specimens from the plains of the Saskatchewan in the Zoological museum, and found them all belonging to the same species.

SCIURUS DOUGLASSII.—BACH.

DOUGLASS' SQUIRREL.

PLATE XLVIII.—Male and Female.

S. Hudsonio quarta parte major; cauda corpore curtiore; supra subniger, infra flavus.

CHARACTERS.

About one-fourth larger than the chickaree (S. Hudsonius); tail, shorter than the body; colour, dark-brown above, and bright-buff beneath.

SYNONYMES.

SCIURUS DOUGLASSII, Gray, Proceedings Zool. Society, London, 1836, p. 88, named, but apparently not described.
" " Bachman, monograph of the Genus Sciurus, Proceedings Zool. Soc., London, 1838.

DESCRIPTION.

Incisors, a little smaller than those of *Sciurus Hudsonius;* in the upper jaw the anterior molar, which is the smallest, has a single rounded eminence on the inner side; on the outer edge of the tooth there are two acute points, and there is one in front; the next two grinders, which are of equal size, have each a similar eminence on the inner side, with a pair of points externally; the posterior grinder, although larger, is not unlike the anterior one. In the lower jaw the bounding ridge of enamel in each tooth forms an anterior and posterior pair of points. The molars increase gradually in size from the first, which is the smallest, to the posterior one, which is the largest.

This species, in the form of its body, is not very unlike *Sciurus Hudsonius*; its ears and tail, however, are much shorter in proportion, and in other respects, as well as in size, it differs widely from *Hudsonius.*

Head, considerably broader; and nose, less elongated and blunter than in the latter; body, long and slender; ears, rather small, nearly rounded, slightly tufted posteriorly. As usual in this genus, the third inner toe is the longest, and not the second, as in the spermophiles.

On Stone by Wm H. Hitchcock

Douglass Squirrel.

Drawn from Nature by J.J. Audubon, F.R.S.F.L.S.

Lith Printed & Col.d by J.T. Bowen, Phila

COLOUR.

The whiskers, which are longer than the head, are black; hair from the roots to near the points, plumbeous, tipped with brownish-gray, a few lighter-coloured hairs interspersed, giving it a dark-brown appearance. When closely examined it has the appearance of being thickly sprinkled with minute points of rust-colour on a black ground. The tail, which is distichous but not broad, is for three-fourths of its length the colour of the back; in the middle the fur is plumbeous at the roots, then irregularly marked with brown and black, and is tipped with dull white, giving it a hoary appearance; on the extremity of the tail the hairs are black from the roots and are tipped with light brown; the belly, the inner sides of the extremities and the outer surfaces of the feet, together with the throat and mouth and a line above and under the eyes, are bright-buff. The colours on the upper and under parts are separated by a line of black, commencing at the shoulders, and running along the flanks to the thighs; this line is broadest in the middle of the body and is there about three lines wide, narrowing from thence to a point. The hairs, which project beyond the outer margins of the ears and form a slight tuft, are dark-brown, and in some specimens black.

DIMENSIONS.

	Inches.	Lines.
Length from point of nose to insertion of tail - -	8	4
Tail (vertebræ) - - - - - - - - -	4	6
Tail, including fur - - - - - - - -	6	4
Height of ear posteriorly - - - - - - -	0	6
Palm to end of middle fore- claw - - - - - -	1	4
Heel and middle hind-claw - - - - - - -	1	10

HABITS.

Our specimens of Douglass' Squirrel were procured by Mr. Townsend. He remarks in his notes:—"This is a very plentiful species, inhabits the pine trees along the shores of the Columbia River, and like our common Carolina squirrel lays in a great quantity of food for consumption during the winter months. This food consists of the cones of the pine, with a few acorns. Late in autumn it may be seen very busy in the tops of the trees, throwing down its winter stock; after which, assisted by its mate, it gathers in and stows away its store, in readiness for its long incarceration."

GEOGRAPHICAL DISTRIBUTION.

Douglass obtained his specimens of this Squirrel on the Rocky Mountains, and Townsend found it on the Columbia River.

GENERAL REMARKS.

This species was found by Douglass and by Townsend about the same time. These gentlemen, if we have been rightly informed, met together in the far West. We drew up a description from specimens sent us by Mr. Townsend, and used the grateful privilege of a describer, in naming it (*S. Townsendii*) after the individual who we supposed had been the first discoverer. Under this name we sent our description to the Acad. of Nat. Sciences of Philadelphia, which was read Aug, 7th, 1838. After arriving in England, however, the same year, we saw a similar specimen in the Museum of the Zool. Society, and heard that it had been named by Gray, on the 11th October, 1836, who had called it after Douglass, (*S. Douglassii.*) He had not, as far as we have been able to ascertain, published any description of it. All that we can find in reference to this species is the following: " Mr. Gray gave a description of two foxes, a squirrel (*Sciurus Douglassii*), and three hares." The foxes and hares were described by him in the Magazine of Nat. Hist., (new series,) Nov., 1837, vol. i., p. 578, but for some reason he appears never to have published a description of this species.

We, however, supposing that he had described it, immediately changed our name to that proposed by Gray, and in our monograph of the genus assigned to him the credit of having been the first describer, although he had, it appears, only named the animal.

No. 10
Plate XLIX
On Stone
Douglasses Spermophile
Drawn from Nature by J. J. Audubon, F.R.S.F.L.S.
Lithd. Printed & Cold. by J. T. Bowen, Philada.

SPERMOPHILUS DOUGLASSII.—Richardson.

Douglass' Spermophile.

PLATE XLIX.

Auribus insignibus; versus humeros canescens; corpore dilute fusco, striis multis indistinctis transversis fuscis et albis, linea nigra inter humeros; cauda, longa, cylindrica, pilis albo nigroque annulatis.

CHARACTERS.

Ears, conspicuous; hoary on the shoulders, with a black stripe between them; general colour of the body, pale-brown, with many indistinct transverse marks of dark-brown and white. Tail, long and cylindrical, hairs annulated with white and black.

SYNONYMES.

Arctomys (Spermophilus ?) Douglassii, Richardson, F. B. A., p. 172.
Arctomys (Spermophilus) Beecheyi, Richardson, F. B. A., p. 170. ?

DESCRIPTION.

In the general form of the body Douglass' Spermophile bears a strong resemblance to several species of squirrel. Its rather slender shape and long ears and tail, together with its large eyes and the form of its head, assimilate it to the Northern gray squirrel, (*S. migratorius.*) Its coarser fur, hovever, cheek-pouches, rounder tail, and the shape of its claws, clearly designate the genus to which it belongs.

Head, rather short, broad and depressed; nose, obtuse; ears, long, semi-oval, covered on both surfaces with short hairs, which in winter specimens extend a line beyond the margins at their extremities; cheek-pouches of moderate size.

The longer hairs of the body are rather coarse, they are slender at their roots, gradually enlarge as they ascend, and suddenly taper off to a point at the tips. The fur beneath is on the back and sides soft and dense; on the under surface, however, the longer hairs predominate, and the animal is in those parts but thinly clothed.

There are on the fore-feet four toes with a blunt nail in place of a thumb. The second toe is longest; the nails are of moderate size, and slightly hooked. The feet are covered with short adpressed hairs to the roots of the nails; the tail is long and cylindrical, the longest hairs two inches in length. Mammæ ten, four pectoral and six abdominal.

COLOUR.

Incisors, dark orange; moustaches, black; on the nose and forehead, a tinge of reddish-brown; around the eyes, white; inner surface of ear, dull yellowish-brown; outer surface, dark-brown, becoming nearly black at the tips; sides of the face, yellowish white. The sides of the neck and shoulders have a hoary appearance. There is a broad, dark-brown stripe commencing on the neck, widening in its descent, and continuing along the centre of the back for about half the length of the body, when it gradually blends with the colours on the sides and hips, which are irregularly speckled with white and black on a yellowish-brown ground. Nails, black; inner surface of legs, and whole under surface of body, dull yellowish-white. All the feet are grayish-brown.

The under-fur on every part of the back is dark-brown; the longer hairs are brown at their roots, then yellowish; those on the dorsal line are broadly tipped with black, whilst on the shoulders the tips are white. The spots on the back and hips are formed by some of the hairs being tipped with white, others with black. The hairs on the tail are at their roots white, then three times annulated with black and white, and are tipped with white; thus when distichously arranged, (which, however, does not seem natural to the animal,) the tail presents three narrow longitudinal black stripes, and four white ones. Under-surface of tail, dull yellowish-gray.

There are some variations in the colour of different specimens. An old female that was suckling her young at the time she was caught had the dark dorsal line on the shoulders very indistinctly visible, and her feet were much lighter coloured than in younger specimens.

DIMENSIONS.

An old female.

Length of head and body - - - - -	13½	inches.
Tail (vertebræ) - - - - - - - - -	7½	do.
Tail, to end of fur - - - - - - - -	9	do.
Height of ear - - - - - - - - -	½	do.
From heel to longest nail - - - - - -	2¼	do.
From eye to point of nose - - - - - -	1¼	do.

An old male.

Length of head and body - - - - -	13¾	inches.
Tail (vertebræ) - - - - - - - -	8	do.
Tail, to end of fur - - - - - - -	9⅓	do.
Height of ear - - - - - - - -	½	do.
From heel to point of nail - - - - -	2¼	do.

Young.

Length of head and body - - - - -	9	do.
Tail (vertebræ) - - - - - - - -	5½	do.
Tail, to end of fur - - - - - - -	6⅓	do.
Height of ear - - - - - - - - -	½	do.
Tarsus - - - - - - - - - -	2	do.

HABITS.

We regret to state, that with the habits of this species we are wholly unacquainted. Mr. TOWNSEND, who kindly loaned us four specimens, from which we made our drawing and prepared our description, did not furnish us with any account of them.

Of *Spermophilus Beecheyi*, which we have supposed might be found identical with this species, Dr. RICHARDSON states that, "Mr. COLLIE, surgeon of his majesty's ship Blossom, informs me that this kind of Spermophile burrows in great numbers in the sandy declivities and dry plains in the neighbourhood of San Francisco and Monterey, in California, close to the houses. They frequently stand upon their hind-legs when looking round about them. In running they carry the tail generally straight out, but when passing over any little inequality, it is raised as if to prevent its being soiled. In rainy weather, and when the fields are wet and dirty, they come but little above ground. They take the alarm when any one passes within twenty or thirty yards of them, and run off at full speed till they can reach the mouth of their hole, where they stop a little and then enter it; they soon come out again, but with caution, and if not molested, will proceed to their usual occupation of playing or feeding. *Artemesias* and other vegetable matters were found in their stomachs."

GEOGRAPHICAL DISTRIBUTION.

One of the specimens obtained by Mr. TOWNSEND is marked "Falls of the Columbia River," another "Walla-walla;" the specimen procured by DOUGLASS was obtained on the banks of the Columbia River, and if our conjectures are correct, that *S. Beecheyi* is the same as the present species, it exists also in considerable numbers in California.

GENERAL REMARKS.

The first description of this species was given by Dr. RICHARDSON, who received from DOUGLASS a hunter's skin, which, containing no skull, he was prevented from deciding on the genus. We have ascertained that in its dentition it is a true Spermophile, and in all other respects possesses the characteristics of that genus.

In the valuable collection of the London Zoological Society we examined a specimen of *S. Beecheyi*, brought by Mr. COLLIE, which so strikingly resembles this species, that we are greatly inclined to think they will yet be found identical; we have, therefore, quoted it for the present as a synonyme, but marked it with a doubt, as an examination of a greater number of specimens might probably change our views.

[illegible]

Richardson's Spermophile.

Drawn from Nature by J. J. Audubon F.R.S. F.L.S.

Lith. Printed & Col. by J. T. Bowen, Philada.

SPERMOPHILUS RICHARDSONII.—Sab.

RICHARDSON'S SPERMOPHILE.

PLATE L.

Sciuro Hudsonio aliquantulum major; dorso fulvescente, pilis nigris mixtis; ventre fusco-rufescens; cauda mediocri, ad extremum nigra apice fulva; auriculis brevissimis.

CHARACTERS.

*A little larger than the **Hudson's Bay** squirrel; back, yellowish-gray, interspersed with black hair; belly, pale grayish-orange; tail, rather short, black at the extremity, tipped with fawn colour; ears, very short.*

SYNONYMES.

ARCTOMYS RICHARDSONII, Sabine, Linn. Trans,, vol. xiii., p. 589, t. 28.
" " Idem, Franklin's Jour., p. 662.
" " Griffith's An. Kingd., vol. v., p. 246.
TAWNEY AMERICAN MARMOT, Godm., Nat. Hist., vol. ii., p. 111.
ARCTOMYS (SPERMOPHILUS) RICHARDSONII, Rich., F. B. A., p. 164, pl. 11.

DESCRIPTION.

Body, rather short and thick; forehead, arched; nose, blunt, covered with short hairs; margins of the nostrils, and septum, naked; whiskers, few, and shorter than the head; eyes, large; ears, small, rounded, clothed with short hairs on both surfaces; cheek-pouches, of moderate size. The fur on the whole body is short and fine.

Legs, rather short; nails, long, weak, compressed, and slightly arched. On the fore-feet there are four toes and a minute thumb; the toes are covered on the upper surface with short hairs which reach the root of the nails. Palms, naked, containing five callosities. The thumb has a very short joint and is covered by a convex nail. Middle toe longest; the first and third are of equal length, and the outer one is shortest and farthest back.

On the hind-feet there are five toes. The three middle ones are nearly of equal length, the other two are smaller, and are situated farther back:

the claws are shorter than those of the fore-feet; the soles are naked, but the heel is covered with hairs along the edges which curve over it. The tail is not very bushy and is about the size of that of the chipping-squirrel, (*Tamias Lysteri.*)

COLOUR.

Teeth, light orange; whiskers, black; nails, dark-brown; the back is yellowish-brown, intermixed with a few blackish hairs; on the sides, this colour is a shade lighter; on the nose, there is a slight tinge of chesnut-brown. The cheeks, throat, and inside of the thighs, are dull white; belly, brownish-gray. The tail is of the colour of the back; the hairs on the margins, near the end, are dark-brown tipped with yellowish-white.

DIMENSIONS.

Adult female.

From point of nose to root of tail	9¼	inches.
Head	2	do.
Tail (vertebræ)	2½	do.
Tail, to end of hair	3½	do.
From heel to end of middle claw	1½	do.
Height of ear	0¼	do.

HABITS.

We possess no personal knowledge of this species, never having met with it in a living state. The specimens from which our figures and descriptions were made, were obtained by Mr. Townsend, and we are indebted to the excellent work of Richardson for the following account of its habits: "This animal inhabits sandy prairies, and is not found in thickly wooded parts. It is one of the animals known to the residents of the fur countries by the name of Ground-squirrel, and to Canadian voyagers by that of Siffleur. It has considerable resemblance to the squirrels, but is less active, and has less sprightliness and elegance in its attitudes.

"It can scarcely be said to live in villages, though there are sometimes three and four of its burrows on a sandy hummock or other favourable spot. The burrows generally fork or branch off near the surface, and descend obliquely downwards to a considerable depth; some few of them have more than one entrance. The earth scraped out, in forming them, is thrown up in a small mound at the mouth of the hole, and on it the animal seats itself on its hind-legs, to overlook the short grass, and reconnoitre before it ventures to make an excursion. In the spring, there are seldom more than two, and most frequently only one individual seen at a time at the mouth of a hole; and, although I have captured many of them at that season, by pouring water into their burrows, and compelling them to come out, I have never obtained more than one from the same hole, unless when a stranger has been chased into a burrow already occupied by another. There are many little well-worn pathways diverging from each burrow, and some of these roads are observed, in the spring, to lead directly to the neighbouring holes, being most probably formed by the males going in quest of a mate. They place no sentinels, and there appears to be no concert between the Tawny Marmots residing in the neighbourhood, every individual looking out for himself. They never quit their holes in the winter; and I believe they pass the greater part of that season in a torpid state. The ground not being thawed when I was at Carlton House, I had not an opportunity of ascertaining how their sleeping apartments were constructed, nor whether they lay up stores of food or not. About the end of the first week of April, or as soon as a considerable portion of the ground is bare of snow, they come forth, and when caught on their little excursions, their cheek-pouches generally contain the tender buds of the *Anemone Nuttalliana*, which is very abundant, and the earliest plant on the plains. They are fat when they first appear, and their fur is in good condition; but the males immediately go in quest of the females, and in the course of a fortnight they become lean and the hair begins to fall off. They run pretty quick, but clumsily, and their tails at the same time move up and down with a jerking motion. They dive into their burrows on the approach of danger, but soon venture out again if they hear no noise, and may be easily shot with bow and arrow, or even knocked down with a stick, by any one who will take the trouble to lie quietly on the grass near their burrow for a few minutes. Their curiosity is so great that they are sure to come out to look around.

"As far as I could ascertain, they feed entirely on vegetable matter, eating in the spring the young buds and tender sprouts of herbaceous plants, and in the autumn the seeds of grasses and leguminous plants.

"Their cry when in danger, or when angry, so nearly resembles that of *Arctomys Parryi*, that I am unable to express the difference in letters.

"Several species of falcon that frequent the plains of the Saskatchewan, prey much on these Marmots; but their principal enemy is the American badger, which, by enlarging their burrows, pursues them to their inmost retreats. Considerable parties of Indians have also been known to subsist for a time on them when large game is scarce, and their flesh is palatable when they are fat."

GEOGRAPHICAL DISTRIBUTION.

This species has not been observed further north than latitude 55°. In the appendix to FRANKLIN'S Journey, it was said to inhabit the shores of the Arctic Sea, but it appears that another species had been mistaken for it. It is found in the grassy plains that lie between the north and south branches of the Saskatchewan River. It is very common in the neighbourhood of Carlton House, its burrows being scattered at short distances over the whole plain. TOWNSEND obtained his specimens in the Rocky Mountains, (about latitude 45°,) and we have traced it as far south as latitude 38°.

GENERAL REMARKS.

"The Tawny Marmot Squirrel is most readily distinguished from the true squirrels by the smallness of its ears, the shape of its incisors, which are larger but not so strong and much less compressed; the second and not the third toe being the largest, and its comparatively long claws and less bushy tail. It seems to be the American representative of *A. concolor* or the *Jevraska* of Siberia."—(RICHARDSON.) The males of this species are represented as very pugnacious in their habits, and we have represented one in our plate that has lost the end of its tail, the figure being taken from one of the specimens sent to us.

INDEX

[A complete List of the Subscribers will be given at the end of the Work.]

TABLE OF CONTENTS

TABLE OF GENERA DESCRIBED IN THIS VOLUME.